# Advances in Material Forming

## Esaform 10 years on

HF348394

**Springer**
*Paris*
*Berlin*
*Heidelberg*
*New York*
*Hong Kong*
*Londres*
*Milan*
*Tokyo*

Francisco Chinesta
Elias Cueto

# Advances in Material Forming

Esaform 10 years on

Springer

**Francisco Chinesta**
LMSP UMR CNRS-ENSAM
151, boulevard de l'Hôpital
75013 Paris
France

**Elias Cueto**
I3A, University of Zaragoza
Maria de Luna, 3
50018 Zaragoza
Spain

ISBN-13 : 978-2-287-72142-7 Paris Berlin Heidelberg New York

© Springer-Verlag France, Paris 2007

Springer-Verlag France est membre du groupe Springer Science + Business Media

Cet ouvrage est soumis au copyright. Tous droits réservés, notamment la reproduction et la représentation, la traduction, la réimpression, l'exposé, la reproduction des illustrations et des tableaux, la transmission par voie d'enregistrement sonore ou visuel, la reproduction par microfilm ou tout autre moyen ainsi que la conservation des banques de données. La loi française sur le copyright du 9 septembre 1965 dans la version en vigueur n'autorise une reproduction intégrale ou partielle que dans certains cas, et en principe moyennant les paiements des droits. Toute représentation, reproduction, contrefaçon ou conservation dans une banque de données par quelque procédé que ce soit est sanctionnée par la loi pénale sur le copyright.
L'utilisation dans cet ouvrage de désignations, dénominations commerciales, marques de fabrique, etc., même sans spécification ne signifie pas que ces termes soient libres de la législation sur les marques de fabrique et la protection des marques et qu'ils puissent être utilisés par chacun.
La maison d'édition décline toute responsabilité quant à l'exactitude des indications de dosage et des modes d'emplois. Dans chaque cas il incombe à l'usager de vérifier les informations données par comparaison à la littérature existante.

SPIN : 12055838

*Maquette de couverture : Jean-François MONTMARCHÉ*

# Foreword

This book groups the main advances in material forming, considering different processes (conventional and non-conventional) focusing in polymers, composites and metals, that are analyzed from the state of the art, describing the most significant recent advances and identifying the present challenges from the experimental, modeling and numerical points of view. Chapters include a large list of references and have been written by recognized specialists.

Special emphasis is devoted to the contributions of the European Scientific Association on Material Forming (ESAFORM) during the last 10 years (1998-2007) and in particular the ones coming from its annual international conference. The first chapter includes an excellent introduction to the Esaform association (please visit www.esaform.org for further information).

We hope that this book will be valuable for all the readers, and it is specially addressed to young researchers trying to define the state of the art or identifying the open problems in the different areas covered by this book.

The editors,
Francisco Chinesta (ENSAM-Paris)
Elias Cueto (I3A-Zaragoza)

# Table of Contents

# A Brief Historical Review of the ESAFORM Association

Jean-Loup Chenot and Jean-Marc Haudin

CEMEF (Center for Material Forming)
Ecole des Mines de Paris and UMR CNRS 7635, BP 207, F06904 Sophia Antipolis Cedex
Jean-Loup.Chenot@ensmp.fr, Jean-Marc.Haudin@ensmp.fr

**Abstract.** The paper is an attempt to recall the principal events during the first ten years of the ESAFORM association. Nuphymat, a European network on material forming, was a first trial to coordinate research in the field and was soon followed by the creation of the ESAFORM association. The main "philosophical principles" and general scientific goals of the association are briefly recalled. Some examples of achievements are listed and their impact is discussed.

**Keywords:** Scientific Association, Conference, Material Forming.

## 1  Introduction

In 2007, the ESAFORM Association is ten-year old and organizes its tenth conference in Zaragoza (Spain), which is chaired by Professor Chinesta and Professor Cueto.

This anniversary gives us the opportunity to have a reflection on the short, but quite rich, history of our scientific association. The objective of this paper is to analyze briefly the accomplishments of ESAFORM and stress its contributions in scientific progress in the field of material forming processes, as well as it success in enhancing exchanges between scientists of different laboratories and different countries and also exchanges between scientists and engineers working on industrial applications.

## 2  Nuphymat, the prehistory of ESAFORM

The necessity of assessing the importance of a scientific approach in the field of material forming, and strengthening the research effort, was clearly recognized in the early 90's. A significant initiative in this direction was launched in 93 by submitting the collaborative project:

NUPHYMAT: Numerical and Physical Study of Material Forming Processes
to the Commission of the European Community within the program HCM (Human Capital and Mobility). The project was defined by J.F. Agassant and J.L. Chenot, in close collaboration with 13 responsibles of renown laboratories (see Table 1). The

project was coordinated by E. Massoni and covered most of the scientific topics related to metal, polymer or composite forming: mechanics and computational mechanics, rheology, tribology, and micro-structure modeling.

**Table 1.** Participants to the Nuphymat network.

| Responsible | Organization | Country |
|---|---|---|
| J.L. Chenot<br>J.F. Agassant | CEMEF – Ecole des Mines de Paris | France |
| S. Cescotto | Université de Liège – Département M.S.M. | Belgium |
| J. Covas | Universidade do Minho – Dept. Engenharia de Polimeros | Portugal |
| F. Dupret | Université Catholique de Louvain – CESAME | Belgium |
| B. Kroëplin | Universität Stuttgart – Institut für Statik und Dynamik der Luft und Raumfarhrtkonstruktionen | Germany |
| H.G. Fritz | Universität Stuttgart – Institut für Kunststofftechnologie | Germany |
| R. Kopp | Rheinisch Westfälische Technische Hochschule Aachen – Institut für Bildsame Formgebung | Germany |
| P. Hartley | University College of Swansea – School of Manufacturing and Mechanical Engineering | U K |
| R. Owen | University College of Swansea – Rockfield Software Limited | U K |
| J. Whiteman | Brunel University – Institute of Computational Mathematics | U K |
| H. Huétink | University of Twente | Holland |
| H. Meijer | Eindhoven University of Technology – Faculty of Mechanical Engineering, Dept. WFW | Holland |
| S. Piccarolo | Universita Degli Studi di Palermo – Dipartemento di Ingegneria Chimica dei Processi e dei Materiali | Italy |
| T. Wanheim | Technical University of Denmark – Laboratory for Mechanical Processing of Materials | Denmark |

The project started officially in 94 and before the end, 4 additional Eastern Europe laboratories from Russia, Romania, Slovenia and Hungary were selected and encouraged to join the project.

Many fruitful exchanges between scientists were organized within the frame of the HCM project. At the end of the collaboration project it was decided to organize a workshop which gathered 38 scientists in June 1996 in Sophia Antipolis and was opened to a few other distinguished speakers. As a conclusion of the meeting, Professor J.L. Chenot proposed to create a new European Scientific Association for Material Forming.

## 3  Launching of the Association

After many discussions with European Colleagues, the CEMEF laboratory took a major part in creating the European Scientific Association in Material Forming in 1997. The French law for non-profit associations was seen as a convenient frame for creating ESAFORM. The first Board of Directors is presented in Table 2 and in Fig. 1.

**Table 2.** List of the first ESAFORM Board of Directors.

| Name | Function | Country |
| --- | --- | --- |
| J.L. Chenot | President | France |
| J.F. Agassant | Vice President | France |
| P. Coels | Treasurer | France |
| J.M. Haudin | Secretary | France |
| F. Delamare | Deputy-Secretary | France |
| J. Covas | Member | Portugal |
| F. Dupret | Member | Belgium |
| H.G. Fritz | Member | Germany |
| J.C. Gélin | Member | France |
| H. Huétink | Member | The Netherlands |
| E. Oñate | Member | Spain |
| S. Piccarolo | Member | Italy |
| M. Pietrzyk | Member | Poland |
| T. Wanheim | Member | Denmark |
| J. Whiteman | Member | United Kingdom |

**Fig. 1.** A photo of the first ESAFORM Board of Directors.

The main general goals of ESAFORM were clearly stated:
- reinforcement of the importance of Material Forming in Sciences and in Industry,
- enhancement of Teaching and Research in Manufacturing processes,
- improvement of exchanges between the different approaches: continuum mechanics, thermal analysis, rheology, tribology, physics of materials at different scales and numerical analysis,

- increase of the links with industrial partners from which new research topics arise, and to which methods of analysis are transferred from academic groups.

ESAFORM was created as an open structure and after the first year, all of the members of the Board of Directors had to be elected by the scientists registered in the Association. The election took place in 1998 and it was also decided to create a Scientific Committee, chaired by Professor H. Meijer [1].

## 4  General philosophy of ESAFORM

The most important principle for ESAFORM was and remains *opening*, in order to be able to welcome any scientist or engineer who is contributing to the advance of knowledge in the wide field of material forming processes. At the beginning of the association, participants were mainly from Europe, but the Board of Directors decided to open more and more ESAFORM to other continents.

The second objective of the association is to encourage strongly multi-disciplinary studies, involving continuum mechanics, thermal analysis, physical and experimental approaches together with computational mechanics in order to try and solve more and more complex problems occurring in real industrial practice. Also by associating scientists and engineers working on different materials, it is hoped to favor cross fertilization.

The last idea in ESAFORM Community is that all the exchanges and work which is done in common necessitates a friendly atmosphere and, even when scientific competition is present, relations between colleagues and discussions should remain fair and constructive. This is one the major the key for building new collaborative projects involving different teams, with different approaches and complementary skills.

These guidelines were shared not only by the successive Presidents of ESAFORM, but also by all the members of the Board of Directors and the Scientific Committee as well as by most of the Members.

The Board of Directors was renewed every two years according to the statutes of the association, i.e., in 1998, 2000 [2], 2002 [3], 2004 [4] and 2006. Professor J.M. Haudin was elected President in 2000 and kept this responsibility until 2004. Then a new elected President in 2004, A.M. Habraken moved smoothly the ESAFORM Chair, Secretary and Web site from France, where it stayed from the origin, to Belgium.

# 5  Major achievements

## 5.1  ESAFORM Conferences

The first major achievement of ESAFORM to spread scientific information and promote mutual exchanges is the ESAFORM Annual Conference on Materials Forming, devoted to all material forming processes and to all types of materials. The first one took place in Sophia Antipolis in March 1998 and a complete list is given in Table 1, conferences being scheduled until 2009. Serious proposals exist for 2010 and 2011.

Since 2001, minisymposia have been organized on specific topics, e.g., applications of inverse analysis. They have greatly contributed to the success of the conferences, which usually gather more than 200 participants (Table 1). Therefore, the current format for an ESAFORM Conference is now: three days, minisymposia and plenary lectures describing the recent advances in some important processes with a balance between metals, polymers, composites, etc…

**Table 3.** The ESAFORM Annual Conferences on Material Forming

|  | Location | Country | Dates | Chair persons | Attendees |
|---|---|---|---|---|---|
| 1 | Sophia Antipolis | France | March 17-20, 98 | J.L. Chenot J.F.Agassant | 137 |
| 2 | Guimarães | Portugal | April 13-17, 99 | J. Covas | 155 |
| 3 | Stuttgart | Germany | April 11-14, 00 | H.G. Fritz M.H. Wagner | 121 |
| 4 | Liège | Belgium | April 23-25, 01 | A.M. Habraken | 280 |
| 5 | Krakow | Poland | April 14-17, 02 | M. Pietrzyck | 193 |
| 6 | Salerno | Italy | April 28-30, 03 | V. Brucato | 226 |
| 7 | Trondheim | Norway | April 28-30, 04 | S. Stören | 206 |
| 8 | Cluj Napoca | Romania | April 27-29, 05 | D. Banabic | 260 |
| 9 | Glasgow | U.K. | April 26-28, 06 | N. Juster A. Rosochowski | 224 |
| 10 | Zaragoza | Spain | April 18-20, 07 | F. Chinesta, E. Cueto | |
| 11 | Lyon | France | April 23-25, 08 | P. Boisse | |
| 12 | Twente | The Netherlands | April 27-29, 09 | H. Huétink | |

For all the conferences, proceedings are available. Moreover, for several ones, selected papers have been published in special issues of the International Journal of Forming Processes [5-8].

Furthermore, short courses have been delivered within the framework of the ESAFORM Conference. Let us mention:
- from Eshelby to Slender Bodies Theory (1999) [9],
- Microstructural Aspects and Constitutive Modelling of Plastic Deformation of Metals (2002),
- Modelling and Simulation of Flow Localisation in Material Processing. Physical, Theoretical and Numerical Fundamentals (2004).

During the conferences, several social events are offered to the participants to support the communication and to generate a relaxed atmosphere. In some cases, post-conference tours have been organized, e.g., the famous "Dracula tour" after the conference in Cluj Napoca (Fig. 2).

**Fig. 2.** Special certificate delivered to the participants to the "Dracula tour" (2005).

## 5.2  Scientific and industrial prizes

Since its creation, the Association has the idea of awarding research work of young scientists. As soon as it has been possible, the ESAFORM Scientific Prize was launched and the first one was given during the second ESAFORM Conference, in 1999. The ESAFORM Scientific Prize is awarded annually. It intends to distinguish a young confirmed scientist, 35-year old or less, who has brought an outstanding contribution in the field of materials forming.

The scientists awarded since 1999 are the following:
1999    Luis Filipe Menezes,
        University of Coimbra, Portugal [10]
2000    Francisco Chinesta, CNAM, France [11]
        Jean-Marie Drezet, EPFL, Switzerland, 2nd Prize [12]
2001    Rui L. Reis, University of Minho, Portugal [13]
        André Luciani, Péchiney, France, 2nd Prize [14]
2002    Eduardo Car, CIMNE, Spain [15]
2003    Andrew Long, University of Nottingham, UK [16]
2004    Thierry Barrière, University of Besançon, France [17]
2005    Pierre-Olivier Bouchard, Ecole des Mines de Paris, France [18]
2006    Elias Cueto, University of Zaragoza, Spain
        José Luis de Carvalho Martins Alves, University of Minho, Portugal

In the same way, the ESAFORM Industrial Prize was instituted in 2000 by the Board of Directors, in order to attract candidates from Industry and to distinguish outstanding industrial research or design work. At the beginning, it seemed difficult to attract good applications. After modifications of the selection rules taking into account the specificity of industrial research and allowing the application of industrial research teams without limitations for age and number of participants, the Industrial Prize could be awarded for the first time in 2004.

The research teams awarded since 2004 are the following:
2004    Martin Rohleder, Tim Lemke and Klaus Wiegand, Daimler-Chrysler, Germany
        [19]
2005    David Silagy, Johann Laffargue and Damien Rauline, Arkema, France [20]
2006    Lucia Garcia Aranda, Tommaso dal Negro, Audrey Marty and Hatem Khanfir,
        Renault Ingénierie Process and Prototypes, France

The awardees of the Scientific and Industrial Prizes have delivered a lecture at the annual conference. All of them, including the second prizes, have published a paper on their work in the ESAFORM Bulletin [10-20].

## 5.3  Communication tools

To reach its objectives, ESAFORM also needed communication tools. Therefore were created:

- a web server (http://www.ESAFORM.org). It currently offers a number of services to the members: advertisement for conferences and courses, post-doc and job opportunities, minutes of the Board of Directors meetings, access to members address via personal password;
- the International Journal of Forming Processes (IJFP), the official journal of ESAFORM, which covers all the scientific and technical domains concerning the Association. The first issue appeared in March 1998 with Hermès as publisher and Prof. M. Touratier as editor-in chief. Since 2002, the journal is published by Hermès-Lavoisier, and recently, Prof. M. Touratier has been replaced by Prof. F. Chinesta as editor-in-chief. It is currently issued both in electronic and printed versions and has appeared continuously (4 issues per year). Special issues have been dedicated to ESAFORM conferences [5-8].
- The ESAFORM Bulletin, which was launched in 1998, with J.M. Haudin as editor-in-chief and F. Morcamp, as managing editor. 8 issues have been published from 1999 to 2005 by this editorial team, the ninth 2006 and the following ones being under the responsibility of Prof. D. Banabic and D.S. Comsa. The bulletin intends to provide information on the life of the Association (Board of Directors, ESAFORM conferences, ESAFORM prizes, IJFP), on European laboratories and on specific events (conferences, courses, PhD defences). It also publishes scientific and technical papers: contributions of the prize winners [10-20] or papers on some new aspects, e.g., [21-22].

# 6  Some consequences of the Association

More and more objectives established at the foundation of the Association tend to be fulfilled. An example is the organization and the development of European research networks such as PIAM (Polymer Injection Advanced Moulding) and VIF (Virtual Intelligent Forging). Both are Coordination Actions of the sixth Framework Program (FP6).

## 6.1  The PIAM network

PIAM gathers 23 partners from 10 countries. It objectives are to:

- create a wide consortium of European experts involved in all the aspects of polymer injection-moulding (process, materials science, physical and numerical modelling, mechanical properties);
- test existing software packages (benchmark);

- identify new numerical developments to overcome software limitations;
- analyse reliable methods for material data identification (thermoplastic polymers and more complex fluids);
- promote the development of advanced injection-moulding technologies (e.g., micromoulding).

## 6.2 The VIF network

The objectives of VIF (54 partners from 17 countries) are to :
- identify current industrial and societal needs;
- define, validate and use reference benchmarks for virtual process simulations and materials testing;
- create an e-Forging environment;
- determinate the needs for material data (e-Database);
- design a structure for the virtual integration of process simulations (from raw materials to product design);
- promote transverse educational programs (e-Learning platform on forging);
- organize workshops for results dissemination and promote programs for mobility of researchers students and industrial staff.

Many partners of these projects are ESAFORM members. Secondly, ESAFORM is explicitly mentioned as a dissemination tool. It is the reason why the PIAM 12-month meeting took place in Cluj-Napoca, just before the ESAFORM Conference (2005). This offered an opportunity to the PIAM partners to contribute to the conference minisymposia.

## 7 Conclusions

The evolution of the material forming community is quite impressive. Not only this field if now widely recognized as a major issue in the material science, but at least for most metals, alloys, polymers and composites, conditions of forming and heat treatments and their optimization are considered as the most important issues for predicting, and possibly optimizing, the final properties of the work-pieces. In many ambitious projects, the goal is not only to assess the feasibility of a sequence of manufacturing but also to predict quantitatively the (heterogeneous) mechanical and physical properties.

One of the most important contributions of the ESAFORM Association was, and will be, to contribute to the diffusion of a multi-disciplinary approach of material forming and to introduce advanced scientific approaches and numerical simulations for solving industrial problems.

In the future, it is anticipated that ESAFORM association will gather more and more scientists from other continents and become really international.

# References


1. Haudin, J.M.: Prehistory and History of ESAFORM. ESAFORM Bulletin Vol. 1, n° 1 (1999) 2-3
2. Haudin, J.M.: ESAFORM Life. New ESAFORM Board of Directors. ESAFORM Bulletin Vol. 1, n° 3 (2000) 2
3. Haudin, J.M.: ESAFORM Life. The renewal of the Board of Directors. ESAFORM Bulletin Vol. 3, n° 1 (2003) 2
4. Haudin, J.M.: ESAFORM Life. The new Board of Directors. ESAFORM Bulletin Vol. 5, n° 1 (2005) 2
5. Habraken, A.M. (ed.): Material Forming. International Journal of Forming Processes Vol. 4, n° 3-4 (2001)
6. Pietrzyk, M. (ed.): Material Forming, 1. Bulk, Sheet, Composite Forming and Cutting. International Journal of Forming Processes Vol. 6, n° 3-4 (2003)
7. Pietrzyk, M. (ed.): Material Forming, 2. Modeling, Simulation, Optimization and Material Testing. International Journal of Forming Processes Vol. 7, n° 1-2 (2004)
8. Habraken, A.M., Stören, S. (eds): Multiscale Simulations and Experiments to Optimize Material Forming Processes. International Journal of Forming Processes Vol. 8, Special Issue (2005)
9. Poitou, A.: From Eshelby to the Slender Bodies Theory. Rheology of Short Fibers Composites. ESAFORM Bulletin Vol. 1, n° 2 (2000) 6-7
10. Menezes, L.F.: Implicit Algorithm for the Numerical Simulation of the Deep-Drawing Process. ESAFORM Bulletin Vol. 1, n° 2 (2000) 4-5
11. Chinesta, F.: Short Fiber Reinforced Thermoplastics: Rheology and Forming Processes. ESAFORM Bulletin Vol. 1, n° 3 (2000) 5-6
12. Drezet, J.M.: Predicting Hot Tearing in Aluminium Castings. ESAFORM Bulletin Vol. 1, n° 3 (2000) 7-8
13. Reis, R.L.: Research on Biomaterials, Biodegradables and Biomimetics. ESAFORM Bulletin Vol. 2, n° 2 (2002) 8-9
14. Luciani, A.: Polymer Blends Mixing. ESAFORM Bulletin Vol. 2, n° 2 (2002) 5-7
15. Car, E., Oller, S., Oñate, E.: Numerical Simulation of Composite Materials. Two Constitutive Models. ESAFORM Bulletin Vol. 3, n° 1 (2003) 3-6
16. Long, A.: Forming Analysis for Textile Composites. ESAFORM Bulletin Vol. 4, n° 1 (2004) 7-10
17. Barrière, T.: Metal Injection Molding (MIM). ESAFORM Bulletin Vol. 5, n° 1, 15 (2005) 6-10, 15
18. Bouchard, P.O.: Numerical Modeling of Self-Pierce Riveting – From Riveting Process Modeling down to Structural Analysis. ESAFORM Bulletin, Vol. 6, n° 1 (2006) 4-7
19. Rohleder, M., Linke, T., Wiegand K.: Numerically Based Compensation of Springback Deviations during the Die Development Process of Complex Car Parts. ESAFORM Bulletin Vol. 5, n° 1 (2005) 3-5
20. Silagy, D., Laffargue J., Rauline D.: Drawability Limits in Elongational Processes, Theoretical Approach Efficiently Applied to Develop ARKEMA's Technical Polymers. ESAFORM Bulletin Vol. 6, n° 1 (2006) 8-10
21. Hauger, A.: Flexible Rolling, a New Technology for Light Weight Constructions. ESAFORM Bulletin, Vol. 3, n° 1 (2003) 10
22. Tillier, Y., Paccini, A., Chenot, J.L., Durand-Reville, M.: Three-Dimensional Finite Element Simulation of Surgery: from Material Forming to Biomechanics. ESAFORM Bulletin Vol. 4, n° 1 (2004) 3-6

# New and Advanced Numerical Strategies for the Simulation of Material Forming

Francisco Chinesta[1], Elias Cueto[2], and Thierry Coupez[3]

[1] LMSP, UMR CNRS-ENSAM, 151 Boulevard de l'Hopital, F-75013 Paris, France
francisco.chinesta@paris.ensam.fr,
[2] I3A, University of Zaragoza, Maria de Luna 3, E-50018 Zaragoza, Spain
ecueto@unizar.es,
[3] CEMEF, UMR CNRS-ENSMP, BP 207, F-06904 Sophia-Antipolis Cedex, France
Thierry.Coupez@ensmp.fr,

**Abstract.** In recent years new and advanced numerical strategies have opened new possibilities in the simulation of forming processes. Multiscale descriptions, meshless methods and enhanced finite element approaches are some techniques that have contributed to the enhancement of forming process simulations. These approaches will be revisited in this chapter.

## 1 Microscopic approaches

### 1.1 Solid mechanics framework

Materials can be described at different scales. The finest level of description consists of the atomic level and the coarsest one concerns the scale at which the conformed part is defined. Molecular dynamics simulation works at the atomic level and allows to account for complex physics in a simple and natural way. Thus, knowing at a certain time the position of the atoms, the resultant force applying at each atom can be easily computed from a semi-empirical atomic potential, and from it the atoms acceleration computed and the velocities and atomic positions updated. The main drawback of this approach lies in the extremely large computing time required to perform realistic simulations even when small domains and time periods are considered,as well as the semi-empirical interatomic potentials usually considered in such approaches.

The establishment of more accurate atomic potentials require the solution of the Schrodinger equation in the quantum mechanics framework, whose main difficulty lies in the curse of dimensionality that we consider later, and that constitutes today a real challenge. In recent years molecular dynamics approaches were considered in the framework of forming processes involving cutting [39] or contact with friction [12], both presented during the recent Esaform conferences.

The micro-macro approach when both descriptions coexist in the physical space as well as the definition of efficient bridges between both descriptions defined in contiguous regions are topics in active development nowadays [24] [43] [23].

In an intermediate scale other approach based on the discrete finite element was successfully applied for treating granular media, being [40] or [21] some examples of works presented during the Esaform conferences in recent years.


## 1.2  Fluid mechanics framework

This section concerns the liquid state of different materials involved in forming processes. In some cases, as in casting, the resulting constitutive equation of involved materials results simple, but the high Reynolds number involved in the forming processes induce numerous numerical difficulties. For other kind of materials, the ones involving microstructure, the main difficulty is coming from the inherent multiscale character of its mechanical behavior. This section focuses on this complex fluids (polymer melts or particle suspensions).

Many natural and synthetic fluids are viscoelastic materials, in the sense that the stress endured by a macroscopic fluid element depends upon the history of the deformation experienced by that element. Notable examples include polymer solutions and melts, liquid crystalline polymers and fibre suspensions. Rheologists thus face a challenging non-linear coupling between flow-induced evolution of molecular configurations, macroscopic rheological response, flow parameters (such as the geometry and boundary conditions) and final properties. Theoretical modelling and methods of computational rheology have an important role to play in elucidating this coupling.

Atomistic modelling is the most detailed level of description that can be applied today in rheological studies, using techniques of non equilibrium molecular dynamics. Such calculations require enormous computer resources, and then they are currently limited to flow geometries of molecular dimensions. Consideration of macroscopic flows found in processing applications calls for less detailed mesoscopic models, such as those of kinetic theory.

Kinetic theory models can be very complicated mathematical objects. It is usually not easy to compute their rheological response in rheometric flows, and their use in numerical simulations of complex flows has long been thought impossible. The traditional approach has been to derive from a particular kinetic theory model a macroscopic constitutive equation that relates the viscoelastic stress to the deformation history. The majority of constitutive equations used in continuum numerical simulations are indeed derived (or at least very much inspired) from kinetic theory. Indeed, derivation of a constitutive equation from a model of kinetic theory usually involves closure approximations of a purely mathematical nature such as decoupling or pre-averaging. It is now widely accepted that closure approximations have a significant impact on rheological predictions for dilute polymer, solutions, or fiber suspensions.

Since the early 1990's the field has developed considerably following the introduction of the CONNFFESSIT method by Ottinger and Laso [37]. Kinetic theory provides two basic building blocks: the diffusion or Fokker-Planck equation that governs the evolution of the distribution function (giving the probability distribution of configurations) and an expression relating the viscoelastic stress to the distribution function. The Fokker-Planck equation has the general form:

$$\frac{d\psi}{dt} + \frac{\partial}{\partial \mathbf{X}}\left(\mathbf{A}\psi\right) = \frac{\partial}{\partial \mathbf{X}}\left(\mathbf{D}\frac{\partial \psi}{\partial \mathbf{x}}\right) \tag{1}$$

where $\frac{d\psi}{dt}$ is the material derivative, vector $\mathbf{X}$ defines the coarse-grained configuration and has dimensions $N$. Factor $\mathbf{A}$ is a $N$-dimensional vector that defines the drift or deterministic component of the molecular model. Finally $\mathbf{D}$ is a symmetric, positive definite $N \times N$ matrix that embodies the diffusive or stochastic component of molecular model. In general both $\mathbf{A}$ and $\mathbf{D}$ (and in consequence the distribution function $\psi$ ) depend on the physical coordinates $\mathbf{x}$, on the configuration coordinates $\mathbf{X}$ and on the time $t$.

The second building block of a kinetic theory model is an expression relating the distribution function and the stress. It takes the form:

$$\tau_p = \int_C \mathbf{g}(\mathbf{X})\psi d\mathbf{X} \tag{2}$$

where $C$ represents the configuration space and $\mathbf{g}()$ is a model-dependent tensorial function of configuration. In a complex flow, the velocity field is a priori unknown and stress fields are coupled through the conservation laws. In the isothermal and incompressible case the conservation of mass and momentum balance are then expressed (neglecting the body forces) by:

$$\begin{cases} \nabla \cdot \mathbf{v} = 0 \\ \rho\frac{d\mathbf{v}}{dt} = \nabla \cdot (-p\mathbf{I} + \tau_p + \eta_s\mathbf{d}) \end{cases} \tag{3}$$

where $\rho$ is the fluid density, $p$ the pressure and $\eta_s\mathbf{d}$ a purely viscous component. The set of coupled equations (1)-(3), supplemented with suitable initial and boundary conditions in both physical and configuration spaces, is the generic multiscale formulation. Three basic approaches have been adopted for exploiting the generic multiscale model:

1. *The continuum approach* wherein a constitutive equation of continuum mechanics that relates the viscoelastic stress to the deformation history is derived from, and replaces altogether, the kinetic theory model (1) and (2). The derivation process usually involves closure approximations. The resulting constitutive model takes the form of a differential, integral or integro-differential equation.
2. *The Fokker-Planck approach* wherein one solves the generic problem (1) to (3) as such, in both configuration and physical spaces. The distribution function is thus computed explicitly as a solution of the Fokker-Planck equation (1). The viscoelastic stress is computed from (2).
3. *The Stochastic approach* which draws on the mathematical equivalence between the Fokker-Planck equation (1) and the following Ito stochastic differential equation:

$$\mathbf{dX} = \mathbf{A}\, dt + \mathbf{B}\, \mathbf{dW} \tag{4}$$

where $\mathbf{D} = \mathbf{B}\,\mathbf{B}^T$ and $\mathbf{W}$ is a Wiener stochastic process of dimension $N$. In a complex flow, the stochastic differential equation (4) applies along individual flow trajectories, the time derivative is thus a material derivation. Instead of solving the deterministic Fokker-Planck equation (1), one solves the associated stochastic differential equation (4) for a large ensemble of realizations of the stochastic process $\mathbf{X}$ by means of a suitable numerical technique. The distribution function is not computed explicitly, and the viscoelastic stress (2) is readily obtained as an ensemble average.

The control of the statistical noise is a major issue in stochastic micro-macro simulations based on the stochastic approach (for more details concerning the micro-macro approach reader can refers to the excellent review paper [28] and the references therein). Some stochastic simulations of muti-bead-spring (MBS) models have been successfully carried out, see for example [44]. These problems do not arise at all in the Fokker-Planck approach. The difficulty, however, is that the Fokker-Planck equation (1) must be solved for the distribution function in both physical and configuration spaces. This necessitates a suitable discretization procedure for all relevant variables, namely position $\mathbf{x}$, configuration $\mathbf{X}$ and time $t$. Until now, the dimensionality of the problem could be daunting and consideration of molecular models with many configurational degrees of freedom did not appear feasible. This probably explains why relatively few studies based of the Fokker-Planck approach have appeared in the literature until very recently at least. In [14] [30] the resolution of the Fokker-Planck equation involving a moderate number of dimensions is considered. Another deterministic particle approach, very close to that proposed in [13], was analyzed in [3] using the SPH meshless approach.

An appealing strategy that allows alleviating the computational effort is based on the use of reduced approximation bases obtained by applying the Karhunen-Loève decomposition, succesfully applied in complex fluid simulation in [41] and [2], however high dimensional models are out of its applicability, because in this case the definition of a mesh results simply prohibitory.

Some attempts exist concerning the treatment of multidimensional problems. The interested reader can refer to [11] for a review on sparse grids methods involving sparse tensor product spaces, but despite of its optimality, the interpolation is defined in the whole multidimensional domain, and consequently only problems defined in spaces of dimension of the order of tens can be treated [1]. In [10] multidimensional problems are revisited and deeply analyzed, and for this purpose new mathematical entities are introduced. In [4] we considered the steady state solution of some classes of multidimensional partial differential equations by using a separated representation. In [5] this technique was extended for solving accurately and efficiently multidimensional transient kinetic theory models.

Some works focussing in the solution of the multidimensional Fokker-Planck equation defining the micro-macro description were presented during the recent Esaform conferences (see for example [6]) however these models only concerned simple rheological flows. The real challenge for the next 10 years will be the extension of these procedures for treating the complex flows encountered in polymer forming processes.

## 2 Meshless methods: the possibility for Lagrangian simulations and much more

### 2.1 Introduction

It is not easy to explain what a meshless method is in few words. Since the meshless irruption after the pioneer work by Touzot and Villon [36], many different meshless methods have arisen, with many different characteristics and names. Just to cite a few, the Element-Free Galerkin [9], the Reproducing Kernel Particle Method [29], the Natural Element Method [46] [20], or the Generalized Finite Element Method [45], are examples of these different methods. They are based either in Galerkin or collocation approaches, employ different kinds of approximation for the essential variables (moving least squares, natural neighbour approximants, wavelets, etc.) and also employ —when talking about Galerkin strategies— different numerical integration procedures.

What is essential then in a numerical method to be considered as *meshless* is its ability to maintain the accuracy despite the distortion of the mesh (or, more properly, the cloud of points). It is well-known that the Finite Element method suffers from lack of accuracy if the mesh becomes distorted [7]. Meshless methods, however, adapt the connectivity of the "elements" as the cloud of points evolves, in a process transparent to the user. Many meshless methods use radially-supported shape functions (or employ tensor-product shape functions, thus giving a rectangular support), see Fig. 1. Thus, the connectivity of each element (defined as the list of nodes influencing the portion of the space in which numerical integration is to be performed) changes as the cloud evolves.

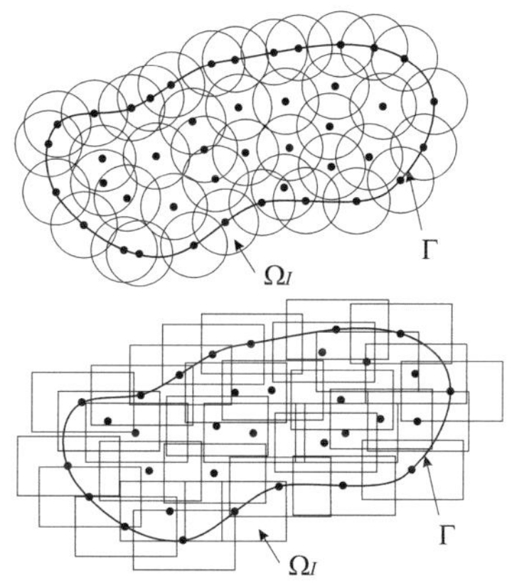


**Fig. 1.** Covering of a two-dimensional domain $\Omega$ by the shape functions' support, $\Omega_I$.

Element-Free Galerkin methods (EFGM) fall within the class of meshless methods that employ cobertures of the domain such as that in Fig. 1. In this case, EFGM employ a Galerkin perspective and Moving Least Squares interpolation to construct shape functions with arbitrary degree of reproducibility. Reproducing Kernel Particle Methods, although originally developed from Smooth Particle Hydrodynamics approaches, are entirely equivalent to EFGM.

However, many meshless methods with radially-supported shape functions lack of an appropriate interpolation along the boundaries (which is readily seen from Fig. 1 if we note that interior nodes influence on the boundary values). This is a particularly important problem in the simulation of forming processes, where issues related to friction, for instance, are frequently noteworthy. The Natural Element Method (NEM) [46] [20] solves this problem in a very elegant way, by utilizing natural neighbour interpolation, instead of Moving Least Squares, in a Galerkin framework. Natural neighbour interpolation functions do not have circular support (instead, they cover the union of circumcircles of each Delaunay triangle containing the node) and this fact makes possible the exact imposition of essential boundary conditions, up to the degree of consistency of the method.

Other, similar, methods, that also employ natural neighbour-based inter-polants have been developed in recent years and applied to the simulation of forming processes, see, for instance, [25] [26].

Many other meshless methods exist, based on collocation as well as Galerkin approaches, but probably the EFGM and the NEM have been the most popular ones within the forming processes community.

## 2.2   Application of meshless methods to the simulation of forming processes.

The fact that meshless methods do not lack accuracy as mesh distorts opens the possibility to perform Lagrangian simulations instead of Eulerian or Arbitrary Lagrangian-Eulerian ones, which had been the most employed ones. This is especially challenging in fields where traditional Finite Element procedures fail or present difficulties. Some examples developed by the authors follow. They are only included intending show how, qualitatively, meshless methods can help in the simulation of very complex forming processes.

**Simulation of injection moulding of short fiber reinforced thermoplastics.** Free-surface flows is a typical example of this kind of problems. The location of the free surface had been traditionally done by means of Volume of Fluid (VoF) or similar techniques, in which a variable representing the portion of an Eulerian element which is filled by a liquid must be advected with the material velocities.

Mechanical modelling of short fibers suspensions flows is usually achieved in the framework of dilute or semi-dilute suspensions of non-spherical particles in a Newtonian fluid. The resulting system of equations involves the coupling of an elliptic problem with an advection problem related to the fluid history. The elliptic problem is associated with the equations of motion whereas the advection equation describes the time evolution of the anisotropic viscosity tensor (fiber orientation) or more generally the microstructural state. The second problem presents two difficulties: it is non-linear and hyperbolic.

Coupled models take into account both the dependence of the kinematics with the fiber orientation and the orientation induced by the flow kinematics. Usually the coupled models are solved by means of a fixed point strategy. In this case, at each iteration the flow kinematics results from the solution of motion and

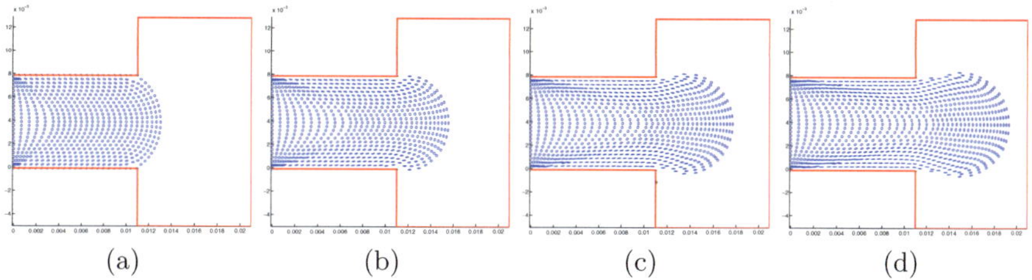

(a)      (b)      (c)      (d)

**Fig. 2.** Evolution of the orientation field in the simulation of an injection moulding.

mass conservation equations, using the fiber orientation field from the previous iteration. From the kinematics just computed, the fiber orientation is updated solving the advection equation governing its evolution. In Fig. 2 four snapshots of the evolution of the orientation field are shown. The orientation is represented by elipses indicating the probability of finding a fiber oriented in each direction.

In [31] a deeper insight on the constitutive modelling of such flows can be found. The accuracy in the numerical treatment of the free surface flow is also noteworthy [32].

**Simulation of orthogonal cutting.** One of the very first applications of the NEM to the field of forming processes was made towards the simulation of cutting [17]. The extremely large deformations appearing in such a process make meshless methods an interesting approach to be considered.

Essentially, in this first applications a very simple visco-plasticity model based on a Norton-Hoff law and very simple contact detection algorithms, that considered rigid tool surfaces, was employed. The main purpose was, however, to demonstrate that such a method can easily suffer these high levels of strain without lack of accuracy, see Fig. 3.

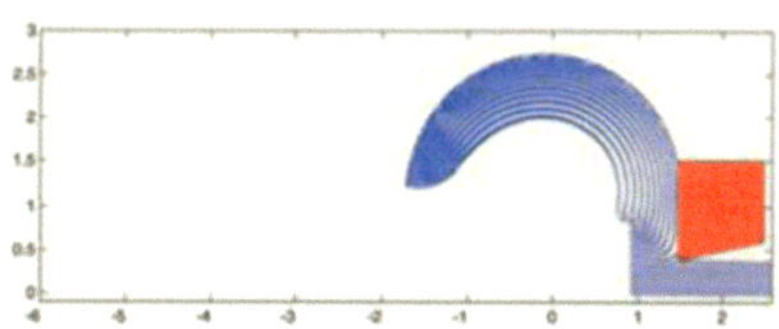

**Fig. 3.** Simulation of the orthogonal cutting process.

**Three-dimensional simulation of the extrusion of a cross-shaped profile.** In order to show the capabilities of the technique presented before, we analyze now the simulation of a cross-shaped aluminium profile.

Nodes located on the upper side of the billet were forced to move with a speed of $2\frac{mm}{s}$, in order to obtain an exit velocity of $1\frac{m}{min}$ approximately. Initially, slipping boundary conditions were considered between the billet and the die and the container. The initial temperature was set to $723K$. The whole model was considered adiabatic, including the profile surface in contact with air. In this case a rigid-plastic Sellars-Tegart material model was used. The simulation ran over 42 time steps of $0.025s$. The obtained evolution for the equivalent strain rate is depicted in Fig. 4.

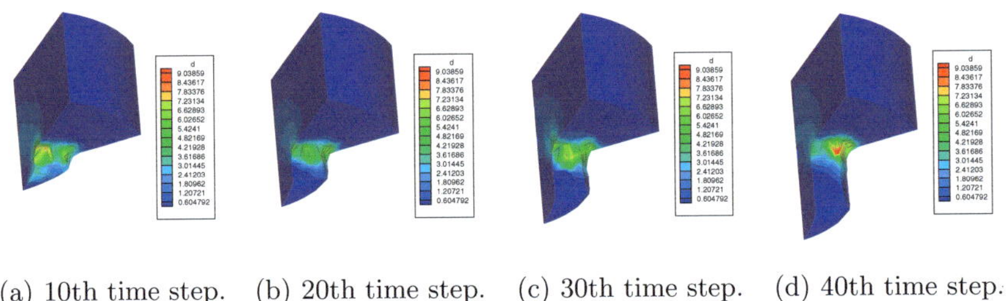

(a) 10th time step.    (b) 20th time step.    (c) 30th time step.    (d) 40th time step.

**Fig. 4.** Equivalent strain rate $(s^{-1})$ for the extrusion simulation.

## 2.3   Meshless methods in the ESAFORM conferences: past, present and future

These examples are intended for suggesting only the wide range of potential application that meshless methods can have in the field of forming processes. But the applications are by no means limited to the before presented. Other examples include the simulation of expanding foams [16], Stefan problems, etc.

In all these cases, meshless methods opened the possibility for a Lagrangian procedure, which is in some cases a very convenient way of overcoming the before-mentioned problems of numerical diffusion in the results due to extensive remeshing.

Concerning the ESAFORM life, meshless methods have been present, up to our knowledge, since the conference held in Liege in 2001 [33] and their contribution extends up to the ninth ESAFORM conference's scientific prize plenary lecture [19]. In between, many papers have been presented in ESAFORM conferences employing meshless method to simulate forming processes. For instance, there is a very active group in Portugal applying these methods to the simulation of various forming proceses [51]. Applications include forging, sheet metal forming, etc. Other contributions include applications of the so-called eXtended Finite Element Method (X-FEM) [22] [35], or the excellent keynote given by N. Sukumar at Salerno in 2003 [47]. Concerning RKP methods, Joyot has made some interesting contributions along the years, see for instance [27] and references therein. Of course, Prof. Villon has been one of the most active researchers in this field and has presented many works on the topic along the years, see for instance [48].

There remain some interesting challenges concerning meshless methods (both within and out of ESAFORM life). There is a more or less unanimous opinion that the issue of numerical integration is crucial in this field, and many groups are now focusing their efforts towards this end. Within ESAFORM life, there remains some problems where meshless methods have much to say. In general, these problems involve very large strains, as mentioned before. Friction stir welding is an example of candidate processes. The extremely large speed of rotation, the coupled thermo-mechanical problem and the extremely large deformations appearing make this process an ideal candidate to benchmark the behaviour of meshless methods.

# References


1. Y. Achdou, O. Pironneau, *"Computational methods for option pricing"*, Siam Frontiers in Applied Mathematics, 2005.
2. A. Ammar, D. Ryckelynck, F. Chinesta, R. Keunings, *"On the Reduction of Kinetic Theory Models Related to Finitely Extensible Dumbbells"*, J. Non-Newtonian Fluid Mech., In press.
3. A. Ammar, F. Chinesta, D. Ryckelynck, *"Deterministic particle approach of multibead-spring polymer models"*, European Journal of Computational Mechanics, In press.
4. A. Ammar, B. Mokdad, F. Chinesta, R. Keunings, *"A fast solver for some classes of multidimensional partial differential equations encountered in kinetic theory modelling of complex fluids"*. Journal of Non-Newtonian Fluid Mechanics, Submitted.
5. A. Ammar, B. Mokdad, F. Chinesta, *"A fast solver for some classes of multidimensional partial differential equations encountered in kinetic theory modelling of complex fluids. Part 2: Transient simulation using space-time separated representations"*. Journal of Non-Newtonian Fluid Mechanics, Submitted.
6. A. Ammar, F. Chinesta, D. Ryckelynck *"Deterministic approach of the multi-bead-spring polymer models"*, Proceedings of the 8th Esaform Conference on Material Forming, Cluj-Napoca, Romania.
7. I. Babuška and A. Aziz. On the angle condition in the finite element method. *SIAM J. Numer. Anal.*, 13:214–227, 1976.
8. T. Belytschko, Y. Krongauz, M. Fleming, D. Organ, and W. K. Liu. Smoothing and accelerated computations in the element free galerkin method. *Journal of Computational and Applied Mathematics*, 1995.
9. T. Belytschko, Y. Y. Lu, and L. Gu. Element-Free Galerkin Methods. *International Journal for Numerical Methods in Engineering*, 37:229–256, 1994.
10. G. Beylkin, M. Mohlenkamp, *"Algorithms for numerical analysis in high dimensions"*, SIAM J. Sci. Com., 26/6, 2133-2159, 2005.
11. H.J. Bungartz, M. Griebel, *"Sparse grids"*, Acta Numerica, 13, 1-123, 2004.
12. P. Chantrenne, *"Atomic scale simulations: use of molecular dynamics"*, Proceedings of the 9th Esaform Conference on Material Forming, Glasgow, UK, 123-126, 2006.
13. C.V. Chaubal, A. Srinivasan, O. Egecioglu, L.G. Leal, *"Smoothed particle hydrodynamics techniques for the solution of kinetic theory problems"*, J. Non-Newtonian Fluid Mech., 70, 125-154, 1997.
14. C. Chauviere, A. Lozinski, *"Simulation of dilute polymer solutions using a Fokker-Planck equation"*, Computer and Fluids, 33, 687-696, 2004.

15. J. S. Chen, C. M. O. L. Roque, C. H. Pan, and S. T. Button. Analysis of metal forming process based on meshless method. *JOURNAL OF MATERIALS PROCESSING TECHNOLOGY*, pages 642–646, 1998.

16. F. Chinesta, E. Cueto, P. Quintela, and J. Paredes. Induced anisotropy of foams forming processes: modelling and simulation. *Journal of Materials Processing Technology*, 155:1482–1488, 2004.

17. F. Chinesta, Ph. Lorong, D. Ryckelynck, M. A. Martinez, E. Cueto, M. Doblaré, G. Coffignal, M. Touratier, and J. Yvonnet. Thermomechanical cutting model discretisation: Eulerian or Lagrangian, mesh or meshless? *International Journal of Forming Processes*, 7:83–98, 2004.

18. E. Cueto, M. Doblaré, and L. Gracia. Imposing essential boundary conditions in the Natural Element Method by means of density-scaled $\alpha$-shapes. *International Journal for Numerical Methods in Engineering*, 49-4:519–546, 2000.

19. E. Cueto. The Natural Element Method for the simulation of forming processes. *Plenary conference. Ninth ESAFORM conference on Material Forming, Glasgow, U. K.*, 2006.

20. E. Cueto, N. Sukumar, B. Calvo, M. A. Martínez, J. Cegoñino, and M. Doblaré. Overview and recent advances in Natural Neighbour Galerkin methods. *Archives of Computational Methods in Engineering*, 10(4):307–384, 2003.

21. A. Delaplace, C. Rey, *"Numerical strategies for describing material failure with discrete element model"*, Proceedings of the 8th Esaform Conference on Material Forming, Cluj-Napoca, Romania, 101-104, 2005.

22. A. Devan and J. E. Dolbow. An enhanced assumed strain method with discontinuous enrichment. In V. Brucato, editor, *ESAFORM Conference on Material Forming*, Salerno University, Salerno, Italy, 2003.

23. F. Feyel, *"Some application of parallel computing for multiscale and multiphysics problems"*, Proceedings of the 7th Esaform Conference on Maaterial Forming, Trondheim, Norway, 85-88, 2004.

24. T. Hughes, G. Feijoo, L. Mazzei, J.B. Quincy, *"The variational multiscale method - a paradigm for computational mechanics"*, Computer Methods in Applied Mechanics and Engineering, 166, 3-24, 1998.

25. S. R. Idelsohn, E. Oñate, N. Calvo, and F. del Pin. The meshless finite element method. *International Journal for Numerical Methods in Engineering*, 58:893–912, 2003.

26. S. R. Idelsohn, E. Oñate, and F. del Pin. The particle finite element method: a powerful tool to solve incompressible flows with free-surfaces and breaking waves. *International Journal for Numerical Methods in Engineering*, 61 (7):964–989, 2004.

27. P. Joyot, J. Trunzler and F. Chinesta. Accounting for incompressibility in Reproducing Kernel Particle meshless approximations. In A. Rosochowsky, editor, *ESAFORM Conference on Material Forming*, Strathclyde University, Glasgow, UK, 2006.

28. R. Keunings, *"Micro-macro methods for the multiscale simulation viscoelastic flow using molecular models of kinetic theory"*, Rheology Reviews, D.M. Binding and K. Walters (Edts.), British Society of Rheology, 67-98, 2004.

29. W. K. Liu, S. Jun, S. Li, J. Adee, and T. Belytschko. Reproducing kernel particle methods. *International Journal for Numerical Methods in Engineering*, 38:1655–1679, 1995.

30. A. Lozinski, C. Chauviere, *"A fast solver for Fokker-Planck equation applied to viscoelastic flows calculations: 2D FENE model"*, Journal of Computational Physics, 189, 607-625, 2003.

31. M. A. Martínez, E. Cueto, M. Doblaré, and F. Chinesta. Fixed mesh and meshfree techniques in the numerical simulation of injection processes involving short fiber suspensions. *Journal of Non-Newtonian Fluid Mechanics*, 115:51–78, 2003.
32. M. A. Martinez, E. Cueto, I. Alfaro, M. Doblare, and F. Chinesta. Updated Lagrangian free surface flow simulations with Natural Neighbour Galerkin methods. *International Journal for Numerical Methods in Engineering*, 60(13):2105–2129, 2004.
33. M.A. Martinez, E. Cueto, M. Doblare and F. Chinesta. A meshless simulation of injection processes involving short fibers molten composites. In A. M. Habraken, editor, *ESAFORM Conference on Material Forming*, Université de Liege, Belgium,2001.
34. N. Moes, J. Dolbow, and T. Belytschko. A Finite Element Method for Crack Growth without Remeshing. *International Journal for Numerical Methods in Engineering*, 46:131–150, 1999.
35. G. Legrain, N. Moës and E. Verron. Fracture with large deformation using X-FEM. In S. Støren, editor, *ESAFORM Conference on Material Forming*, Trondheim, Norway, 2004.
36. B. Nayroles, G. Touzot, and P. Villon. Generalizing the finite element method: Diffuse approximation and diffuse elements. *Computational Mechanics*, 10:307–318, 1992.
37. H.C. Öttinger, M. Laso, *"Smart polymers in finite element calculation"*, Int Congr. on Rheology, Brussel, Belguim, 1992.
38. J.A. Sethian. *Level Set Methods and Fast Marching Methods Evolving Interfaces in Computational Geometry, Fluid Mechanics, Computer Vision, and Materials Science*. Cambridge University Press, 1999.
39. R. Rentsch, *"Molecular dynamics simulation of microscopic phenomena in cutting and forming process"*, Proceedings of the 7th Esaform Conference on Material Forming, Trondheim, Norway, 93-96, 2004.
40. J. Rojek, F. Zarate, C. Agelet de Saracibar, M. Chiumenti, M. Cervera, P.M. Haigh, C. Gilbourne, P. Verdor, *"Numerical simulation of sand mould manufacture for lost foam casting process"*, Proceedings of the 6th Esaform Conference on Material Forming, Salerno, Italy, 655-658, 2003.
41. D. Ryckelynck, F. Chinesta, E. Cueto, A. Ammar, *"On the a priori model reduction: overview and recent developments"*, Archives of Computational Methods in Engineering, 13, 91-128, 2006.
42. D. Ryckelynck, *"A priori hyperreduction method: an adaptive approach"*, J. Computational Physics, 202, 346-366, 2005.
43. V.B. Shenoy, R. Miller, E.B. Tadmor, D. Rodney, R. Phillips, M. Ortiz, *"An adaptive finite element approach to atomic scale mechanics - the quasicontinuum methdos"*, Journal of the Mechanics and Physics of Solids, 47, 611-642, 1999.
44. M. Somasi, B. Khomami, N.J. Woo, J.S. Hur, E.S.G. Shaqfeh, *"Brownian dynamics simulations of bead-rod and bead-spring chains: numerical algorithms and coarse-graining issues"*, J. Non-Newtonian Fluid Mech., 108/1-3, 227-255, 2002.
45. T. Strouboulis, K. Copps, and I. Babuška. The generalized finite element method. *Computer Methods in Applied Mechanics and Engineering*, 190:4081–4193, 2001.
46. N. Sukumar, B. Moran, and T. Belytschko. The Natural Element Method in Solid Mechanics. *International Journal for Numerical Methods in Engineering*, 43(5):839–887, 1998.
47. N. Sukumar. Meshless methods and Partition of Unity Finite Elements. In V. Brucato, editor, *ESAFORM Conference on Material Forming*, Salerno University, Italy, 2003.

48. D. Chamoret, A. Rassineux, J.-M. Bergheau, P. Villon, Modelling of contact surface by local hermite diffuse interpolation. *Proceedings of ESAFORM 2001, The 4th international ESAFORM conference on material forming*, Liège, Belgique, 1, p. 179-182, 2001.

49. S. W. Xiong, C. S. Li, J. M. C. Rodrigues, and P. A. F. Martins. Steady and non-steady state analysis of bulk forming processes by the reproducing kernel particle method. *FINITE ELEMENTS IN ANALYSIS AND DESIGN*, pages 599–614, 2005.

50. J. Yvonnet, D. Ryckelynck, P. Lorong, and F. Chinesta. A new extension of the Natural Element method for non-convex and discontnuous problems: the Constrained Natural Element method. *International Journal for Numerical Methods in Enginering*, 60(8):1452–1474, 2004.

51. Cai Zheng and J. M. A. Cesar de Sa. A meshless method application on the localization phenomena of ductile damage in metal forming. In A. Rosochowsky, editor, *ESAFORM Conference on Material Forming*, Strathclyde University, Glasgow, UK, 2006.

52. J. Zhou, L. Li, and J. Duszczyk. 3D FEM simulation of the whole cycle of aluminium extrusion throughout the transient state and the steady state using the updated Lagrangian approach. *Journal of Materials Processing Technology*, 134:383–397, 2003.

# Flow-Induced Crystallization in Polymer Processing

Jean-Marc Haudin

CEMEF (Center for Material Forming)
Ecole des Mines de Paris and UMR CNRS 7635, BP 207, F06904 Sophia Antipolis Cedex
Jean-Marc.Haudin@ensmp.fr

**Abstract.** The paper first presents the experimental set-ups designed to produce flow-induced crystallization and the techniques used to analyze structure development. Then, the general effects of flow are discussed in terms of crystallization thermodynamics (nucleation, growth, overall kinetics), with the consequences on morphologies (from spherulites to shish-kebab morphologies). Finally, some open questions are presented, which shows that many things are not yet explained, especially the nucleation event.

**Keywords:** Polymer Crystallization, Nucleation, Growth, Overall Kinetics, Morphologies.

## 1 Introduction

Structure development is more and more a key issue in polymer processing, with a view to master the final properties of the products (mechanical, optical, etc.). Among the phenomena involved, crystallization plays a major role. It generally occurs under complex, inhomogeneous and coupled mechanical (flow, pressure), thermal (cooling rates, temperature gradients) and geometrical (contact with processing tools) conditions. According to the process, it takes place after cessation of flow (cast-film extrusion) or under flow (film blowing). If the conditions of crystallization are not homogeneous, both may occur as a function of the location in the part (injection molding). The type of flow may be shear, elongation or mixed.

Many papers have been devoted to the relationships between processing conditions, microstructure and properties. In parallel, many laboratory experiments have been designed to capture the specific influence of flow under well-defined conditions. Most of them concern shear. Conversely, little work has been dedicated to elongation, mainly because of experimental difficulties. In a first step, the devices used to produce flow-induced crystallization will be described together with the techniques used to follow structure development. Then, the state of art will be presented, i.e., what is currently known on the effects of flow on crystallization. This knowledge is based both on studies of processes and on laboratory experiments. Finally, some open questions will be addressed, which leads to the conclusion that a lot of things remain unknown or at least unexplained.

## 2 Experimental

Keller reports in [1] that in his first experiments, he melted an isotropic polyethylene film above a Bunsen flame and stretched it by hand. Since this pioneering work, whose major results are still considered as a reference [1,2], many devices have been proposed to control both flow and temperature. Most of them concern shear.

A shear flow can be produced in commercial rheometers with parallel-plate [3-5] or cone-and-plate [6,7] geometry. Commercial rheometers generally ensure a good mechanical but a rather poor thermal control. In fact, in most cases, rheometers have been modified or equivalent set-ups have been built with different geometries: parallel disks [5,7-16] including the very popular Linkam CSS 450 shearing hot stage (Cambridge Optical Shearing Device) [5,7,11-16], Couette [8,17,18], biconical [19,20].

A simple-shear flow is easily generated by the translation of a plate with respect to another parallel to it [16,21-25]. This can be achieved in a modified hot stage [16,23,24], which ensures a good temperature control, but only a limited strain can be applied. More recently, this type of flow has been used in a sandwich construction to impose very high shear rates (up to 1000 $s^{-1}$) during a short time [25].

Many authors have used the fiber pull-out device [26-32]. The fiber, usually a glass fiber, does not induce surface crystallization under quiescent conditions. It is moved along its axis in the polymer melt, at constant velocity. Due to shear, a layer of solidified polymer forms at the fiber surface.

Most of the above-mentioned set-ups generally apply a small or moderate shear rate during a rather long time at low crystallization temperatures. On the contrary, Janeschitz-Kriegl and his co-workers proposed to impose a high shear rate during a short time at high temperatures. The objective was to separate nucleation and growth, and also to be closer to industrial conditions. In their system [33], a box-like pressure-time profile creates a Poiseuille flow in a duct of rectangular section and large aspect ratio. The same type of apparatus has been by Kumaraswamy et al. [34-36] and adapted to small quantities of polymer. Note that the same concept of short-term shearing is applied in [25].

Compared to shear, less work has been devoted to crystallization from the melt under elongation, mainly because of experimental difficulties. Among the systems used to produce an elongational flow, let us mention:
- the cross-slot flow cell of Eindhoven University [37];
- the modified Minimat tensile machine (PolymerLab) of CEMEF [38];
- the Windbix extensional rheometer of Linz University [39,40].

Structure development has been followed by numerous experimental techniques, many authors combining several of them: rheological measurements, optical microscopy, optical measurements (birefringence, dichroism, turbidity, depolarized light intensity (DLI), small-angle light scattering (SALS), small-angle X-ray scattering (SAXS), wide-angle X-ray diffraction (WAXD), etc.

# 3  General Effects of Flow

Thermoplastics are composed of flexible macromolecules, which adopt a random-coil conformation at rest, and can be easily deformed by flow. Under flow, they also tend to be oriented along the flow direction. Stretching and orientation of macromolecules in the melt has consequences on crystallization thermodynamics and kinetics, as well as on the subsequent morphologies.

## 3.1 Thermodynamics and Kinetics

First of all, crystallization and melting are first-order *thermodynamical transitions*, characterized by an equilibrium melting temperature. Then, the crystallization of a polymer consists of two stages: (i) a *nucleation stage*, i.e., the formation, within the liquid phase, of entities, called active nuclei, from which crystals can appear and (ii) a *growth stage*, i.e.; the development of nuclei into observable crystals. Crystallization can also be considered in an *overall approach* integrating nucleation and growth mechanisms as well as growth geometry. In this approach, the transformed volume fraction is described by a function which varies between 0 and 1.

Flow increases the equilibrium melting temperature and enhances all the kinetics (nucleation, growth, overall kinetics). These major results will be illustrated by some comments and selected examples.

The increase of the equilibrium melting temperature can be explained by a simple model. Under quiescent conditions, the equilibrium melting temperature is defined by:

$$T_0 = \frac{\Delta h}{\Delta s}, \tag{1}$$

where $\Delta h$ and $\Delta s$ are the enthalpy and the entropy of fusion per unit volume, respectively. Due to the reduction of the number of the conformations, flow mainly affects the entropy of fusion, which decreases. The equilibrium melting temperature becomes:

$$T_0' = \frac{\Delta h}{\Delta s'} > T_0, \tag{2}$$

since $\Delta s' < \Delta s$. It means that under isothermal conditions, the undercooling is increased and that during cooling, crystallization occurs at higher temperature, which is observed experimentally. This could simply explain why the nucleation and growth kinetics are enhanced. Nevertheless, the melting point elevation theory cannot predict the rheological behavior of the crystallizing system, due to complex coupling effects. More sophisticated thermodynamic approaches are necessary (e.g., McHugh et al. [41]).

A large number of experimental studies (e.g., [24]) have shown that flow increases both the total number of activated nuclei (Fig. 1) and the nucleation rate. Concerning the effect of flow on growth rate, there is still some debate. However, Monasse [23] has clearly shown that growth is enhanced in the three directions of space: along the shear-flow direction, but much more perpendicularly to it (Fig. 2). This anisotropic growth has a direct consequence on morphologies (see 3.2). Furthermore, the flow effect on growth rate is greater when molecular increases [30-32].

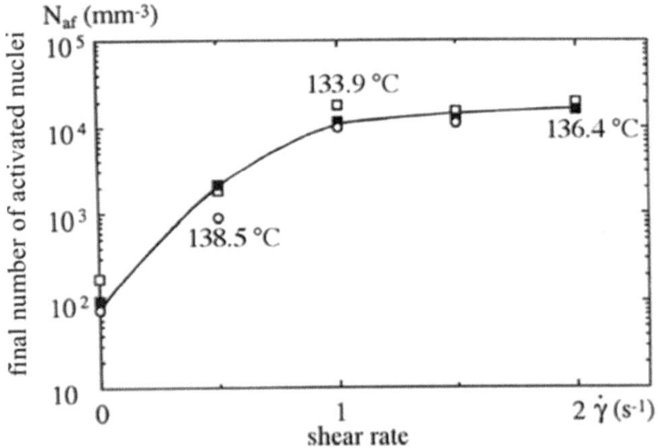


**Fig. 1.** Final number of activated nuclei vs. shear rate. Isotactic polypropylene (iPP) sheared at different temperatures [24].

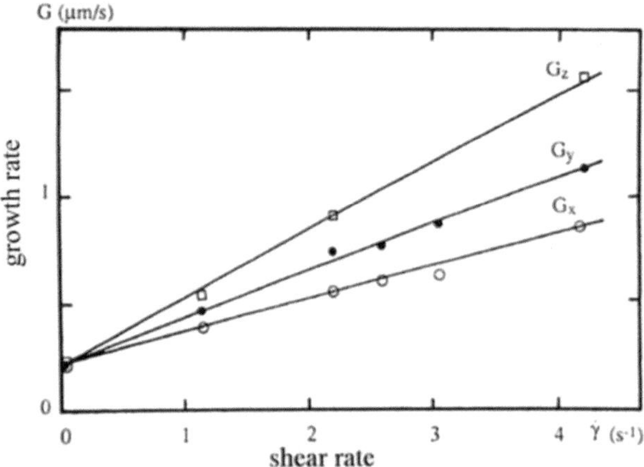


**Fig. 2.** Growth rates of polyethylene vs. shear rate. (x) longitudinal (flow), (y) transverse and (z) thickness direction [23].

The enhancement of overall kinetics can be characterized:
- by the decrease of characteristics times defined experimentally: induction time [17,22], half-transformation time [7,15];
- by the evolution of the transformed volume fraction $\alpha(t)$. From DLI measurements, Devaux et al. [16] define $\alpha$ as the ratio between the

transmitted intensity $I(t)$, corrected for the intensity prior to crystallization $I(0)$, to the maximum intensity observed at the end of crystallization $I_\infty$, also corrected for $I(0)$:

$$\alpha(t) = \frac{I(t) - I(0)}{I_\infty - I(0)} \, .$$  (3)

Fig. 3 demonstrates the acceleration of the crystallization kinetics with increasing shear rate.

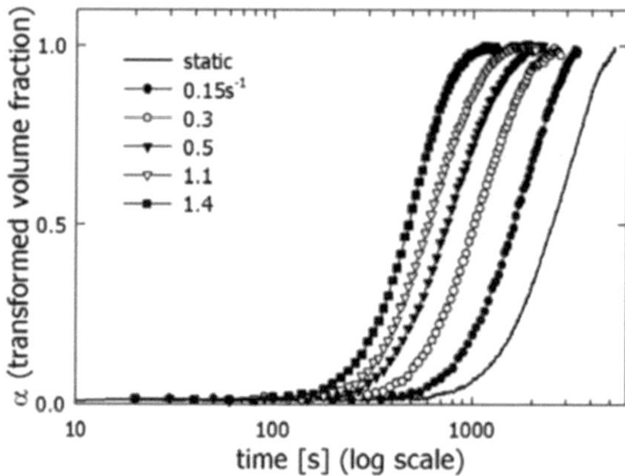


**Fig. 3.** Evolution of the transformed volume fraction with shear rate (s⁻¹); iPP at 136 °C [16].

## 3.2 Morphologies

The effects of flow on the development of morphologies are to:
- decrease the size of the crystalline entities, which results from the increase of the number of nuclei (see Fig. 1);
- arrange nuclei in lines parallel to the flow direction (row-nucleation). Simultaneously, the shape of the nuclei changes: from point-like to thread like precursors;
- induce an anisotropic growth of lamellar crystals. As growth proceeds by the deposit of molecular stems on the growth surface, orientation of macromolecules in the melt favors growth in the direction perpendicular to flow (see Fig. 2).

Taking into account these effects, the morphologies which can be expected by increasing flow intensity are spherulites (Fig. 4a), spherulites deformed into ellipsoids (Fig. 4b), sheaf-like morphologies (Fig. 4c), and row-nucleated morphologies or cylindrites [1,2]. At low stress (Fig. 4d), the columnar morphology can be described as a stack of thin spherulitic slices consisting of twisted radial crystallites. Under higher stress, crystalline lamellae are flat (Fig. 4e). It is now admitted that the nucleating line forming in the melt can be identified with the fibrous, partly extended-chain backbone of the solution-crystallized shish-kebabs [42]. The latter morphologies consist of a central thread that is covered with folded-chain platelets at regular intervals [43]. At very high stress, smooth microfibrils are obtained (Fig. 4f).

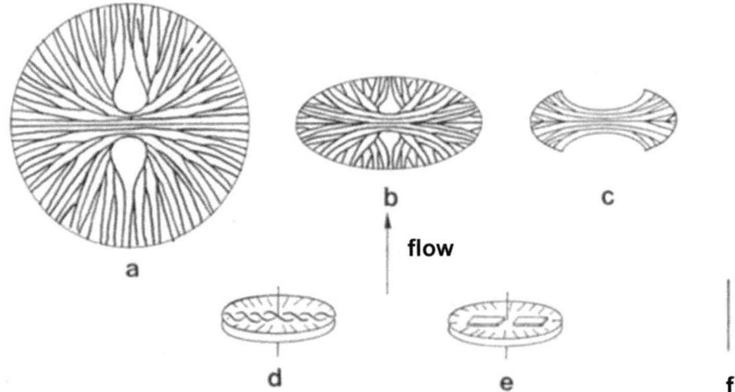


**Fig. 4.** Morphological change with increasing flow intensity: (a) spherulite; (b) ellipsoidal spherulite; (c) sheaf-like texture; (d) and (e) row-nucleated morphologies; (f) smooth microfibrils.

Spherulites are usually observed at the core of injected parts. Ellipsoidal spherulites (Fig. 4b) are observed in 300 µm-thick polypropylene films prepared by cast-film extrusion [44]. Sheaf-like textures (Fig. 4c) and row-nucleated morphologies (Figs 4d and 4e) are usually observed in polyethylene blown films [45]. Polyethylene, spinning at take-up velocities lower than 600m/min also leads to cylindritic morphologies (Figs 4d and 4e) [46]. In industrial spinning processes, the morphology is microfibrillar (Fig. 4f) and the basic crystalline entity is a parallelepipedic extended-chain crystallite, whose dimension along the flow direction increases with take-up velocity. Smooth polyethylene microfibrils have been also obtained from flowing solutions [47]. Dark-field electron microscopy has shown that the crystallites within the microfibrils had lateral dimensions of 15 nm and that their lengths might amount to over 100 nm.

## 4   Some open problems

### 4.1 What are the Relevant Parameters?

The experiments with the shear devices described in 2 are performed in isothermal conditions. Two parameters are generally varied: the shear rate $\dot{\gamma}$ (see Fig. 3) and the shearing time $t_s$. The shear strain $\dot{\gamma}\,t_s$ can also be considered as an operating parameter. It appears that the crystallization kinetics is more sensitive to $\dot{\gamma}$ than to $t_s$. For Liedauer et al. [33] a measure for the intensity of shearing is $\dot{\gamma}^4 t_s^2$ or $\dot{\gamma}^2 t_s$ according to the type of nucleation. In our experiments on polypropylene [16], the strong acceleration of the crystallization kinetics has been quantified using the Avrami equation [48]:

$$\alpha(t) = 1 - \exp[-kt^n] , \tag{4}$$

with $n$ equal to 2.5 for quiescent as well as for shear-induced crystallizations at 136°C. Kinetics is characterized by the rate constant $k$, with the following dependencies on shear rate and shearing time:

$$k(\dot{\gamma}, t_s) \propto \dot{\gamma}^{1.22} t_s^{0.77} , \tag{5}$$

obtained by varying $\dot{\gamma}$ at constant $t_s$=30s (Fig. 3) and varying $t_s$ at constant $\dot{\gamma}$=0.4 s$^{-1}$. In fact, the effects of shear rate and shearing time are coupled, which leads to a more complicated expression:

$$k(\dot{\gamma}, t_s) \propto \dot{\gamma}^{f(t_s)} t_s^{g(\dot{\gamma})} . \tag{6}$$

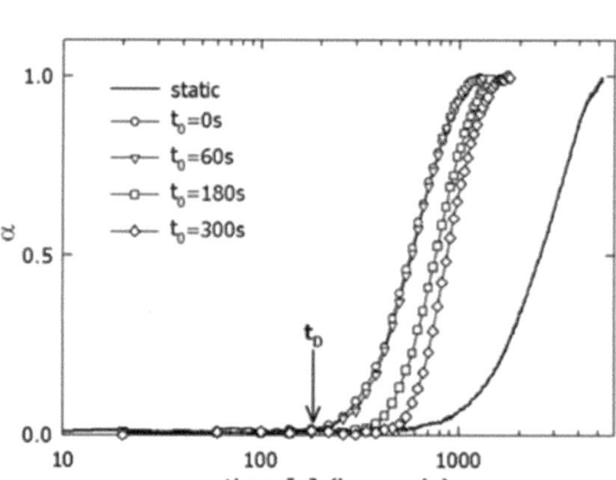


**Fig. 5.** Effect of the instant of the start of shear flow on the evolution of $\alpha$ using a fixed shear rate of 1s$^{-1}$ and a constant shearing time of 30 s [16].

Another important parameter is $t_0$, the time elapsed between the instant at which the isothermal conditions are reached and the inception of the shear flow. Its role is generally ignored in the literature. It has been shown [16] that this dwell time has no influence on the overall kinetics as long it remains smaller than a critical value $t_D$ (Fig. 5). Beyond $t_D$, an enhancement of the kinetics is still observed, but the onset of crystallization is delayed.

The kinematical parameters discussed above do not take into account the effect of temperature, nor the influence of the molecular architecture, which conditions the polymer rheology. They can be introduced in different ways: shear stress, dimensionless numbers based on rheological parameters, invariants of mechanical tensors, etc. For instance, Acierno et al. [15] use a Weissenberg (or Deborah) number, which is the product of the shear rate by the longest relaxation time. This parameter is used to plot master curves for flow-induced overall kinetics and to characterize the morphological transition from spherulites to shish-kebabs.

The most elegant proposal, emanating from Janeschitz-Kriegl and co-workers (e.g., [49]), seems to use the specific work of deformation, defined as:

$$W = \int_0^t \boldsymbol{\sigma}(u) : \mathbf{D}(u)\,du \, , \qquad (7)$$

where $\boldsymbol{\sigma}$ and $\mathbf{D}$ denote the stress and strain-rate tensors, respectively. It makes it possible to compare results obtained:
-   for different shear histories. If the total specific work is the same, a forceful shearing for a short time is equivalent to less forceful shearing for a longer period;
-   in shear and elongation conditions. A common statement is that flow is more efficient than elongation. It does seem to be true if the results are plotted in terms of specific work.


## 4.2 Early Stages of Crystallization

As recalled in 3.1, static polymer crystallization from the melt is supposed to proceed through nucleation, mainly heterogeneous, and growth. The classical nucleation theory has been adapted by Binsbergen [50] to the heterogeneous case. When the size of the nucleus exceeds a critical value, the growth of the crystalline phase becomes energetically favorable. However, these critical sizes have not yet been measured for polymers. Small-Angle X-ray Scattering (SAXS) and Wide-Angle X-ray Diffraction (WAXD) even seem to reveal a more complex behavior (long-range order in the melt before crystal growth), which has been interpreted in terms of spinodal decomposition [51], although other researchers disagree on this point [52]. Okada et al. used Small-Angle Light Scattering (SALS) on polypropylene samples. Their experiments show that large-scale density fluctuations develop in the melt and then relax before a crystalline structure appears [53]. Pogodina et al. [54] extended their work and combined the SALS experiments with rheological measurements. Based on the evolution of the linear viscolelastic moduli with time and frequency, they suggested that during the early stages of crystallization a gel-like structure develops, considerably before crystalline structures appear. SALS gives correlation lengths exceeding those from SAXS by two orders of magnitude, i.e., 1 µm as opposed to 10 nm [51].

Similar SALS results have been obtained by Devaux et al. [16]. These authors also used the two SALS invariants:

$$Q_\eta = \int_0^\infty (I_{VV} - \frac{4}{3} I_{HV}) q^2 dq ,$$

(8)

$$Q_\delta = \int_0^\infty I_{HV} q^2 dq ,$$

(9)

where $I_{VV}$ and $I_{HV}$ are the scattered intensities for the VV (parallel polarizers) and HV (crossed polarizers) geometries, respectively. Under quiescent conditions and in agreement with previous work [53], density fluctuations ($Q_\eta$) are detected at the very early stages and reach a significant level before orientation fluctuations ($Q_\delta$), associated with crystallization, occur. Both density and orientation fluctuations are observed to occur earlier when the crystallization temperature is decreased or when a shear flow is applied (Fig. 6).

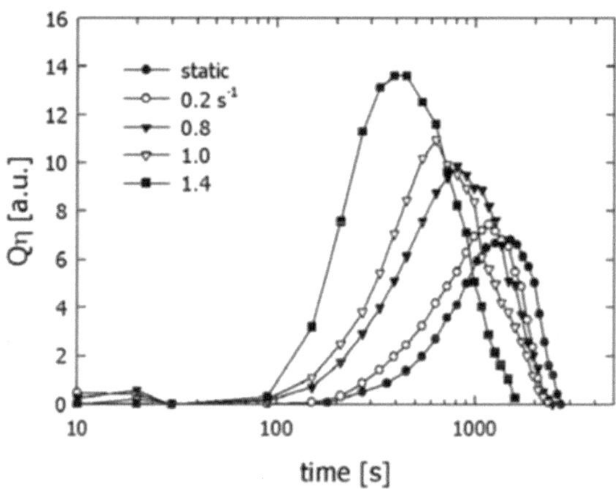


**Fig. 6.** Influence of shear rate on density fluctuations [16].

As long as flow is applied before the critical time $t_D$ introduced in 4.1, no effect of $t_0$ is observed on the evolution of the scattering invariants (Fig. 7); $t_D$ corresponds to the point where the density invariant $Q_\eta$ starts to increase.

A complementary study by optical microscopy [55] has shown that the time at which spherulites appear under quiescent conditions is close to $t_D$. At some crystallization temperatures, spherulites may have appeared before the onset of density fluctuations revealed by SALS. Then, these density fluctuations could be also related to the covering of space by spherulites, as suggested recently by Janeschitz-Kriegl from an independent calculation [49]. Therefore, a clear picture of nucleation is still lacking, even in the case of static crystallization.

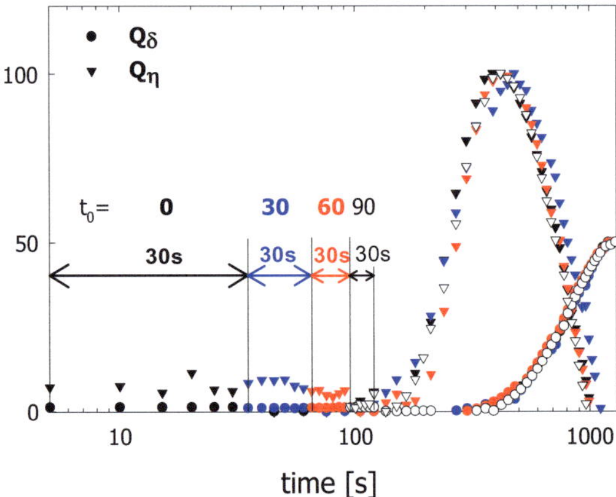


**Fig. 7.** SALS invariants as a function of time for the same shear strain and different $t_0$ [16].

## 5  Conclusions

There exist a huge number of experimental results concerning flow-induced crystallization from the melt under shear conditions. Most of them concern polypropylene. Nevertheless, it is not easy to compare them due to the use of different polymers, experimental conditions and methods of investigation. Comparison of elongation and shear remains an open problem. The concept of specific work of deformation could provide a consistent framework for such evaluations.

The basic mechanisms of nucleation remain unknown, both in quiescent and flow conditions. As an interpretation of their results, many authors invoke a pre-nucleation stage: "aggregates" or "precursors" which become active nuclei by the flow treatment. The "aggregates" or "dormant nuclei" could be at some junctions of the network detected. In the same way, the enhancement of growth by flow is still unexplained.

However, this type of study is of prime importance for deriving nucleation laws to be introduced in simulation models for polymer processing. Avrami's theory [56] gives a good general framework for the treatment of overall crystallization kinetics. The equations can be cast into a system of differential equations, more suitable for numerical integration. Flow introduces additional nuclei. Two types of morphologies can be considered spherulites (point-like nuclei) or shish-kebabs (or thread like nuclei). The number density of flow-induced nuclei is generally given by the following type of equation:

$$\frac{dN}{dt} + \frac{1}{\lambda} N = f \, , \tag{10}$$

where $\lambda$ is a relaxation time, and $f$ a function which describes the effect of flow. The objective is to give a physically-based expression for $f$. Then, such phenomenological laws can be implemented in finite element simulation codes like the ones developed by CEMEF.

# References


1. Keller, A., Machin, M.J.: Oriented Crystallization in Polymers. J. Macromol. Sci. (Phys.) B1(1) (1967) 41-91
2. Hill, M.J., Keller, A.: Direct Evidence for Distinctive, Stress-Induced Nucleus Crystals in the Crystallization of Oriented Polymer Melts. J. Macromol. Sci. (Phys.) B3(1) (1969) 153-159
3. Masubuchi, Y., Watanabe, K., Nagatake, W., Takimoto, J.I., Koyama, K.: Thermal Analysis of Shear Induced Crystallization by the Shear Flow Thermal Rheometer: Isothermal Crystallization of Polypropylene. Polymer 42 (2001) 5023- 5027
4. Nieh, J.Y., Lee, L.J.: Hot Plate Welding of Polypropylene. Part 1: Crystallization Kinetics. Polym. Eng. Sci. 38 (1998) 1121-1132
5. Pogodina, N.V., Lavrenko, V.P., Srinivas, S., Winter, H.H.: Rheology and Structure of Isotactic Polypropylene near the Gel Point: Quiescent and Shear-Induced Crystallization. Polymer 42 (2001) 9031-9043
6. Vleeshouwers, S., Meijer, H.E.H.: A Rheological Study of Shear Induced Crystallization. Rheol. Acta 35 (1996) 391-399
7. Koscher, E., Fulchiron, R.: Influence of Shear on Polypropylene Crystallization: Morphology Development and Kinetics. Polymer 43 (2002) 6931-6942
8. Fritzsche, A.K., Price, F.P., Ulrich, R.D.: Disruptive Processes in the Shear Crystallization of Poly(ethylene oxide). Polym. Eng. Sci. 16 (1976) 182-188
9. Ulrich, R.D., Price, F.P.: Morphology Development during Shearing of Poly(ethylene oxide) Melts. J. Appl. Polym. 20 (1976) 1077-1093
10. Wolkowicz, M.D.: Nucleation and Crystal Growth in Sheared Poly(1-butene) Melts. J. Polym. Sci. Polym. Symp. 63 (1978) 365-382
11. Pople, J.A., Mitchell, G.R., Sutton, S.J., Vaughan, A.S., Chai, C.K.: The Development of Organized Structures in Polyethylene Crystallized from a Sheared Melt, Analyzed by WAXS and TEM. Polymer 40 (1999) 2769-2777
12. Floudas, G. Hilliou, L., Lellinger, D., Alig, I.: Shear-Induced Crystallization of Poly($\varepsilon$-caprolactone). 2. Evolution of Birefringence and Dichroism. Macromolecules 33 (2000) 6466-6472
13. Somani, R.H., Hsiao, B.S., Nogale, A., Srinavas, S., Tsou, A.H., Sics, I., Balta-Calleja, F.J., Ezquerra, T.A.: Structure Development during Shear Flow-Induced Crystallization of iPP: in-Situ Small-Angle X-Ray Scattering Study. Macromolecules 33 (2000) 9385-9394
14. Somani, R.H., Yang, L.Y., Hsiao, B.S.: Precursors of Primary Nucleation Induced by Flow in Isotactic Polypropylene. Physica A 304 (2002) 145-157
15. Acierno, S., Palomba, B., Winter, H.H., Grizzuti, N.: Effect of Molecular Weight on the Flow-Induced Crystallization of Isotactic Poly(1-butene). Rheol. Acta 42 (2003) 243-250
16. Devaux, N., Monasse, B., Haudin, J.M., Moldenaers, P., Vermant, J.: Rheooptical Study of the Early Stages of Flow Enhanced Crystallization in Isotactic Polypropylene. Rheol. Acta 43 (2004) 212-222

17. Sherwood, C.H., Price, F.P., Stein, R.S.: Effect of Shear on the Crystallization Kinetics of Poly(ethylene oxide) and Poly($\varepsilon$-caprolactone) Melts. J. Polym. Sci. Polym. Symp. 63 (1978) 77-94

18. Andersen, P.G., Carr, S.H.: Crystal Nucleation in Sheared Polymer Melts. Polym. Eng. Sci. 18 (1978) 215-221

19. Wereta, A., Gogos, C.G.: Crystallization Studies on Deformed Polybutene-1 Melts. Polym. Eng. Sci. 11 (1971) 19-27

20. Tan, V., Gogos, C.G: Flow-Induced Crystallization of Linear Polyethylene above its Normal Melting Point. Polym. Eng. Sci. 16 (1976) 512-525

21. Haas, T.W., Maxwell, B.: Effects of Shear Stress on the Crystallization of Linear Polyethylene and Polybutene-1. Polym. Eng. Sci. 9 (1969) 225-241

22. Lagasse, R.R., Maxwell, B.: An Experimental Study of the Kinetics of Polymer Crystallization during Shear Flow. Polym. Eng. Sci. 16 (1976) 189-199

23. Monasse, B.: Nucleation and Anisotropic Crystalline Growth of Polyethylene under Shear. J. Mater. Sci. 30 (1995) 5002-5012

24. Tribout, C., Monasse, B., Haudin, J.M.: Experimental Study of Shear-Induced Crystallization of an Impact Polypropylene Copolymer. Colloid Polym. Sci. 274 (1996) 197-208

25. Janeschitz-Kriegl, H., Ratajski, E., Stadlbauer, M.: Flow as an Effective Promotor of Nucleation in Polymer Melts: a Quantitative Evaluation. Rheol. Acta 42 (2003) 355-364

26. Devaux, E., Chabert, B.: Nature and Origin of the Transcrystalline Interphase of Polypropylene/Glass Fibre Composites after a Shear Stress. Polym. Comm. 32 (1991) 464-468

27. Monasse, B.: Polypropylene Nucleation on a Glass Fibre after Melt Shearing. J. Mater. Sci. 27 (1992) 6047-6052

28. Varga, J., Karger-Kocsis, J.: Rules of Supermolecular Structure Formation in Sheared Isotactic Polypropylene Melts. J. Polym. Sci. Polym. Phys. 34 (1996) 657-670

29. Wang, C., Lin, C.R.: Transcrystallization of PTFE Fiber/PP Composites. III. Effect of Fiber Pulling on the Crystallization Kinetics. J. Polym. Sci. Polym. Phys. 36 (1998) 1361-1370

30. Jay, F., Haudin, J.M., Monasse, B.: Shear-Induced Crystallization of Polypropylene: Effect of Molecular Weight. J. Mater. Sci. 34 (1999) 2089-2102

31. Duplay, C., Monasse, B., Haudin, J.M., Costa, J.L.: Shear-Induced Crystallization of Polypropylene: Effect of Molecular Weight. J. Mater. Sci. 35 (2000) 6093-6103

32. Haudin, J.M., Duplay, C., Monasse, B., Costa, J.L.: Shear-Induced Crystallization of Polypropylene. Growth Enhancement and Rheology in the Crystallization Range. Macromol. Symp. 185 (2002) 119-133

33. Liedauer, S., Eder, G., Janeschitz-Kriegl, H., Jerschow, P., Gemayer, W., Ingolic, E.: On the Kinetics of Shear Induced Crystallization in Polypropylene. Intern. Polym. Proc. 8 (1993) 236-244

34. Kumuraswamy, G., Verma, R.K., Kornfield, J.A.: Novel Apparatus for Investigating Shear-Enhanced Crystallization and Structure Development in Semicrystalline Polymers. Rev. Sci. Instrum. 70 (1999) 2097-2104

35. Kumuraswamy, G., Issaian, A.M., Kornfield, J.A.: Shear-Enhanced Crystallization in Isotactic Polypropylene. 1. Correspondence between in Situ Rheo-optics and ex Situ Structure Determination. Macromolecules 32 (1999) 7537-7547

36. Kumuraswamy, G., Verma, R.K., Issaian, A.M., Wang, P., Kornfield, J.A., Yeh, F., Hsiao, B.S., Olley, R.H.: Shear-Enhanced Crystallization in Isotactic Polypropylene. 2. Analysis of the Formation of the Oriented "Skin". Polymer 41 (2000) 8931-8940

37. Swartjes, F.H.M.: Stress Induced Crystallization in Elongational Flow, Thesis, Technical University of Eindhoven (2001)

38. Monasse, B.: Crystallization under Elongation of a Polypropylene, 6th ESAFORM Conference on Material Forming, Salerno, Italy (2003)

39. Stadlbauer, M., Janeschitz-Kriegl, H., Lipp, M., Eder, G., Forstner R.: Extensional Rheometer for Creep Flow at High Tensile Stress. Part I. Description and Validation. J. Rheol. 48 (2004) 611-629

40. Stadlbauer, M., Janeschitz-Kriegl, H., Eder, G., Ratajski, E.: Extensional Rheometer for Creep Flow at High Tensile Stress. Part II. Flow Induced Nucleation for the Crystallization of iPP. J. Rheol. 48 (2004) 631-639

41. Doufas, A.K., Dairanieh, I.S., McHugh, A.J.: A Continuum Model for Flow-Induced Crystallization of Polymer Melts. J. Rheol. 43 (1999) 85-109

42. Keller, A., Kolnaar, J.W.H.: Chain Extension and Orientation: Fundamentals and Relevance to Processing and Products. Prog. Colloid Polym. Sci. 92 (1993) 81-102

43. Pennings, A.J.: Bundle-like Nucleation and Longitudinal Growth of Fibrillar Polymer Crystals from Flowing Solutions. J. Polym. Sci. Polym. Symp. 59 (1977) 55- 86

44. Duffo, P., Monasse, B., Haudin, J.M.: Cast Film Extrusion of Polypropylene. Thermomechanical and Physical Aspects. J. Polym. Eng. 10 (1991) 151-229

45. Haudin, J.M., Piana, A., Monasse, B., Monge, G., Gourdon, B.: Etude des Relations entre Mise en Forme, Orientation et Rétraction dans des Films de Polyéthylène Basse Densité Réalisés par Soufflage de Gaine. II. Orientation de la Phase Cristalline. Ann. Chim. Sci. Mat. 24 (1999) 555-580

46. White, J.L., Cakmak, M.: Orientation Development and Crystallization in Melt Spinning of Fibers. Adv. Polym. Technol. 6 (1986) 295-338

47. Zwijnenburg, A., van Hutten, P.F., Pennings, A.J., Chanzy, H.D.: Longitudinal Growth of Polymer Crystals from Flowing Solutions. V. Structure and Morphology of Fibrillar Polyethylene Crystals. Kolloid Z.Z. Polym. 256 (1978) 729-740

48. Avrami, M.: Kinetics of Phase Change. II. Transformation-Time Relations for Random Distributions of Nuclei. J. Chem. Phys. 8 (1940) 212-224

49. Janeschitz-Kriegl, H.: Phases of Flow-Induced Crystallization of i-PP: How Remote Pieces of the Puzzle Appear to Fit. Macromolecules 39 (2006) 4448-4454

50. Binsbergen, F.L.: Heterogeneous Nucleation in Crystallization of Polyolefins. III. Theory and Mechanism. J. Polym. Sci. Polym. Phys. 11 (1973) 117-135

51. Terrill, N.J., Fairclough, P.A., Towns-Andrews, E., Komanshek, B.U., Young, R.J., Ryan, A.J.: Density Fluctuations: the Nucleation Event in Isotactic Polypropylene Crystallization. Polymer 39 (1998) 2381-2385

52. Wang, Z.G., Hsiao, B.S., Sirota, E.B., Srinivas, S.: A Simultaneous Small- and Wide-Angle X-ray Scattering Study of the Early Stages of Melt Crystallization in Polyethylene. Polymer 41 (2000) 8825-8832

53. Okada, T., Saito, H., Inoue, T.: Time Resolved Light-Scattering Studies on the Early Stages of Crystallization in Isotactic Polypropylene. Macromolecules 25 (1992) 1908-1911

54. Pogodina, N.V., Siddique, S.K., van Egmond, J.W., Winter, H.H.: Correlation of Rheology and Light Scattering in Isotatic Polypropylene during Early Stages of Crystallization. Macromolecules 32 (1999) 1167-1174

55. Devaux, N.: Influence d'un Cisaillement sur les Premiers Stades de la Cristallisation du Polypropylène. Thesis , Ecole des Mines de Paris (2003)

56. Haudin, J.M., Chenot, J.L.: Numerical and Physical Modelling of Polymer Crystallization. Part I. Theoretical and Numerical Analysis. Intern. Polym. Proc. 19 (2004) 267-274

# An Overview of Polymer Processing Modelling

Jean-François Agassant [1] and José António Covas [2]

[1] Ecole des Mines de Paris, CEMEF, UMR CNRS 7635, BP 207, 06560 Sophia Antipolis Cedex, France - jean-francois.agassant@ensmp.fr

[2] Dept. Polymer Engineering, University of Minho, 4800-058 Guimarães, Portugal jcovas@dep.uminho.pt

**Abstract**

Polymer processing modelling offers new opportunities for die or mould optimization.

We first present a short state of the art of polymer processing technology, followed by a discussion on how to develop reasonable processing models, some illustrations being given. We then deal with the question of constitutive equations in polymer processing modelling, both for classical thermoplastics and for more sophisticated (fibre reinforced) polymer systems. Finally, we revise flow instabilities that limit processability both in confined flows and in free surface extensional flows.

**Key Words:** Polymer Processing, Rheology, Modelling, Instability

## Introduction

Until the late eighties, the development of new polymers (or polymer blends) was predominantly driven by target end-use properties. Polymer producers and compounders assumed that "processability would follow up", i.e., that in order to process the new materials only small trial and error adjustments would be needed to existing processing conditions. In fact, they often discovered that some of these new materials were very difficult to process, as they exhibited extrusion / drawing instabilities and high torque or pressure requirements. This is, for example, the story of the development of new metallocene Polyethylene based compounds.

Currently, polymer producers account for processability at an early development stage of their products. To assist such studies, a variety of small-scale processing devices (needing a few grams of material) is now available in addition to the more conventional laboratory processing equipment. However, one conjectures whether processing with small/lab scale machines can be representative of industrial scale processing. While this remains an open issue, process modelling is now well accepted as a valuable engineering tool, particularly if it complies with the following pre-requisites:

- Provide an appropriate process depiction in terms of geometry, kinematic and heat transfer boundary conditions, solid/liquid transition, etc;
- Use accurate, but reasonable, constitutive equations to describe rheology both in the molten and in the solid state and, which is more difficult, in the

liquid-solid transition temperature range. The term *accurate* refers to a constitutive equation that is able to capture the main response features, while *reasonable* means that the parameters of these constitutive equations can be determined from a limited set of well defined experiments;

- Ensure a user friendly interface and yield representative results with known precision after a reasonable computation time.

What will the future be? Presently, models can be used successfully to forecast the processability of a given polymeric system in most processing tools. Attempts have been made to predict the morphology of the finished part (two-phase morphology, macromolecular/fibre orientation, crystallization rate, size of the spherulites, etc) from knowledge of the thermomechanical process parameters (temperature and temperature gradient, strain and stress rates along the flow path until solidification), but the connection between structural parameters and end-use properties (not only tensile or impact resistance, but also long term properties such as ageing and fatigue) remains an open research field, especially for semi-crystalline polymers. A continuous chain of models (also known as multi-scale modelling) relating structure, rheology, process, morphology and final performance (e.g., mechanical, optical, barrier properties) needs to be put together. Although a dream today, in future such a tool will allow the manufacture of a specific polymer architecture (e.g., molecular weight and molecular weight distribution, degree of chain branching) for target product properties and optimized processability.

In the following sections, we will first present a short state of the art of polymer processing technology, followed by a discussion on how to develop reasonable processing models, some illustrations being given. Then, we will deal with the question of constitutive equations in polymer processing modelling, both for classical thermoplastics and for more sophisticated (fibre reinforced) polymer systems. Finally, we will revise the flow instabilities that limit processability both in confined flows (die extrusion, injection moulding) and in free surface extensional flows (fibre spinning and cast film).

## Processing technology

Most of the existing industrial polymer forming processes (single and twin screw extrusion, injection moulding, calendering, film blowing, blow moulding) emerged after the second world war, simultaneously with the development of the most important commodity thermoplastics. Since then one has observed a continuous but spectacular stream of technological development, which has allowed, for example, doubling the output rate of extruders every five to ten years. Some examples include:

- Single screw extruders: invented by Maillefer [1], barrier screws separate the solid bed from the liquid pool during plastication, enhancing melting rate and process stability; mixing sections improve distributive / dispersive mixing; grooved barrels in the feeding zone induce more efficient solid drag and pressure generation; two stage screws allow for efficient devolatilization;
- Twin screw extruders: counter-rotating solutions became progressively more efficient to process temperature sensitive materials such as PVC dry-blends,

while the excellent mixing capabilities of co-rotating machines (complemented by an array of devices, such as left handed elements, staggered kneading blocks, mixing disks) made them suitable for compounding and reactive extrusion (where the machine becomes simultaneously a chemical reactor and an extrusion device);

- Extrusion lines: for example, in the case of tubular film blowing, very sophisticated air rings and internal bubble cooling ensure uniform film characteristics; double-bubble technology enables better controlled biaxial orientation;
- Injection moulding: production cells with increasing reproducibility and degree of automation, development of variations to the basic technique (yielding parts with innovative geometrical/material structure features) and more sophisticated mould technology (e.g., hot runners, quick mould change systems);
- Thermoforming: variations to the conventional vacuum technique, capable of producing more complex / deeper parts with relatively uniform thickness.

Although it is difficult to distinguish between a continuing development and singular innovations, it seems worth identifying some of the latter:
- New twin screw technology capable of attaining very high screw speeds (>1500 rpm), or outputs (up to 100 tons/hour!);
- Adjustable feed blocks for film/sheet or coating co-extrusion, much more flexible than traditional multilayer systems;
- All-electrical injection moulding machines, capable of huge energy savings in relation to conventional equipment, with higher repeatability and suitable for operation in clean rooms (important for electronic and medical applications);
- Unconventional injection moulding techniques, such as gas and water assisted injection, allow to overcome shrinkage problems and to shorten cycle times without altering mechanical properties. They represent a huge step forward to mould parts with very irregular thicknesses;
- Micro-injection moulding can produce parts with wall thicknesses of $10\mu$m, surface roughness of about $0.05\mu$m, and structural details in the range of $0.2\mu$m;
- Micro-foaming technology creates small closed cells that improve many properties (such as impact, ductility and insulation) compared to standard foaming techniques. The well-known MuCell Process® is an injection moulding/extrusion foaming system that produces microcellular foamed structures exclusively through the injection of either Nitrogen or Carbon Dioxide in its supercritical state;
- Stretch blow moulding allows the manufacture of PET bottles with improved mechanical and barrier properties.

Very recent trends in polymer processing technology involve hybridization (e.g. combination of twin screw extrusion and injection moulding to manufacture long-fibre reinforced parts for the automotive industry) and the emergence of new processing routines for specific materials (e.g., manufacture of towpregs followed by filament winding and melting/cooling for continuous glass-fibre thermoplastic composites).

## Thermomechanical modelling

Thermomechanical modelling is now recognized by the plastics industry as an useful tool to optimize process conditions and design processing tools. However, as mentioned above, adequate predictions require a good model, based on a pertinent representation of the underlying physics.

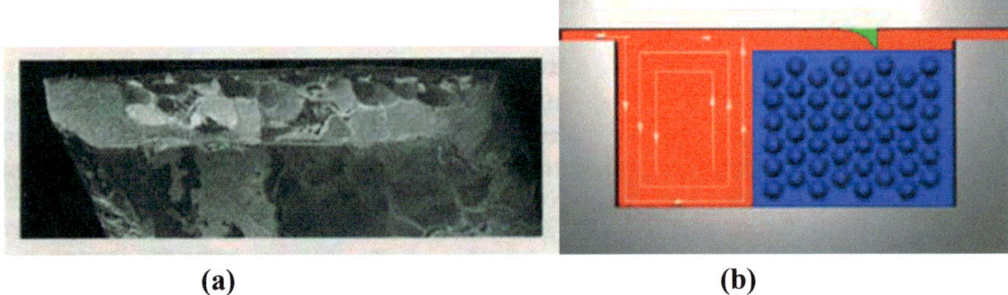

(a)                                                    (b)

**Figure 1** - Rear flight plastication mechanism: (a) experimental observation, (b) Tadmor model

As an example, take the very popular single screw extruder, which combines various complex mechanisms in the same machine: conveying of polymer pellets, plastication of solid polymer and metering of molten polymer, which is forced into a die to shape the product. Initially, extrusioneers believed that plastication was somewhat chaotic, located arbitrarily in the machine and driven by localized friction between pellets, or between pellets, screw and barrel. As a consequence, for several years they tried to improve the process by designing more or less sophisticated mixing elements positioned along the screw, in order to favour friction. In 1959, Maddock [2] decided to study the process. For that purpose, he stopped the extruder once operating steady state had been reached, cooled down the barrel very rapidly, extracted the screw, removed and unrolled the polymer helix from the screw channel and cut cross-sections at regular intervals, to identify the so called *rear flight plastication mechanism* (Figure 1a): instead of the anticipated chaotic melting process at pellets interfaces, he observed a compacted solid bed along the front flight, segregated from a liquid pool along the rear flight. Later, Tadmor [3] proposed a model (Figure 1b) based on simple mass and temperature balance equations that can be solved analytically and provides very interesting qualitative results, which have driven technological improvements for decades. Since then, more sophisticated models have been put forward [4-6], namely taking into account for solid polymer compressibility and more physical and accurate rheological descriptions of the liquid pool - thus requiring numerical resolution - but remain based on the early experiments of Maddock and on the initial Tadmor's representation.

A good model must rely on a well adapted rheology. This does not necessarily mean the use of a sophisticated constitutive equation, but of an equation whose parameters can be determined from a reasonable number of experiments involving flow situations that are qualitatively equivalent to those encountered in the process (e.g. shear vs. elongation dominant flows). This problem will be addressed in the next section.

A good model requires also accurate boundary conditions, particularly for the kinematics and heat transfer. An advanced 3D finite element computation, with a refined mesh and accurate constitutive equations, may generate poor results if non realistic boundary conditions are imposed at the wall:

- *kinematic boundary conditions*: generally, a sticking contact is assumed at the wall. However, this is not true for PVC compounds, elastomers or even for polyolefines at very high rates (when extrusion instabilities are also observed - this will be addressed in the last section - or are equivalent to those often reached in injection moulding);

- *heat transfer conditions*: most computations prescribe a fixed temperature at the die or mould wall. This may be not too far from reality if the die/part shape is not too complex, as the very distributed polymer and metal thermal diffusivity values induce an interface temperature that is near to the imposed tool temperature. Otherwise, important temperature gradients may be measured along the tool periphery, the simplest solution being to account simultaneously for polymer flow in the die / cavity and for thermal interchanges in the surrounding metal (for example, assuming more realistic boundary conditions along the cooling channels).

In order to develop a realistic model, it is very useful to identify *a priori* the order of magnitude of the phenomena, this being achievable via adimensionnal numbers. We present briefly and comment several of these numbers:

- The *Reynolds number* balances inertia and viscous terms in the force balance equation:

$$R_e = \frac{\rho U e}{\eta} \qquad (1)$$

where $\rho$ is fluid density, $\eta$ viscosity, $U$ a characteristic flow velocity and $e$ a representative dimension. Generally, in polymer processing $R_e$ is very small (in the region of $10^{-3}$ or less) and inertia terms may be neglected. Occasionally, $R_e$ may be around 1 (e.g. pin point gate in injection moulding, fibre spinning), but it should be calculated using the right characteristic dimension (e = h, the flow gap, in shear dominant flows, or e = L, the flow length, in extensional flows);

- The *Stokes number* balances mass and viscous forces, where $g$ is mass acceleration; it should be expressed differently for a confined flow geometry (die or mould),

$$S_t = \frac{\rho g L h}{\eta V} \qquad (2a)$$

or, for a free surface flow:

$$S_t = \frac{\rho g L^2}{\eta V} \qquad (2b)$$

Consequently, in confined flows mass forces may be neglected when the vertical distance $L$ is less than 1 m, while in free surface flows they have to be accounted for if the vertical distance is of the order of 0.1m.

- The *Cameron number* [7,8] (or its inverse, the *Graetz number*) is very convenient to estimate the extent of temperature rise in a confined flow. It weights the convection term in the flow direction against the diffusion term in the transversal direction:

$$C_a = \frac{aL}{Vh^2} \tag{3}$$

where $a$ is thermal diffusivity. If $C_a$ is less than $10^{-2}$, flow can be considered as adiabatic, i.e., thermal conduction towards the tool wall is negligible, hence the precise determination of the thermal boundary condition is unimportant. If $C_a$ is higher than one (this is very seldom the case in polymer processing), the thermal regime is fully developed, i.e. the temperature profile will remain constant along the flow geometry. Therefore, since any additional viscous dissipation will be balanced by heat conduction to the die/mould wall, an accurate thermal boundary condition is required;

- The *Peclet number* balances the convection and diffusion terms in the flow direction.

$$P_e = \frac{UL}{a} \tag{4}$$

Since it is always very high, conduction may be neglected in the flow direction;

- The *Brinkman number* compares viscous dissipation with heat transfer to the wall:

$$B_r = \frac{\eta V^2}{k\left(\overline{T} - T_w\right)} \tag{5}$$

where $k$ is thermal conductivity, $\overline{T}$ average polymer temperature and $T_w$ wall temperature. If $B_r$ is higher than 1, viscous dissipation will govern polymer temperature and, again, accurate thermal regulation (and precise definition of corresponding boundary condition) is not mandatory. Conversely, thermal regulation will control flow temperature.

- The *Nahme number* writes as:

$$N_a = \frac{\eta b V^2}{k} \tag{6}$$

where $b$ is the temperature dependence viscosity coefficient (given by $E/R$, where $E$ is activation energy for viscosity and $R$ is the perfect gas constant, assuming that the polymer obeys an Arrhenius law). $N_a$ weights the viscosity temperature sensitivity against temperature rise due to viscous dissipation. When $N_a < 1$, mechanical and thermal calculations may be decoupled;

- The *Deborah number* balances the polymer characteristic time, $\lambda$ (mean relaxation time? longest relaxation time?) with the process characteristic time (typically, residence time $U/L$):

$$D_e = \frac{\lambda L}{U} \tag{7}$$

If it is smaller than one, transient viscoelastic effects may be neglected. $D_e$ is especially relevant for extensional free surface flows, where the residence time is clearly defined.

- The *Weissenberg number* weights the first normal stress difference $N_1$ against shear stress $\tau$ in simple shear flow. So, it is well suited to confined flows where shear rate is dominant:

$$W_e = \frac{N_1}{2\tau} \tag{8}$$

Contrarily to the *Deborah number* $W_e$ is not ambiguous, since $N_1$ and $\tau$ may be measured independently, for example in a cone and plate rheometer. If $W_e$ is small, viscous effects are dominant and an accurate purely viscous constitutive equation is sufficient. Conversely, viscoelastic effects have to be accounted for, even if they affect pressure and velocity only slightly, as they become significant for stress distribution;

- Other adimensional numbers may be put up, but they result from a combination of the above. In general, such numbers provide a useful guide to adapt the complexity of the numerical method to the problem under consideration.

To obtain an appropriate numerical solution, good meshing is also needed. This does not necessarily mean a very refined mesh, as this would lead to a huge number of elements and unknowns for most 3D parts, which could not be solved at a reasonable cost, but refined where it is required and coarse elsewhere. In addition, we need anisotropic meshing for thin parts, with sufficient number of nodes along the thickness to capture sharp velocity profiles. Although tedious, these meshing procedures may be executed manually for 2D flow geometries, but become totally impractical for 3D channels, where they have to be implemented automatically and governed by a local metric affixed to the local flow geometry dimensions. Sometimes, automatic remeshing techniques are required to capture specific features of unsteady processes (free surface, for example). Figure 2b presents an anisotropic meshing of a profile die geometry, where the mesh size in the flow direction is a hundred times larger than in the thickness direction.

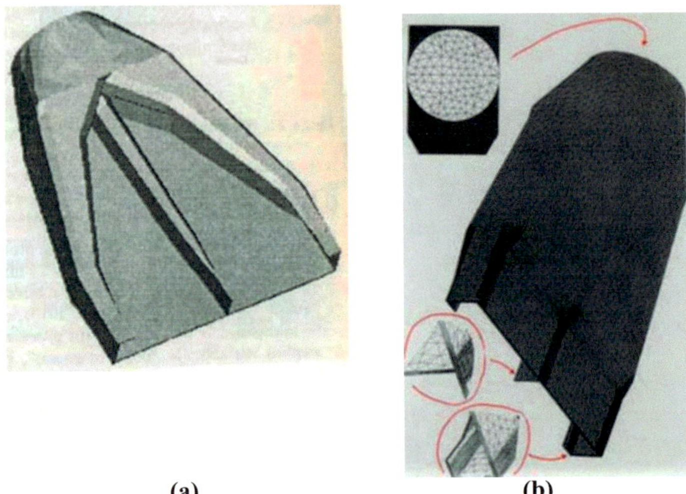

(a)                                                    (b)

**Figure 2 -** Profile die: (a) flow channel geometry; (b) anisotropic meshing

We will now present several examples of 3D computations:

**Computation of flow in a profile extrusion die [9]**

Figure 2a illustrates the flow channel geometry of a profile die, which starts with a circular cross-section at the extruder outlet and ends up with the required profile. The objective of the design is to secure uniform velocity at the die exit and obtain the prescribed thickness distribution.

We use a generalized viscous incompressible temperature dependent constitutive equation, assume sticking contact and a fixed set temperature at the wall. Figure 3 shows that pressure decreases downstream quite regularly, so that one would believe that this would lead to a homogeneous velocity at the exit. However, this is not true, as seen in Figure 4, since velocity is badly distributed.

**Figure 3 -** Profile die: pressure field

**Figure 4 -** Profile die: velocity field at die exit

Computations may avoid tedious and expensive trial and error tests made with dies subjected to successive machined corrections.  In fact, we can carry out "numerical machining" until the velocity field becomes balanced, but this has only practical value if we have confidence on the numerical results. Figure 5 presents the average velocity at two zones of the die exit, as a function of  meshing. While with a rough mesh the predicted mismatching between zones 1 and 2 is of the order of 3, using a refined mesh it reduces to only 30%. Machining a new die using the results provided by a computation using a rough mesh would generate enormous errors.

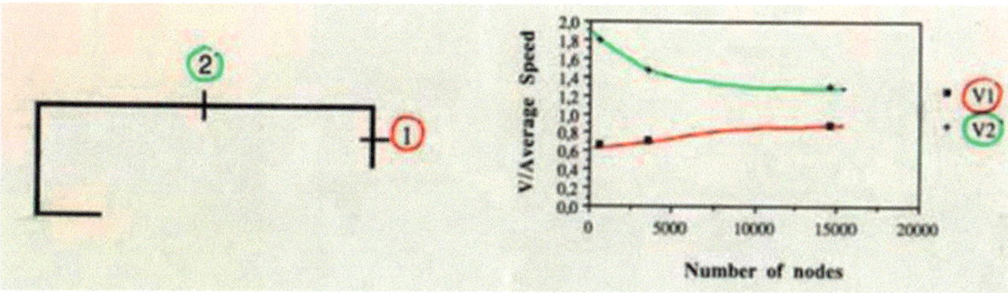


**Figure 5 -** Profile die: velocity as a function of the meshing

Although most of the computational tools used in polymer processing are based on the finite element method, other numerical techniques are available. For example, several authors claim that the finite-volume method (FVM), widely used in traditional Computational Fluid Dynamics, can also capture the details of 3D flows in complex geometries, while requiring less computational resources [10,11]. This feature may be critical in optimization problems, when the searching process of the optimisation algorithm requires the recurring resolution of a flow problem with different sets of input parameters. Figure 6 refers to the automatic optimisation of a profile die [12]. The modelling routine, based on FVM, is coupled to a non-linear SIMPLEX optimization algorithm, which attempts to minimize a penalty function that considers both velocity differences around the die exit contour and differences between a minimum acceptable die land / thickness ratio ($L/e$) and each proposed solution. The velocity contours and extrudate on the left were obtained with a die where the lands of each profile zone have identical $L/e$. The illustrations at the centre result from an optimisation of the individual die lands lengths in order to guarantee uniform average velocity at the die exit, while those at the right refer to the optimisation of individual thicknesses to assure the pre-established local thickness after haul-off.

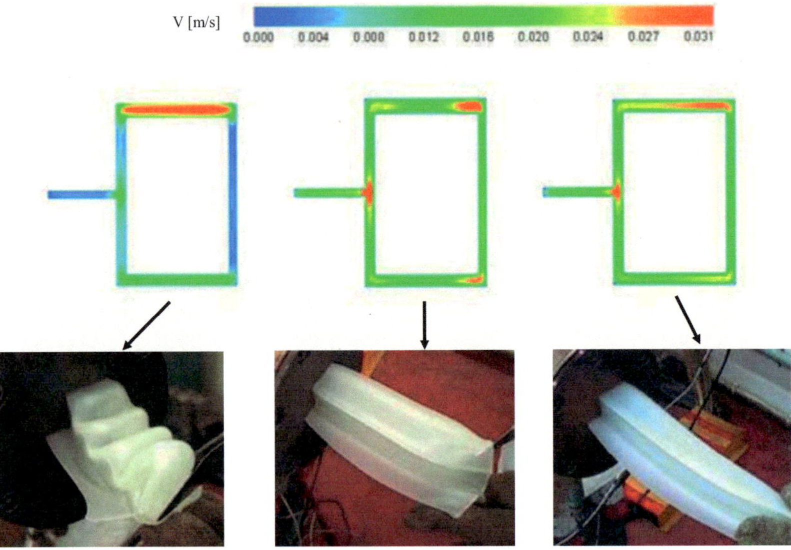

**Figure 6** – Velocity contours and emerging extrudate (see text for details).


**Computation of 3D injection moulding [13]**

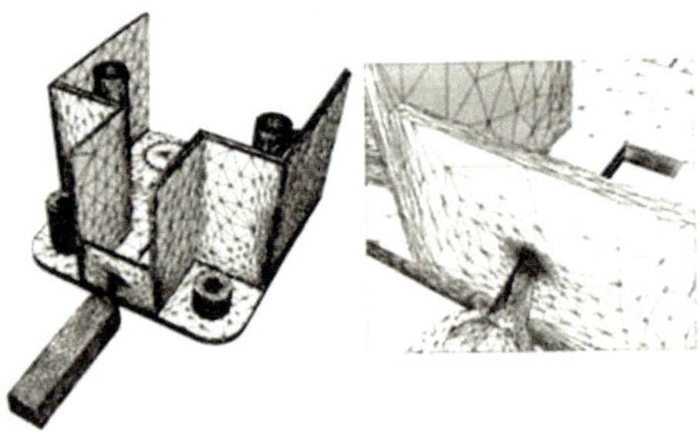

**Figure 7 -** Complex 3D part

We consider a complex 3D part (courtesy of Schneider company) comprising thin and thick regions.  The corresponding mould is equipped with a pressure transducer.
We model flow assuming compressible purely viscous temperature dependent behaviour and fixed temperature at the mould wall. Figure 7 presents the meshing, consisting of refined meshes around the gate and 3D thick regions of the part and coarse and anisotropic meshes in the thin regions. Comparison between experimental (Figure 8, left) and predicted short shots (Figure 8, right) evidences a reasonable agreement. So, it is possible to predict the position of the weld line and change the location of the injection gate in order to shift the weld line to suitable regions.

Figure 9 contrasts the experimental pressure trace of the complete moulding cycle with the one predicted for two imposed packing pressures (Figure 9a) and two different filling rates (Figure 9b).The agreement is fair and the model is able to predict accurately the weight of the part.

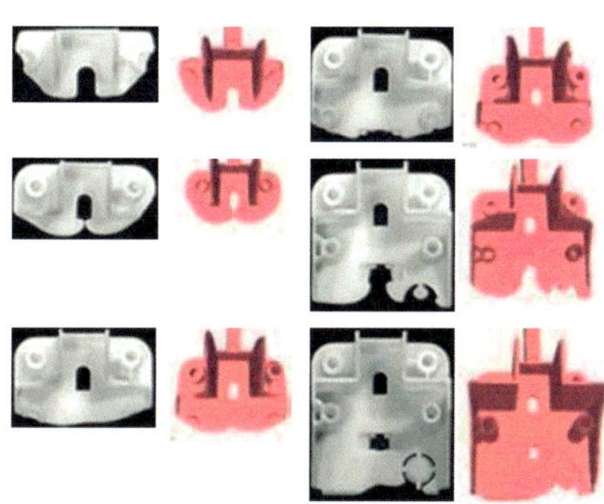

**Figure 8** - Comparison between experimental
(left) and predicted (right) short shots

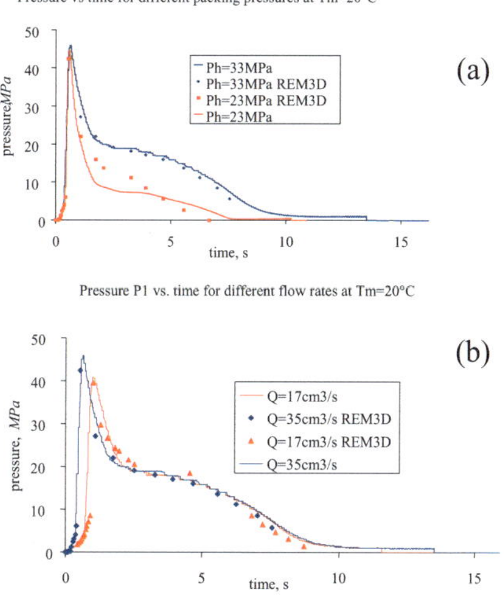


**Figure 9** - Comparison between experimental and computed pressure traces; (a) for varying packing pressures; (b) for varying flow rates

## Computation of blow moulding [14]

We consider here blow moulding of a standard shampoo bottle and assume a membrane hypothesis (in plane deformation, normal stress equal to the inflation pressure), purely viscous polymer and isothermal blowing. The polymer sticks to the mould as soon as contact occurs.

Figure 10 shows the successive membrane meshings during the inflation stage. Automatic remeshing allows for mesh refinement when the membrane becomes highly distorded (high local curvature) and/or in close contact with the mould. Figure 11 presents the isothickness values in the blown bottle. The objective is to obtain a thickness as uniform as possible (for mechanical reasons, the value of thickness around the handle is particularly important). In practice, it is possible to modify a priori the parison thickness distribution, for example via parison programming, in order to obtain a more regular thickness distribution.

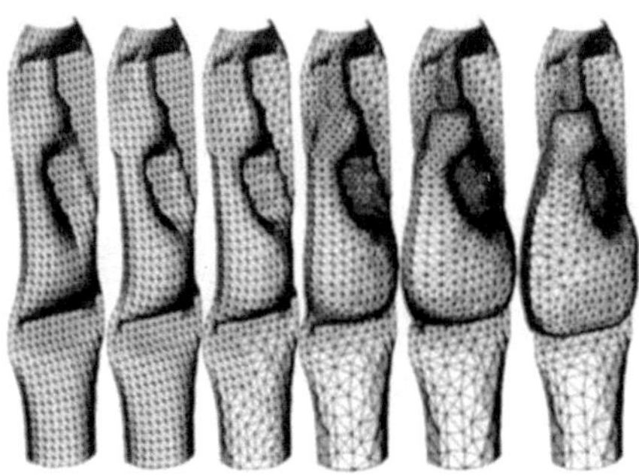

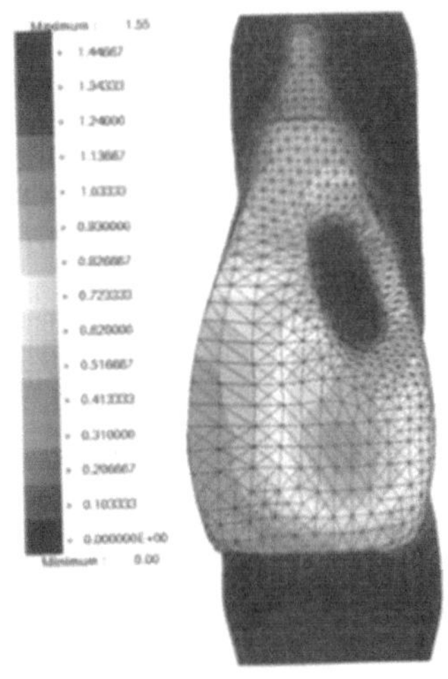

**Figure 10** - Successive membrane meshings during inflation

**Figure 11** - Computed thickness distribution in the final bottle

**Computation of solids conveying in an extruder**

Some specific processing problems may require resorting to particular numerical methods. Solids conveying in an extruder screw is generally considered as equivalent to friction dragging of an isotropic elastic solid plug in perfect contact with the surrounding metallic walls. However, it is not only well known that this assumption is far from the physical reality (the screw is fed with loose pellets, which become progressively compacted), but it also yields poor output predictions. The Discrete Element Method (DEM) is well suited to simulate particle assemblies, since each particle is modelled as a distinct entity, particle interactions are accounted for and the complete dynamics of the system can be characterized [15]. Assuming validity of Newton´s second law and of a linear displacement-force law, we modelled the flow of individual pellets in the initial screw turns, as they fall continuously from the hopper.

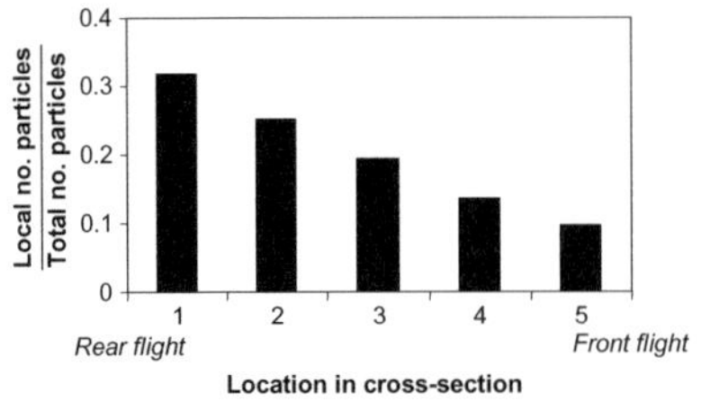


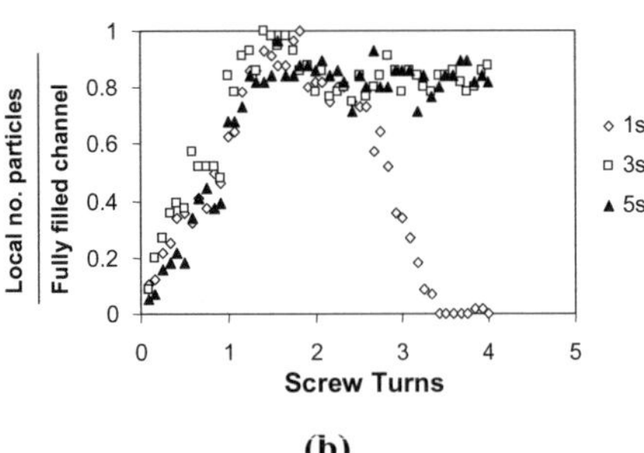


(a)                              (b)

**Figure 12** – Distribution of the number of particles: a) in a channel cross-section; b) along the channel.

Figure 12a shows the distribution of particles in a channel cross-section, thus evidencing that the active flight is a major contributor to their progression. Figure 12b illustrates how the number of particles changes in the down-channel direction. Immediately after flow initiation, a peak value develops close to the hopper throat, but the number of particles remains relatively constant and steady state is reached. From the evolution of density along the channel, one may predict the onset of plug flow [16]. Pellets follow a full 3D pattern, with transverse and backward flow components, the complexity of the trajectory depending on the local degree of filling, screw speed and particle size/channel height ratio.

## Rheology of molten polymer systems

Rheology is commonly considered as a means to investigate the physics of polymers in the molten state, progressively more sophisticated constitutive equations having been developed to account for an array of experimental data. However, until recently polymer processors were still using only very straightforward constitutive equations, assuming purely Newtonian or shear thinning behaviour (power law, Carreau law [17]). In fact, as shown in Figure 13 for a classical coat hanger die, even a simple power law constitutive equation will influence not only pressure drop, which is evident, but also velocity distribution at the die exit (a high value of the power law index will promote higher velocity at the die periphery, a lower value will induce more flow through the centre). At present, given the huge evolution in polymer processing modelling, it is possible to use the same type of constitutive equations used in polymer Physics. Two questions become particularly relevant:

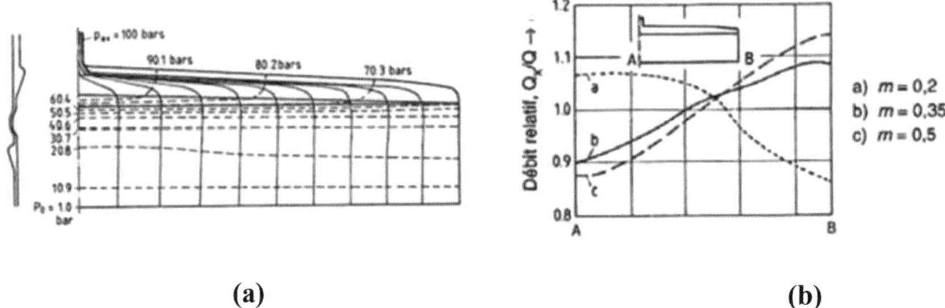


**(a)**                                                       **(b)**

**Figure 13 -** Flat die geometry: (a) Flow lines and pressure field; (b) velocity distribution at die exit as a function of the power law index

- What is the most suitable constitutive equation for a given problem? This will depend on flow type (shear dominant, extensional dominant, mixed) and on the value of the adimensionnal numbers presented above.
- How to determine the parameters of the constitutive equation from a limited series of experiments?

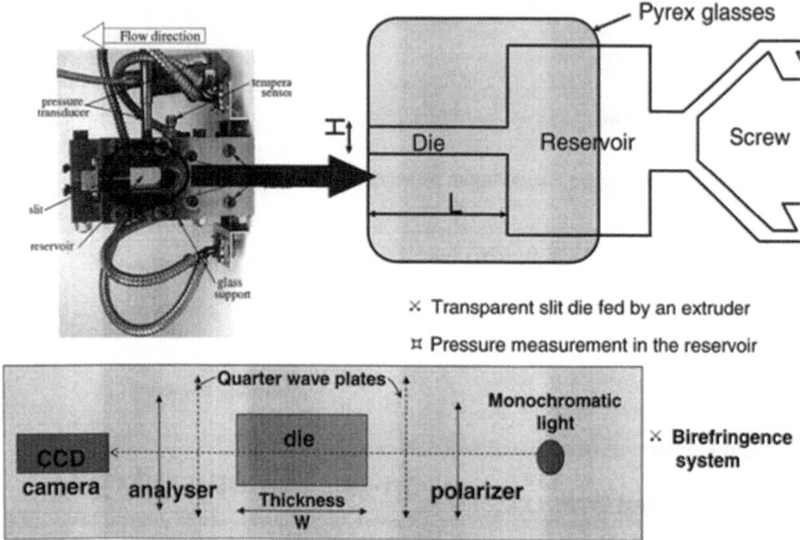


**Figure 14** - Prototype industrial flow instrument

Many commercial rheometers (e.g., cone and plate, parallel-plates, Couette, capillary, extensional) and well established experimental protocols exist (dynamic measurements, step strain tests, steady state flows etc), reference text books [18,19] describing in detail these techniques, their advantages and shortcomings. In general, determination of viscoelastic parameters under conditions relevant to processing (non linear regime) requires a set of time consuming experiments needing the availability of different kinds of rheometers. We suggest an alternative approach, which consists in building a *prototype industrial flow instrument* (figure 14) [20] which allows monitoring the flow of molten polymer with complementary techniques (flow birefringence to identify the main stress difference distribution, laser Doppler anemometry, LDA, to measure the velocity field), followed by computations with 3D finite element softwares using various candidate constitutive equations. Comparison between experimental and numerical results allows fine tuning of the adjusting parameters (especially the non linear viscoelastic parameters), in order to achieve the best fit for each. As this could be tedious if only trial and error adjustments were made, an optimization algorithm can be used to minimize the value of a cost function measuring the difference between numerical and experimental results. The outcome would be the constitutive equation that is able to minimize the cost function and the corresponding set of parameters.

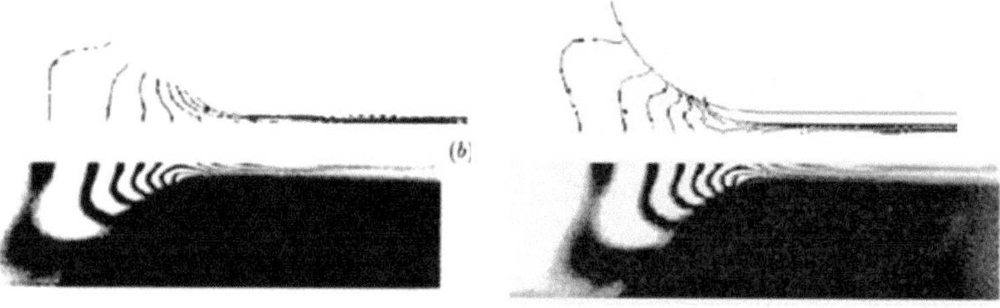


**Figure 15** - Comparison between experimental (bottom) and computed (top) birefringence patterns in half of the die geometry, for the initial (left) and final (right) vicoelastic parameters.

On figure 15 we present the experimental birefringence pattern in the mid plane of a convergent flow and a 2D finite element computation in the same plane. The candidate is the multi-mode Phan-Thien Tanner model [21]:

$$\sigma = -pI + \Sigma \tau_i + 2\eta_s \dot{\varepsilon}$$

$$\exp\left(\frac{\varepsilon \tau_i}{\Sigma \eta_{vi}} tr(\tau_i)\right)\tau_i + \tau_i\left[\left(1 - \frac{a}{2}\right)\overset{\triangledown}{\tau_i} + \frac{a}{2}\overset{\triangle}{\tau_i}\right] = 2\eta_{vi}\dot{\varepsilon}(u) \qquad (9)$$

where $\sigma$ is stress tensor, $\dot{\varepsilon}$ is rate of strain tensor, $p$ is pressure, $\left(\eta_s, \eta_{vi}, \tau_i\right)$ is the distribution of relaxation times and viscosity, deduced by well consolidated dynamic rheology experiments, $a$ and $\varepsilon$ are two non linear viscoelastic parameters to be determined and $\overset{\triangledown}{\tau_i}$ and $\overset{\triangle}{\tau_i}$ are the upper and the lower convective stress derivatives.

The initial choice of the non linear parameters (a, $\varepsilon$) is on the left - the disagreement with the experiment is clear, while the computation with the final set of (a, $\varepsilon$) values after minimisation of the cost function produces a fair agreement, as shown on the right.

This "advanced" rheological technique looks promising, despite a few limitations concerning optical properties: i) it can only be applied to transparent materials (many commercial compounds contain additives, such as pigments and fillers, that have an effect on transparency), ii) the material must possess a sufficient birefringence level (in order to create a minimum number of fringes), but not an exaggerated one (in this case the picture cannot be analysed and the cost function linking computation and experiment cannot be built)  and iii) the existence of a linear relationship between stress and birefringence must be checked [22].

Very often, more complex polymer systems are used, such as polymer blends, or short fibre composites. In this case, the minor or co-continuous phase deformation, or fibre orientation, respectively, will influence the rheology at macroscopic level.

We examine here the rheology and orientation of short fibre reinforced polymer melts. In 1922, Jeffery [23] calculated analytically the orientation $\vec{p}$ of a single fibre under several flow configurations:

$$\frac{Dp}{Dt} = \Omega p + \tau \left( \dot{\varepsilon} p - \dot{\varepsilon} : p \otimes p \right) p \qquad (10)$$

In this equation, $\Omega$ is the rotation tensor and $\tau = \left( \beta^2 - 1 \right)\left( \beta^2 + 1 \right)$, where $\beta$ is the fibre shape factor. In shear flow, the fibre remains most of the time aligned in the flow direction, but rotates suddenly with a well defined period. In a positive extensional flow, the fibre has a stable orientation in the flow direction, while in a negative extensional flow it remains stably aligned in a direction perpendicular to the flow. This very simple model allows a good qualitative understanding of many phenomena, but ignores fibre interaction under realistic reinforcement conditions (up to 50% fibre content). In addition, it is necessary to derive average numbers for the orientation of a fibre population, which was done by introducing an orientation tensor $a_2$ defined as $a_{ij} = <p_i, p_j>$, where $p_i$ and $p_j$ are unit vectors affixed to fibres $i$ and $j$, respectively. Tucker and co-workers [24] introduced the following generalized orientation equation:

$$\frac{da_2}{dt} = \Omega : a_2 - a_2 : \Omega + \tau \left( \dot{\varepsilon} : a_2 + a_2 : \dot{\varepsilon} - 2\dot{\varepsilon} : a_4 \right) + 2CI \left( I - 3a_2 \right) \qquad (11)$$

Here, $CI$ is an interaction coefficient which has to be adjusted and $a_4$ is the fourth order orientation tensor, $a_{ijkl} = <p_i p_j p_k p_l>$. If one writes an evolution equation for $a_4$, a sixth order orientation tensor is introduced, and so on. Therefore, a "closure approximation" is needed, relating the fourth order orientation tensor to $a_2$. Although many such relationships have been proposed, none is able to capture fibre orientation in all flow configurations. In addition, the constitutive equation must be modified to account for the presence of fibre orientation. Tucker [25] proposed a model for a Newtonian matrix with two adjusting parameters $N_s$ and $N_p$:

$$\sigma = -pI + 2\eta \left( \dot{\varepsilon} + N_s \left( \dot{\varepsilon} : a_2 + a_2 : \dot{\varepsilon} \right) + N_p \dot{\varepsilon} : a_4 \right) \qquad (12)$$

Hence, it is a considerable step forward to perform a direct numerical computation involving a large number of fibres and a more general viscous, or even viscoelastic, matrix [26,27]. As an example, Figure 16 presents the first term of the orientation tensor ($a_{11}$) in pure shear flow as a function of time for 60 fibres, with a shape factor equal to 6, and with two different fibre concentrations. The fibres, with initial random orientation, are progressively aligned in the flow direction, though not perfectly. Orientation seems to decrease with increasing fibre concentration.

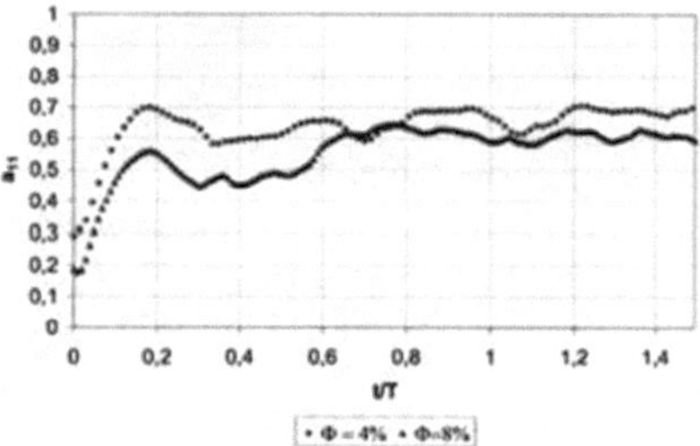

Figure 16 - Direct computation of component $a_{11}$ of the orientation tensor: 60 fibres, shape factor = 6, two different concentrations ( from [27])

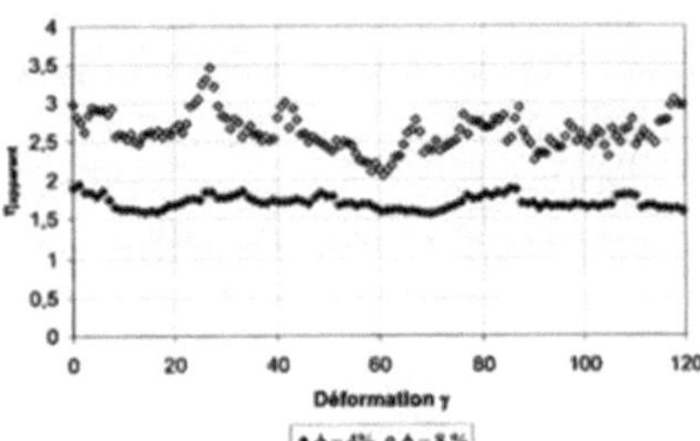

Figure 17 - Direct computation of average shear viscosity; same conditions as in Figure 16 (from [27])

The fibres, with initial random orientation, are progressively aligned in the flow direction, though not perfectly. Orientation seems to decrease with increasing fibre concentration. For the same conditions, Figure 17 shows the evolution of apparent Newtonian viscosity (ratio of imposed stress to apparent shear rate) with deformation (i.e., time), this being quite distinct from Einstein's theory for spheres, but in agreement with Kitano and Kataoka's experiments [28]. We are only at the early stages of this approach: we need to introduce interaction forces between particles, periodic boundary conditions in order to account for more complex 3D flow configurations, and to consider a higher number of fibres.

## Instability phenomena in polymer processing

Instability phenomena are often the real restriction to increasing the productivity of a given polymer processing sequence. Generally, three types of instabilities may be observed, their origin being quite different:

(a)

(b)

(c)

(d)

(e)

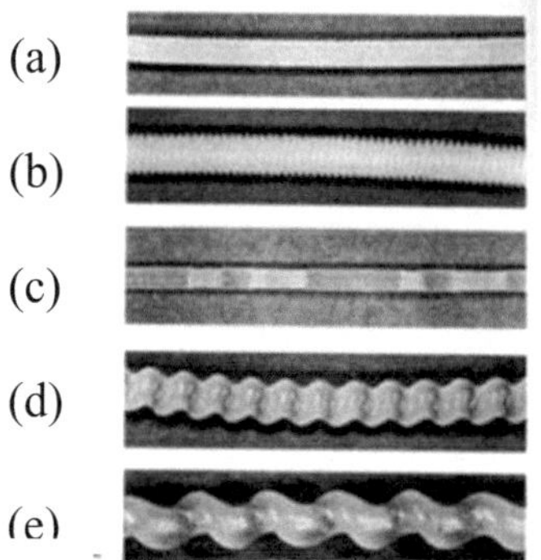

Figure 18 - Extrusion defects observed in a capillary die: (a, b) sharkskin; (c) stick slip; (d,e) helical

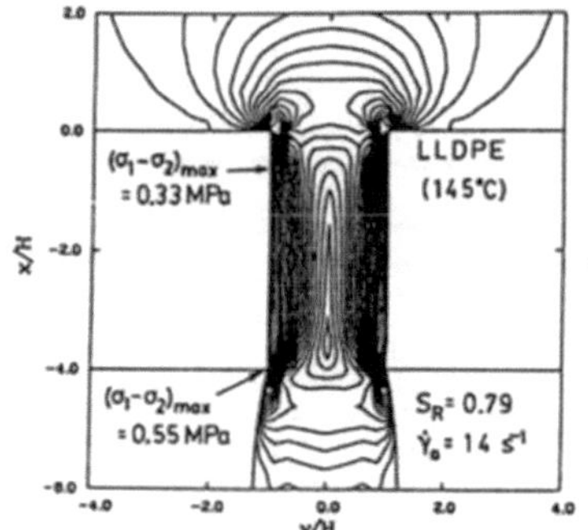

Figure 19 - Computed stress singularity at die exit ( from [34])

*Confined flow instabilities* are commonly observed in most extrusion processes (pipes, cast and blown film, wire coating) and have been extensively studied in capillary geometries [29,30]. They are often called *melt fracture*, even if they encompass different defects, depending on polymer and flow rate:

- For linear polymers such as HDPE or LLDPE, the first defect that appears when increasing flow rate is known as *sharkskin*, when a rough - but often periodic - surface develops on the extrudate (Figure 18a and b). It shows at very low shear rates (a few reciprocal seconds) for some metallocene Polyethylenes. Its occurrence may be linked to a combination of two phenomena: i) a stress singularity at the die exit (Figure 19), ii) a high elongation rate along the surface of the extrudate, as it emerges from the die, that will cause its periodic rupture [31]. At higher flow rates, a stick-slip defect appears, consisting of a succession of smooth and rough extrudate surfaces (Figure 18c), accompanied by a discontinuity on the pressure - flow rate relationship (Figure 20). This defect may be caused by an oscillating relaxation mechanism [32].

- For branched polymers (like LDPE or Polystyrene), a different instability may develop (Figure 18d and e), denoted as helical (in capillary geometry) or volume instability (in more general flow geometries). It may be associated with both stress and velocity fluctuations in the reservoir upstream of the die (these can be measured with flow birefringence and  LDA techniques – see Figure 21) [33].

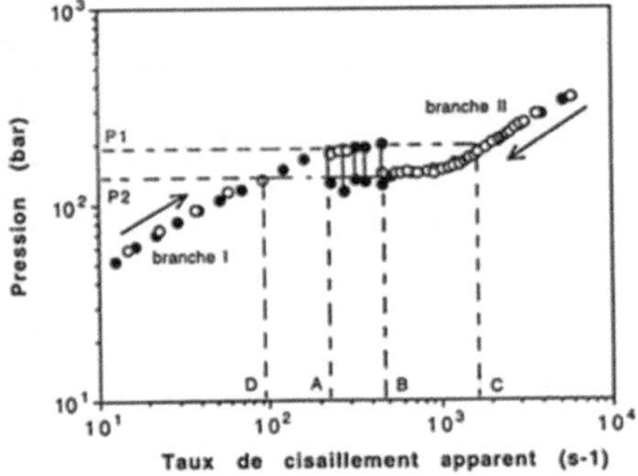


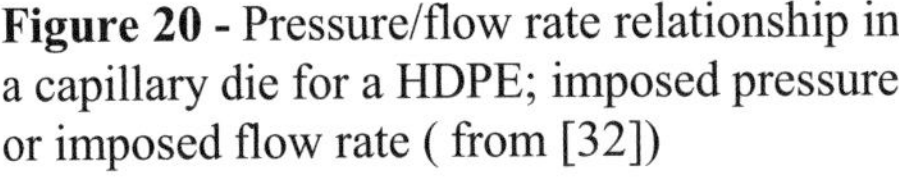


**Figure 20 -** Pressure/flow rate relationship in a capillary die for a HDPE; imposed pressure or imposed flow rate ( from [32])

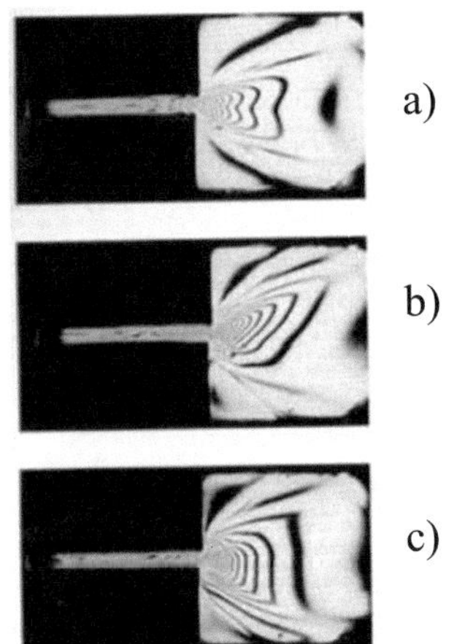


**Figure 21 -** Stable and unstable flow birefringence pattern for a Polystyrene

The process engineer should identify the exact nature of the defect and its origin, if he wants to take appropriate action:

- If it is a sharkskin instability, it is important to decrease the stress singularity at die exit by incorporating processing aids or by smoothing out the die exit geometry.
- If it is a stick-slip defect, melt compressibility upstream must be decreased by reducing, for example, the die land.
- If it is a helical or volume defect, it is best to modify the upstream geometry of the die.

*Interface instabilities* may be observed in coextrusion flows, depending on the relative thickness of each layer, on the relative flow rate of each polymer, and on viscosity and elasticity ratios (Figure 22) [35]. Linear stability analyses [36] and dynamic computations [37] demonstrated that the instability originates at the contact point between polymers and is then convected along the die land. Depending on geometry, processing conditions, and relative rheology of polymers, the initial instability will either decrease or increase, leading to a smooth or disturbed interface, respectively (Figure 23).

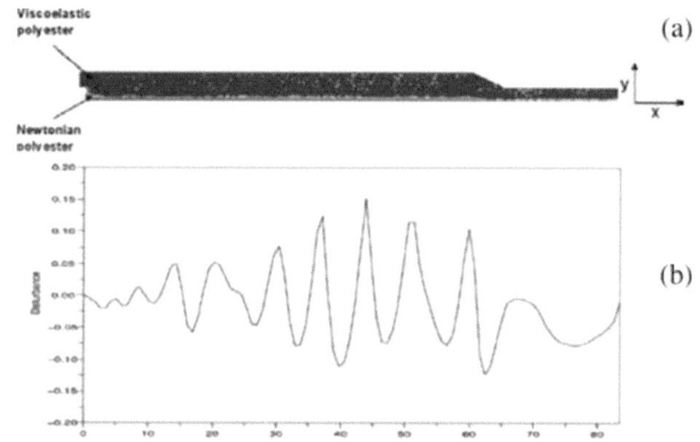


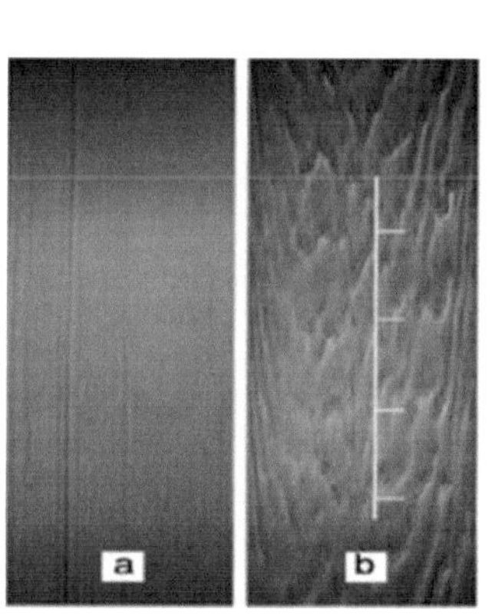


**Figure 22** – Coextrusion of a plaque: (a)smooth surface; (b) interface instability

**Figure 23** – Dynamic computation in a 2D convergent geometry: (a) interface location between the viscoelastic and the Newtonian Polyester; (b) development of the interface instability

*Drawing instabilities* are observed in free surface flows such as fibre spinning, cast and blown film, when the draw ratio (ratio between take-up and extrusion velocities) exceeds a critical value, which depends on polymer rheology (Deborah number) and cooling conditions. The phenomenon has been extensively investigated in fibre spinning and corresponds to a Hopf bifurcation [38]. In cast film, apart from viscoelasticity, draw resonance depends also on a film shape factor (ratio of drawing distance to film width) [39]. In the case of blown film (Figure 24), École

Polytechnique de Montreal developed a device capable of measuring axi-symmetric and non axi-symmetric bubble instabilities (Figure 25) [40]. It was noted that the instability develops for a draw ratio higher than a critical value but also, which is less trivial, that increasing the blow up ratio will stabilize the process (Figure 26). We are able to draw a limit stability curve which is the envelope, for all processing conditions and machine sizes, of stable operating conditions (Figure 27a). A linear stability analysis, developed first for axi-symmetric conditions and then for non axi-symmetric ones was able to capture these features (Figure 27b).

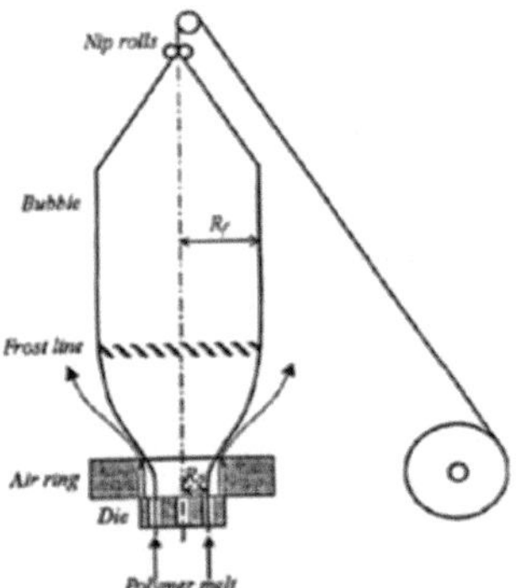


**Fig. 24** – Film blowing line

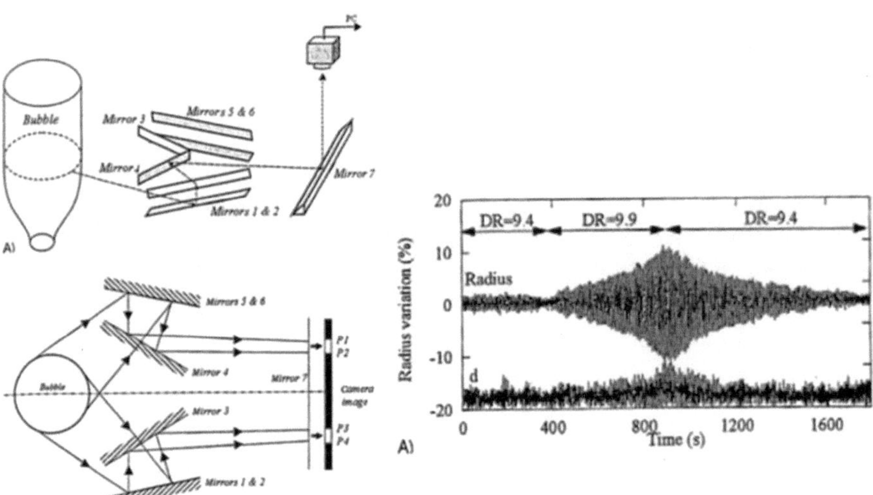


**Figure 25** – Bubble instability measuring technique developed at Ecole Polytechnique de Montréal (from [40])

**Fig. 26** – Bubble radius as a function of draw ratio

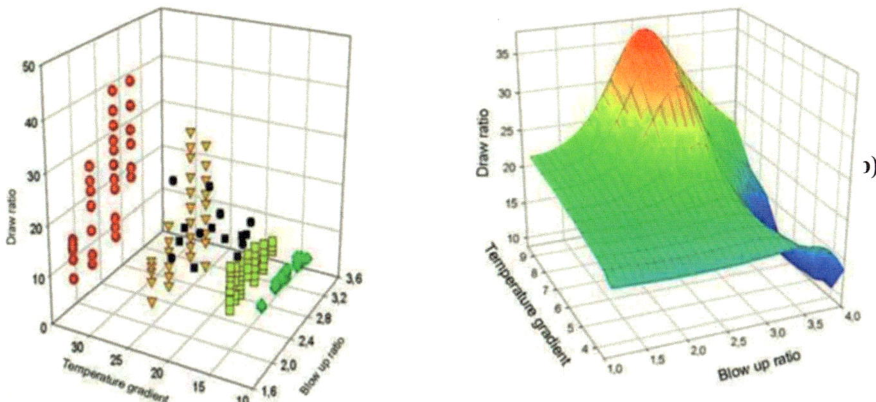

**Figure 27** – Film blowing
stability envelope: (a) experiments, (b) linear stability results

## Conclusions

During the last decade there have been huge progresses in polymer processing modelling: we have now available 3D finite element methods, take into account compressibility, incorporate complex viscoelastic constitutive equations, or perform thermo-mechanical coupling in steady or transient conditions. However, research in this field still contains new challenges, particularly in the following directions:

- reduce the time required before an actual process computation is performed, which involves, for example, developing fully automatic meshing procedures, driven by the local channel/part dimensions;
- reduce the computation time, which means developing adaptative meshing, using iterative solvers with parallel computing, or new space-time finite elements;
- determine the precision of a numerical result via the development of error estimators and remeshing techniques driven by these estimators.

Obviously, these advanced numerical methods have to be based on proper physical models, which should integrate:

- accurate constitutive equations containing parameters that can be determined with reasonable experimental effort;
- realistic boundary conditions;
- all process stages, namely plastication, cooling and crystallisation.

Last but not least, although this chapter focused on the molten state, the big challenge is now to relate processing conditions, induced orientation and final part structure to resulting performance under service conditions.

# References


(1)  C. Maillefer, Swiss Patent n°363149 (1959)
(2)  B.H. Maddock, Soc. Plast. Engs. S. $\underline{15}$, 383 (1959)
(3)  Z. Tadmor and I. Klein, "Engineering Principles of Plastificating extrusion" Van Nostrand Reinhold, New York (1970)
(4)  J.T. Lindt, Polym. Eng. Sci. $\underline{25}$, 585 (1985)
(5)  P.G. Lafleur and K. Amellal, Plast. Rubb. Proc. Appl., $\underline{19}$ (1993)
(6)  B. Vergnes, E. Wey and J.F. Agassant, Caout. Et Plast., 633, $\underline{81}$ (1983)
(7)  A. Cameron, Principles of lubrification, Longsman Green, London (1966)
(8)  J.F. Agassant, P. Avenas, J.Ph. Sergent, B. Vergnes, M. Vincent, La Mise en Forme des Matières Plastiques, Tec. et Doc. (1996)
(9)  J.F.Gobeau, T.Coupez, B.Vergnes and J.F.Agassant, Computations of profiles dies for thermoplastic polymers using anisotropic meshing, in " simulation of materials processing: theory, methods and applications" S.F.Shen and P.Dawson edts, Balkema, Rotterdam, P 59 ( 1995)
(10) S.C.Xue,N. Phan Thien, , and R.I. Tanner., J. Non-Newt. Fluid Mech., $\underline{59}$, 191 (1995)
(11)R.Y. Chang, H.C.Hsu, C.S.Ke, and C.C. HSu, *Polymer Processing Society: Asia/Australia Regional Meeting.* Taipei, Taiwan (2002).
(12) J.M. Nobrega, O.S. Carneiro, F.T. Pinho, and P.J.Oliveira, Intern. Polym. Proc, $\underline{18}$, 298 (2003)
(13) L.Silva, C.Gruau, J.F.Agassant, T.Coupez and J.Mauffrey, Intern.Polym.Proc.$\underline{20}$, 265  (2005)
(14) A.Rodriguez-Villa, J.F.Agassant and M.Bellet, Finite element simulation of the extrusion blow molding process, in" simulation of material processing: theory, methods and applications",S.F.Shen and P.Dawson edts,p 1053 ( 1995)
(15) P.A.Cundall, O.D.L. Strack, *Geotechnique* $\underline{29}$, 47 (1979)
(16) He Hong, J.A. Covas, A. Gaspar-Cunha, *Numiform 2007*, Porto, Portugal (2007)
(17) P.J.Carreau, Trans.Soc.Rheol.$\underline{16}$, 99 ( 1972)
(18) C.W. Macosko, Rheology: Principles, Measurements and applications, VCH Publishers, NY (1994)
(19) T.G. Mezger, The Rheology handbook, Vincentz Verlag, Hannover (2002)
(20) J.F.Agassant, F.Baaijens, H.Bastian, A.Bernnat, T.Coupez, B.Debbaut, A.L.Gavrus, A.Goublomme, M.Van Gurp, R.J.Koopmans, H.M.Laun, K.Lee, O.H.Nouatin, M.R.Mackley, G.W.M.Peters, E.Reckers, V.M.H.Verbeeten, B.Vergnes, M.H.Wagner, E.Wassner, W.F.Zoetelief, Intern.Polym. Proc. $\underline{17}$, 347, ( 2002)
(21) N.Phan Thien and R.I.Tanner, J.Non Newt.Fluid Mech.2, 353, ( 1977)
(22) R.Muller and B.Vergnes, in" rheology for polymer melts processing", J.M.Piau and J.F.Agassant eds.,Elsevier, Amsterdam ( 1996)
(23) G.B.Jeffery, Proceeding Royal Society of London, A 102, 161 ( 1922)
(24) F.Folgar and C.L.Tucker, Orientation, Journal of reinforced Plastics and Composites, $\underline{3}$, 98, ( 1984)
(25) C.L.Tucker, J. Non Newt.fluid Mech. $\underline{39}$, 239 ( 1991)

(26) A.Megally, P.Laure and T.Coupez, Direct Simulation of rigid fibres in viscous fluids, 3rd international Symposium on two-phase flow modelling and experimentation, Pisa, Italy ( 2004)

(27) A.Megally, Etude et modelisation de l orientation des fibres dans des thermoplastiques renforcés, These de doctorat en Sciences et Genie des Materiaux, Ecole des Mines de Paris ( 2005)

(28) T.Kitano and T.Kataoka, Rheologica Acta 20, 390 ( 1981)

(29) M.M.Denn, Ann.Rev.Fluid Mech. 22, 13 ( 1990)

(30) R.G.Larson, Rheologica Acta, 31, 213 ( 1992)

(31) J.F.Agassant, D.R.Arda, C.Combeaud, A.Merten, H.Munstedt, M.R.Mackley, L.Robert, B.Vergnes, Intern.Polym.Proc. 21, 239, 5 2006)

(32) V.Durand, B.Vergnes, J.F.Agassant, E.Benoit , R.J.Koopmans, J.Rheol. 40, 383 (1996)

(33) C.Combeaud , Y.Demay, B.Vergnes, J.Non Newt.Fluid Mech. 121, 175 ( 2004)

(34) D.G.Kiriakidis, H.J.Park, E.Mitsoulis, B.Vergnes and J.F.Agassant, J.Non Newt. Fluid Mech. 47, 339 ( 1993)

(35) R.Valette, P.Laure, Y.Demay and J.F.Agassant, Intern.Polym.Proc.18, 171, (2003)

(36) R.Valette, P.Laure, Y.Demay and J.F.Agassant, J.Non Newt.Fluid Mech. 121, 41 ( 2004)

(37)O.Mahdaoui, J.F.Agassant, P.Laure, R.Valette, Luisa Silva, Esaform Conference, Saragossa (Spain) ( 2006)

(38) Y.Demay and J.F.Agassant, J. Meca. Theorique et Appliquée, 1, 763, (1982)

(39) D.Silagy, Y.Demay , J.F.Agassant, J.Non Newt. Fluid Mech. 79, 563 ( 1998)

(40) J.Laffargue, L.Parent, P.G.Lafleur, P.J.Carreau, Y.Demay, J.F.Agassant, Intern. Polym. Proc. 17, 347 ( 2002)

(41) J.Laffargue, Etude et modélisation des instabilités du procédé de soufflage de gaine, Thèse de doctorat en Sciences et Genie des materiaux, Ecole des Mines de Paris ( 2003)

# Composites Forming

Philippe Boisse[1], Remko Akkerman[2], Jian Cao[3], Julie Chen[4],
Stepan Lomov[5] and Andrew Long [6]

[1] INSA de Lyon, Laboratoire de Mécanique des Contacts et des Solides,
Bâtiment Jacquard, 27 av. Jean Capelle, 69621 Villeurbanne Cedex, France
Philippe.Boisse@insa-lyon.fr

[2] University of Twente, Faculty of Engineering Technology, Composites Group, PO Box
217, 7500 AE Enschede, the Netherlands
r.akkerman@utwente.nl

[3] Northwestern University, Department of Mechanical Engineering,
2145 Sheridan Road, Evanston, IL 60208, USA
jcao@northwestern.edu

[4] University of Massachusetts Lowell, Advanced Composite Materials

and Textile Research Laboratory, One University Ave. Lowell, MA 01854, USA
Julie_Chen@uml.edu

[5] Katholieke Universiteit Leuven, Department MTM
Kasteelpark Arenberg, 44, B-3001 Leuven, Belgium
Stepan.Lomov@mtm.kuleuven.be

[6] University of Nottingham, School of Mechanical, Materials
and Manufacturing Engineering, University Park, Nottingham NG7 2RD, UK
Andrew.Long@nottingham.ac.uk

**Abstract.** The strong development of composite materials in particular in aeronautics leads to a demand for knowledge and simulation codes concerning composites forming. The mini-symposia developed on this theme within the ESAFORM conference are a privileged framework for the researchers in this field to communicate and to exchange ideas. This paper gives a progress report on the advancements achieved during these ten last years and on the research prospects in the various fields of composite material forming.

**Keywords:** Composites Forming, Material characterization, Constitutive modelling, Forming simulation, Mesoscopic analyses, Resin injection, Benchmarks

# 1  Introduction

The use of composite materials is expanding rapidly in various fields, in particular in civil aeronautics. The centre wing box of A380 is made from composite materials; the future aircraft of Boeing and Airbus are planned with composite wings and a composite fuselage.

The use of these composite materials became a major issue in order to decrease the mass and the fuel consumption of the aircraft. These high level applications of composite materials created a significant demand for scientific knowledge and computational tools of composite materials.

The mechanical behaviour of the composite materials in service is dominated by the fibre orientation and density. The fibre distribution, in turn, is determined by the forming process. Hence, not only the in service performance (stiffness, damage, fatigue,…) has to be predicted, but certainly also the complex manufacturing processes, of which there are many. Their knowledge and their modelling are essential for the analysis of the composite structures in service.

Since 2001 and the Liege Esaform conference, an annual "Composite forming" mini symposium gathers researchers from Europe, and also from USA, Asia and Australia, who can present their works (about twenty communications each year) and exchange their points of view concerning research in the field of composite forming. Conference ESAFORM thus became a privileged and single place for this subject. Experimental and numerical "benchmarks" were set up on the initiative of J. Cao and J. Chen. They are discussed within the composites forming mini-symposia.

This chapter was written by the coordinators of the mini-symposia whose teams are strongly involved in the ESAFORM conferences. It reports on the current state of research and the prospects in the field, in particular on materials characterisation, constitutive laws during forming, forming simulations, mesoscopic analysis of the geometry of the textile reinforcements, resin injection and the benchmark efforts.


## 2  Materials Characterisation

### 2.1  Intra-ply shear

This mechanism occurs when the material is subjected to in-plane shear, and is usually considered to be the primary deformation mechanism for aligned fibre based materials. Coupled with low bending resistance, the ability of materials to shear allows them to be formed to three dimensional shapes without folds or wrinkles. Amongst the composites forming community, two widely used test methods to characterise intra-ply shear are the picture frame test [1-4] and the bias extension test [4-5] (see Fig. 1).

The picture frame (PF) test measures shear force versus shear angle (or strain) directly. Cross-shaped test samples are used, mounted in a square frame hinged at each corner. The frame is loaded into a tensile test machine, and two diagonally opposite corners are extended, imparting pure and (ideally) uniform shear in the test specimen on a macroscopic scale. For pre-impregnated (prepreg) materials, tests should be conducted at different rates and temperatures. Care should be taken to ensure that the fibres are perfectly aligned with the edges of the rig.

Typical results from PF experiments are shown in Fig. 2. For clarity these graphs represent single experiments, although it should be noted that scatter of up to ±20% in shear force is typical for a particular angle. It is usual for samples to wrinkle towards the end of the test (between 40° and 60° of shear deformation) and by careful observation the test can be used to estimate the "locking angle" of a material. Whilst graphs like these capture the essential mechanical response of the material, more detailed analysis during testing can provide insights into material deformation behaviour. For example Harrison [5-6] and Zouari [7] have demonstrated that the rate of deformation within the tows is consistently lower than that of the sample as a whole. This is related to the balance between tow shear resistance and inter-tow friction; such observations are helping in development of fully predictive models for textile composites forming [8].

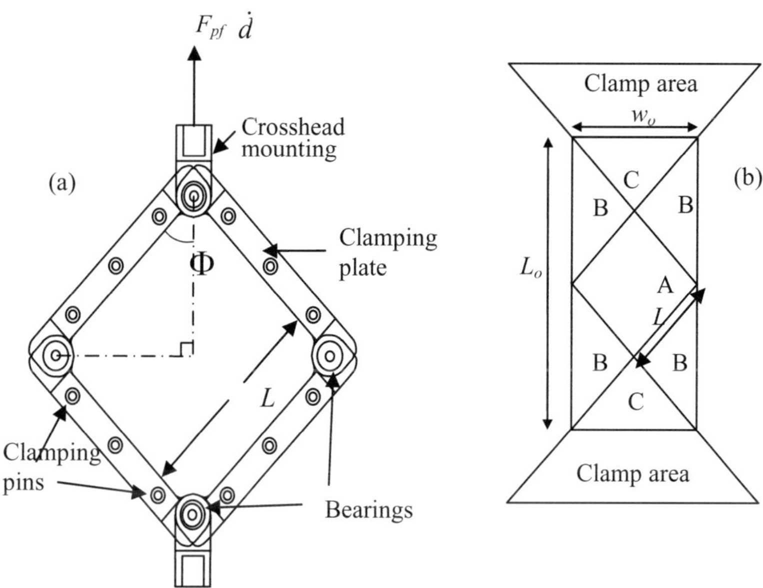


**Fig. 1.** Intra-ply shear test methods. (a) Picture frame. (b) Bias extension.

The bias extension (BE) test involves clamping a rectangular bidirectional sample with tows orientated initially at $\pm$ 45° to the direction of applied force. The material sample can be characterised by the aspect ratio, $\lambda = L_o/w_o$ , where the sample width $w_0$ is usually greater than 100mm. Fig. 1(b) shows an idealised BE test sample with (the minimum value suitable for testing). Classically it is assumed that the shear angle in region A is twice that in regions denoted B, while region C remains un-deformed. The deformation in region A is similar to that produced by a PF test, although for a given sample extension the shear angle is lower in a BE sample as intra-ply slip will occur (ie. tow spacing will increase). Recently Creech [9] has shown that this classical description is a simplification, and that in fact deformation varies considerably over the sample.

BE tests are simple to perform and can provide reasonably repeatable results. In particular the test provides a useful method to estimate the locking angle of a material; once the material in region A reaches the locking angle, it usually ceases to shear. Shear deformation within region A may need to be recorded by visual analysis, which can prove time consuming without an automated image acquisition and analysis approach. Forces from bias extension tests can be normalised by dividing by a characteristic dimension, such as sample width. However this does not allow data from samples with different aspect ratios to be compared directly. Recent work by Harrison4 considered the energy dissipated within regions A, B and C to develop a more sophisticated normalisation technique for bias extension test data. In addition to allowing results from different aspect ratio samples to be compared, this also allows PF and BE test data to be correlated directly.

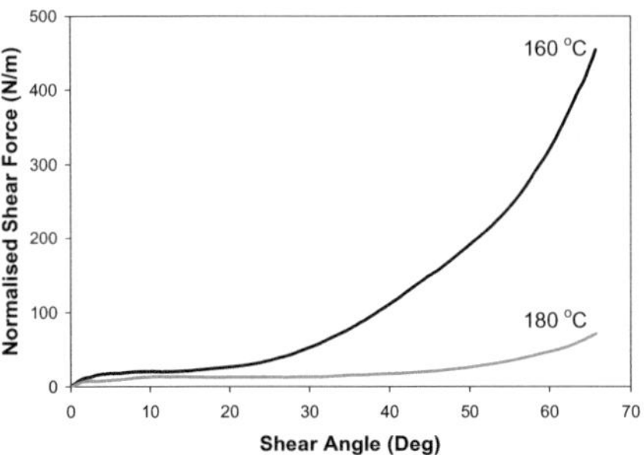


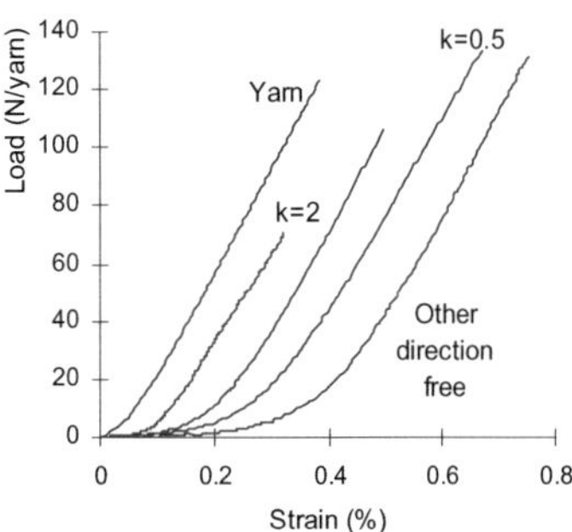


**Fig. 2.** Picture frame shear data for 2:2 twill weave pre-consolidated glass/polypropylene thermoplastic composite at two temperatures.

**Fig. 3.** Results of biaxial tensile tests for a balanced glass plain weave fabric. The load is measured along one tow direction, with the constant $k$ determining the ratio between strains in the loading and transverse directions.

## 2.2  Axial loading

Loading of aligned fibre based materials along the fibre axes typically results in very large forces and very low maximum strains in comparison to intra-ply shear. This might suggest that axial deformation is of secondary importance, although Boisse [10] has long argued that this behaviour cannot be neglected as this high resistance accounts for the majority of energy dissipated during forming. Axial loading of textile composites can be conducted using standard tensile testing equipment using wide test samples. Woven materials exhibit an initial non-linear stiffening due to crimp in the tows. As the fibres become aligned with the load direction, the response becomes linear and is determined by the fibre modulus. The importance of this "de-crimping" depends on the properties of the transverse tows, and in particular their resistance to bending and compaction. If the transverse tows are also loaded, then the de-crimping zone will decrease in magnitude. Boisse [11] has analysed a wide range of fabrics using a specially designed biaxial loading frame. Typical results are given in Fig. 3 for a plain weave fabric. When loaded uniaxially (denoted "other direction free"), the non-linear region extends to a strain of approximately 0.5%, but the force to completely straighten the tows is low. As the ratio between strains in the tested (warp) and transverse (weft) directions increases, the force curve tends towards the behaviour of an individual tow (yarn).

## 2.3  Ply/tool and ply/ply friction

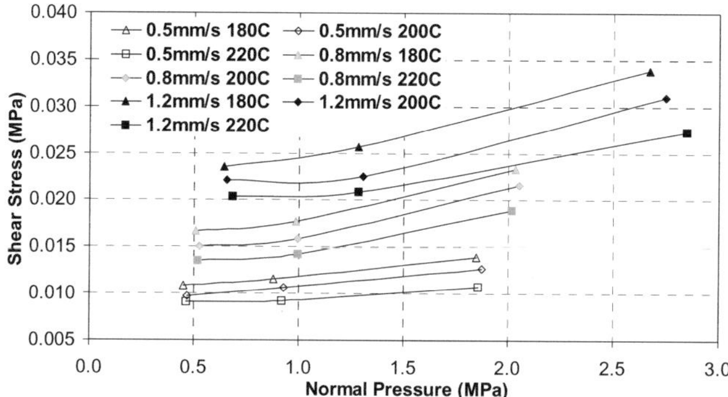


**Fig. 4.** Shear stress at the interface between steel tools and pre-consolidated 2:2 twill weave glass/polypropylene under applied normal pressure. Results are for pull-out velocities of 0.5, 0.8 and 1.2mm/s at 180°C, 200°C and 220°C.

During an automated forming operation, friction between the material and forming tools governs the transfer of loads to the material. In multi-layer forming, friction between individual layers is also of importance. Ply pull out tests have been used to measure ply/ply friction [12]. Murtagh [13] developed a device whereby a layer of thermoplastic prepreg or tooling material was sandwiched between two prepreg layers, held together with a controlled normal pressure. The whole apparatus was heated to evaluate the effect of temperature on behaviour, and the tooling material was withdrawn at various rates with the required force measured using a load cell. Wilks [14] and Gorczyca [15] used apparatus based on a similar principle. Typical results are shown in Fig. 4 for friction between a glass/polypropylene thermoplastic composite and a steel tool. Shear stress increases with normal pressure, although the relationship is not linear. At a given normal pressure, the shear stress increases with increasing rate and with decreasing temperature.

# 3  Constitutive modelling for composites forming

The material properties of the composite sheets during the forming processes are determined by the fibre and resin properties, their respective fractions, the reinforcement structure and orientation. The stiffness of the fibres generally exceeds the resin stiffness by several orders of magnitude. Glass and carbon fibres are usually linearly elastic up to fracture, organic fibres exhibit rate/time dependent behaviour.

The polymer resin, whether thermoplastic or thermoset, typically exhibits viscoelastic behaviour. The composite sheet constitutive behaviour will be an anisotropic function of the constituent properties. Closed form expressions as well as numerical multiscale approaches are in use for composites forming simulations.

## 3.1  The past

The composite material during forming can be interpreted as a single continuum or as a system, composed of several fractions, usually connected in series or in parallel. Elastic, viscous, viscoelastic and plastic types of descriptions have been used to this end. The resulting constitutive equation will be an addition of terms related to the separate constituents and the reinforcement structure [16].

The framework for continuum type constitutive equations of fibre reinforced materials was formulated in Nottingham [17-18], based on continuum mechanics arguments. The simplest equations result for unidirectionally reinforced fluids (assuming transversal isotropy) subject to constraints of incompressibility and fibre-inextensibility,

$$\tau\left(\boldsymbol{D},\boldsymbol{a}\right)=2\eta_{T}\boldsymbol{D}+2\left(\eta_{L}-\eta_{T}\right)\left(\boldsymbol{A}\cdot\boldsymbol{D}+\boldsymbol{D}\cdot\boldsymbol{A}\right)+\gamma\boldsymbol{A}\,\mathrm{tr}\left(\boldsymbol{A}\cdot\boldsymbol{D}\right),\tag{1}$$

with the extra stress $\boldsymbol{\tau}$, the rate of deformation tensor $\boldsymbol{D}$, fibre orientation vector $\boldsymbol{a}$ and its dyadic product $\boldsymbol{A}=\boldsymbol{aa}$, and material dependent constants $\eta_L$, $\eta_T$, $\gamma$. This Ideal Fibre Reinforced Fluid (IFRF) model was also formulated for woven fabrics with biaxial reinforcement in terms of an additional fibre direction $\boldsymbol{b}$,

$$\begin{aligned}\tau\left(\boldsymbol{D},\boldsymbol{a},\boldsymbol{b}\right)=2\eta\boldsymbol{D}+2\eta_{1}\left(\boldsymbol{A}\cdot\boldsymbol{D}+\boldsymbol{D}\cdot\boldsymbol{A}\right)+2\eta_{2}\left(\boldsymbol{B}\cdot\boldsymbol{D}+\boldsymbol{D}\cdot\boldsymbol{B}\right)+\\+2\eta_{3}\left(\boldsymbol{C}\cdot\boldsymbol{D}+\boldsymbol{D}\cdot\boldsymbol{C}^{T}\right)+2\eta_{4}\left(\boldsymbol{C}^{T}\cdot\boldsymbol{D}+\boldsymbol{D}\cdot\boldsymbol{C}\right),\end{aligned}\tag{2}$$

with dyadic products $\boldsymbol{B}=\boldsymbol{bb}$ and $\boldsymbol{C}=\boldsymbol{ab}$ and anisotropic viscosities $\eta_i$. The continuum approach disregards detailed micromechanics. Structure related properties can be included by fitting the material property data to experimental results.

## 3.2  The present

Biaxial woven fabrics are the most investigated materials in the composites forming area. The resistance to intraply shear (Fig.5) is seen as the dominant term in the ply deformation behaviour. Various research groups have used characterisation experiments to find and quantify detailed deformation mechanisms, in both dry fabrics and impregnated plies. Meso-mechanical analyses of the deformations of the fibre tows in Representative Volume Elements have been performed to gain a better understanding of the material's behaviour. Bending, shear, extension, buckling and compaction of the tows have been recognised as important features in the dry fabric deformation characteristics [19-20]. The intratow shear was found to be non-zero during trellis deformation of woven prepreg [21], leading to further complexities in addition to the IFRF model (2).

Implementation of the anisotropic constitutive equations in finite element formulations has been found far from straightforward [22-24]. The high anisotropy with small fibre strains and large shear strains easily leads to erroneous numerical results. Considerable effort has been spent in order to formulate accurate objective constitutive formulations, taking into account the evolution of the non-orthogonal reinforcement structure and e.g. appropriate fibrous derivatives.

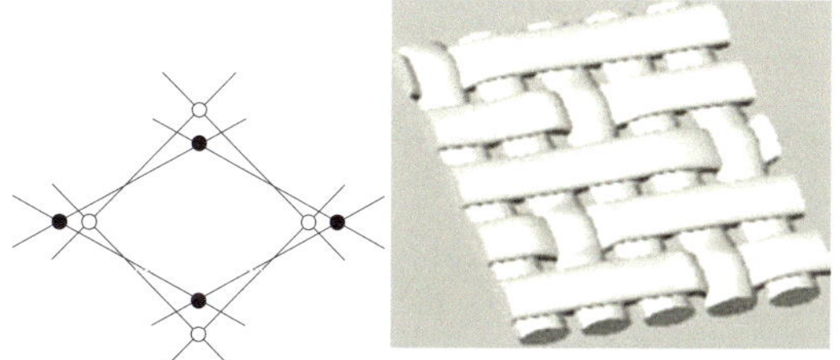

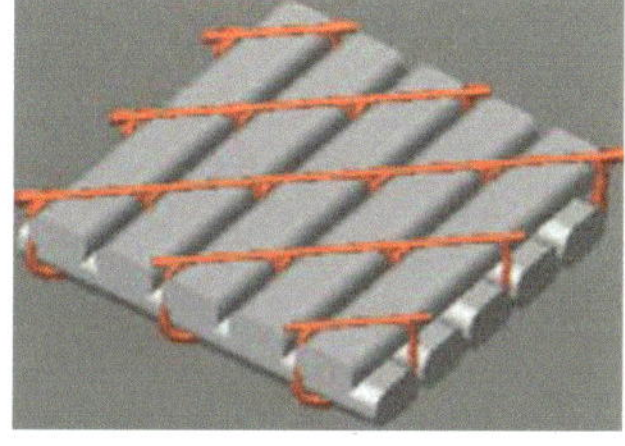

**Fig. 5**. Intraply or 'trellis' shear deformation of biaxial fabrics

**Fig. 6**. Meso-mechanical geometry model of a biaxial woven fabric (left) and a non-crimp fabric (right)

Biaxial relations between the fibre stresses and strains in the two fibre directions were identified by experiments and meso-mechanical analysis (Fig.6). These were implemented in a constitutive model as a separate structure related term. Non-Crimp Fabrics exhibit further structure related phenomena, induced by stitch deformations and fibre slip through the stitches. These are typically modelled as a separate fraction in the total stress-strain relation. The fibre slip adds an extra complexity, e.g. as a convection term to describe the fibres moving through a reference frame formed by the stitches.

Actual products are usually multilayered composite components. The forming process analysis hence has to consider the interply phenomena such as shear, debonding during laminate heating and reconsolidation during the compression stage. Phenomenological models have been used to include these interply deformations, e.g. by assuming an intermediate resin layer of arbitrary thickness. A similar approach is used for the friction of the laminate with the forming tools.

Process induced residual stresses have been subject of investigations. The relevant phenomena have been identified and constitutive equations have been implemented to allow for the phase changes during the forming process. Generic definitive solutions are not available yet. It is clear that fibre stresses originating from the forming process play an important role in the composite product distortions after forming.


## 3.3  The future

The meso-mechanics of the reinforcement structure in the composite material will be investigated further without a doubt. Meso-mechanics can be applied in numerous areas to explain the composite response under forming conditions. Friction between plies and between the laminate and the tooling are the first phenomena that come to mind. Meso-mechanical considerations also indicate that the relatively low shear resistance of biaxial fabrics is affected significantly by the fibre stresses. This defines a whole research area in its own, in terms of experimental characterisation and constitutive modelling on the meso and macro level.

High precision composites forming requires further quantitative analysis. Esaform's composites forming society has the potential to make good progress in this challenging research area, combining the expertises of various research groups on material characterisation as well as on numerical modelling.

# 4   Composite reinforcement forming simulation

There are several reasons for simulating the composite textile reinforcements (or performs) forming process. First, the simulation can give the conditions (for instance initial orientations, type of material, loads on the tools etc) that will make the forming possible. It also can determine possible defects after forming (wrinkles (Fig. 7), porosities, yarn fracture etc). Secondly, and unique to composites forming analysis, is the need to know at any point the positions of the fibres after forming. The directions and densities of fibres are very important for analysing the composite part in use (stiffness, damage, fatigue etc)

## 4.1   The past

The methods initially used for composite reinforcement draping simulations are "kinematical models" [25]. Several packages are commercially available for fifteen years. In these methods, the knowledge of the shape on which the fabric is formed and initial geometrical conditions leads to the positions of the yarns on the final form. The length of the yarns is constant and the rotation of warp and weft yarn at each intersection is free. This leads to a local geodesic problem that is generally non linear but very small and the resolution is very fast. These approaches are fairly efficient especially in the case of hand operated draping of classic fabrics or prepregs. These methods are routinely used today in design offices.

## 4.2   The present

In order to perform physical (or mechanical) analysis of a composites forming process, in contrast with the previous geometrical analyses, the model must include all the equations for the mechanics, especially equilibrium, constitutive equations, and boundary conditions. These equations must be solved numerically, with some approximations. Finite Element analysis of the composites forming process includes modelling the tools, the contact and friction between the different parts, and the mechanical behaviour of the composite during forming. This is highly specific because of the internal structure of the fabrics. The macroscopic behaviour is much dependent of possible motion and interactions of yarns at meso-scale (scale of the woven unit cell) and at the micro-scale (level of the fibres constituting yarns). The main aspect for the FE approach lies in the requirement for accurate models of all the significant aspects of the forming process. There are different ways that are used and still developed today.

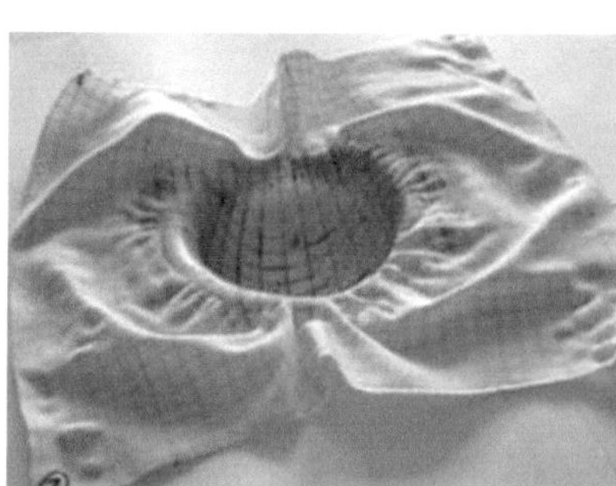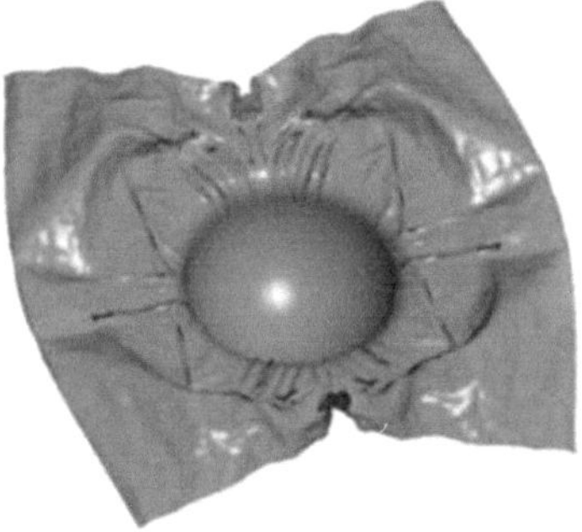

**Fig. 7.** Forming of an unbalanced textile composite. Experiment and FE simulation

### 4.2.1  Continuous approaches

The multiscale nature of the composite and of its fibrous reinforcement permits different possible approaches. The first one considers the fibrous reinforcement as a continuum. The reinforcement is not continuous at lower scales but it is usually continuous in average and a continuous material superimposed to the fibrous material can be considered. The advantage of the continuous approach is that it can be used in standard finite elements. Nevertheless, the constitutive model of this continuum will have to convey the very specific mechanical behaviour of the fibrous reinforcement. Especially this behaviour is mainly depending on the fibre directions that are strongly changing during forming. Most of the proposed continuous approaches for F.E. draping simulations are based on rate constitutive equations (or hypoelastic approaches). This permits to develop the models in user subroutine such as VUMAT in Abaqus code. The stresses are cumulated in a rotated frame. In [26][27] the Green Nagdi frame (which is the frame of Abaqus explicit) is used and the constitutive matrix is deduced in this basis from behaviour in the fibre basis for one single direction of fibre in [26] and two non orthogonal directions in [27]. In [28][29] a specific objective derivative is defined from the fibre rotation.

### 4.2.2  Discrete approaches

The discrete approaches are the opposite of the continuous approach, and consider and use F.E. models of the components of fibrous reinforcement at low scale. [30-32] These components can be yarns, woven cells or stitching, and also sometimes fibres. Because these elements are usually at the mesoscale, the approach is also known as meso-mechanical modelling. Some analyses have been proposed where all fibres are modelled (microscale modelling) [33], but the number of fibres in a composite structure limits these computations to small sub-domains, for instance a woven cell or a few braided or knitted loops. A major difficulty lies in the description of the components at mesoscopic scale, usually the woven yarns. A compromise must be found between a precise description (which will be expensive from the computation time point of view) and a simple description, where it is possible to compute the entire forming process.

### 4.2.3  Semi discrete approach

The semi-discrete approach is a compromise between the above continuous and discrete approaches [34][35] (figure 7).  A finite element method is associated to a mesoscopic analysis of the woven unit cell. Specific finite elements are defined that are made of a discrete number of woven unit cells. The mechanical behaviour of these woven cells is obtained by experimental analyses or from 3D F.E. computations of the woven cell. The nodal interior loads are deduced from this local behaviour and the corresponding strain energy in the element deformation. The objective of the approach is to use a description of the yarns (or woven cells) at mesoscopic level while keeping a limited number of degrees of freedom. This description of the fabric by finite elements needs to assume that two points of a weft and a warp yarns initially superimposed remain superimposed after forming. i.e. there is no translation sliding between the yarns. That has been experimentally shown in most cases.

### 4.3  The future

It is difficult to say today which of the continuous or discrete approach for FE analyses of composite forming reinforcement will establish itself as the approach used in the future. A point in favour of analysis at the mesoscopic scale lies in the strong increase in computer efficiency.

Until now, most of the simulations have concerned one thin reinforcement layer forming. The increase of composite use in aeronautical applications leads to focus on large thickness composites forming. It can be composites with many plies (100 for instance) or interlock thick textile reinforcements. Some work has started in both directions [36] and will be a field where the specificity of composite behaviour will need special simulation tools.

## 5  Mesoscopic geometry and deformation during forming

The deformability of textile fabrics and its dependency on the fabric structure is an important issue for apparel textiles. Kawabata, Niva and Kawai, de Jong and Postle, Hearle and Shanahan have established in 70s an approach to mathematical modelling of deformation of textiles on mesoscopic structural level (unit cell of the fabric), based on descriptions of internal geometry of fabrics by Peirce and others. New challenges for textile mechanics were opened when different processes of draping of composites textile performs came in the order of the day in 90s. Deformability of textile preforms plays a key role in the quality of a composite part formed into a 3D shape. The deformation modes of primary importance are in-plain deformation (tension and shear) and compression of the preform. Comparing with the apparel-oriented models, textile mechanics of composite perform must include description of behaviour under large loads, with deformations close to the jammed, or "locked" yarn structures. Hence the finite element analysis comes in the order of the day together with approximate models, elaborated towards better description of yarn interaction. Deeper understanding of mesoscopic behaviour forms a basis for adequate macro-models, used for simulation and optimisation of draping. Textile architectures of greater complexity need to be modelled, as three-axial and three-dimensional weaves and braids, non-crimp fabrics with complex stitching patterns.

The rise of textile mechanics research in 90s, answering these challenges, coincided with emerging of the ESAFORM community. The textile mechanics research on the mesoscopic level in the past ten years has been recently summarised in two book chapters [37-38].

## 5.1 Internal structure of textile [37, 39]

Modelling of the internal structure of textile can be subdivided into three steps: (1) coding of the interlacing of the yarns (weave, braiding pattern, stitching structure...); (2) reconstruction of the yarns paths; (3) reconstruction of the yarn volumes.

Woven and knitted patterns are coded comprehensively in textile CAD systems, but this is far away from the composite forming models. Simple checkerboard/matrix woven patterns are used widely to describe simple 2D weaves. Lomov has developed a coding for 2D and 3D weaves, two- and three-axial braids, coding for warp- and weft-knitted patterns, which can be created and edited with graphical interactive tools.

Yarns in a relaxed fabric occupy paths determined by equilibrium or minimum energy conditions, which take into account yarn interactions which bend and compress the yarns. This problem is solved for complex textile topologies, to yield models of internal geometry of 2D and 3D woven, two- and three-axially braided, weft-knitted knitted and non-crimp fabrics, covering all the textile architectures used as reinforcement. If parameters of internal geometry, such as dimensions of the yarn cross-section and crimp, are known from experiment or postulated a priori, then the yarn paths can be reconstructed directly, using (in works of different authors) Peirce model, elastica curves, splines or Bezier curves.

Reconstruction of yarn volumes presents a difficult task, as, due to approximate nature of the yarn path models, the yarn volumes, built with a simple cross-section shape tend to interpenetrate, which makes impossible subsequent meshing. Approaches proposed to solve the interpenetration problem use local adjustment of the yarn shape.

## 5.2 Approximate models of compression, biaxial tension and shear [38]

In phenomenological models of compression models the compression curve is broken into three regions (low, medium and high loads), each dominated by different phenomena. More elaborate model uses the balance between the mechanical work of compression and change of yarn crimp and dimensions to calculate compression curve of the fabric based on compressibility of yarns and their bending rigidity. The nesting of layers of the reinforcement is taken into account.

Model of biaxial tension uses calculates equilibrium of the yarns in the unit cell under tension and tension-induced bending and compression. Elongation of the yarn is estimated by the difference between yarn length in the repeat before and after the deformation. Experimental non-linear bending and compression diagrams are used to compute bending energy and resistance to compression. The model stands comparison with experimental data on glass, carbon and commingled yarns, as well as benchmarking by finite element simulations.

Models of shear of woven reinforcements consider very high shear angles (up to 60-70°) occurring in forming of complex 3D parts. This is dealt with by an introduction of models of lateral compression of the yarns, which come into contact when the shear angle reaches and exceeds the locking angle of the fabric "trellis". The simple, but not true-to-life concept of preserving the volume of the unit cell to calculate the change of the thickness of fabric in shear, has been advanced to more correct considerations of yarn compression. More elaborate models take into consideration various sources of shear resistance, as yarn friction, compression, (un) bending and torsion, and coupling between shear and biaxial tension.

When forming thermoplastic performs, the matrix melts and viscous resistance plays an important role. Mesoscopic models of shear deformation of such materials are developed [39].

Finite element analysis of textile deformability [40]. Mesoscopic finite element analysis has been performed for bi-axial tension, shear and compression of woven and tension of knitted reinforcements, aiming at the detailed description of deformation of the textile. Finite element modelling encounters two difficulties. First, the geometry of a textile unit cell is very complex, and creating a solid model manually is not an easy task. The solution is provided by the use of a textile geometry modeller as a preprocessor, creating a finite element model automatically. Second, the description of the material behaviour used in the finite element model must realistically represent the actual behaviour of the fibrous assemblies – yarns. Development of such a library of material models for textiles, available in finite element packages, presents a serious challenge to researchers.


### 5.3   Experimental methods of registration of mesoscopic deformations [41,42]

Validation of mesoscopic models cannot be complete without comparison of the assumed deformation mechanisms with the experimentally observed behaviour on the mesoscopic level. Recent advances in full-filed registration of strain in textiles under shear and biaxial tension have made such comparison possible and created confidence in the models.

Encapsulation of mesoscopic models in constitutive models for macro-simulations [40]. Any model of reinforcement deformation faces serious difficulties if an attempt were made to use it in a draping simulation. The theory would have to give values for the resistance in all the finite elements in the draping model, which means that thousands of cases of unit cell deformation have to be calculated. For this reason, the mechanical calculations for a unit cell are done on beforehand and limiting calculations for each element to simple regression formula or to a fast search in a table [43]. For a similar reinforcements (for example, typical woven glass or carbon fabrics) the regression formulae may be parametrised vis-à-vis fabric parameters (tow linear density, fabric tightness) [44].

## 5.4  Software tools

Mesoscopic modelling of internal structure and deformability of textiles is implemented in a software package WiseTex (K.U.Leuven) [40]. The models calculate internal structure of fabric using minimum energy principle for the relaxed and deformed state, resistance of fabric to compression, biaxial tension and shear. Internal geometry of reinforcements is described using a unified data format, which can be visualised and exported for modules calculating stiffness of impregnated composite, permeability of textile, or to finite element packages (accompanying MeshTex software, University of Osaka).

Software package TexGen (University of Nottingham) [45] creates geometric model of unit cell of textile reinforcement, which can be further exported to finite element package to calculate mechanical response of textile or fluid flow through it.

# 6  Resin infusion and permeability

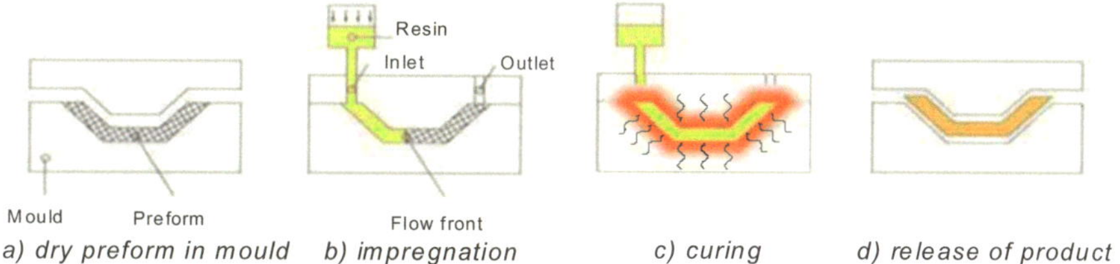


**Fig. 8.** The resin transfer moulding process

Resin transfer moulding (RTM) is a popular method to impregnate dry fibre reinforcement structures [46], see Fig. 8. This process consists of filling a rigid and closed mould cavity by injecting a resin through one or several gates, depending on the size of the component. A subsequent curing reaction leads to a solid composite component. Thermoset resins are generally used, offer a sufficiently low viscosity during infusion. RTM allows the moulding of components with complex shape and large surface area with a good surface finish on both sides. The process is typically suited for small to medium series production.

The 'black box' nature of the process pushes the development of predictive modelling.

## 6.1  The past

The resin infusion process can be interpreted as flow in porous media. Darcy's law describes the relation between the flow rate $q$ and the pressure gradient $\nabla p$ as

$$q = -\frac{1}{\eta} k \cdot \nabla p \tag{3}$$

The flow resistance of the porous medium is represented by the permeability tensor k and $\eta$ is the fluid viscosity. This type of equation is analogous to the Hele Shaw approximation for injection moulding of (also) thin walled products. Earlier extensive work on injection moulding simulations, e.g. at Cornell University developing the C-MOLD program, could be extended to RTM simulations by inclusion of the permeability. The simulation results are particularly sensitive to thickness variations of the cavity (e.g. due to tool deflection) and the permeability values. Permeability measurements typically resulted in large scatter [47,48] in the data obtained by various institutes on similar reinforcements.

## 6.2  The present

Reasonable qualitative predictions can be made of the 'standard' filling process. Various commercial RTM simulation codes are currently available, describing the infusion process in 2D and 3D, and in some cases thermal problems during cure are addressed. Simulations can thus assist in proper positioning of the vents. For constant pressure injection the actual filling time is rarely predicted accurately, however. A wide variety of infusion processes has emerged: vacuum driven, sometimes using flexible tooling, possibly using internal or external flow enhancement layers and infusion through the thickness, in situ polymerising thermoplastics, etc.

The permeability plays an even more important role in these processes, as it varies with fibre compaction. A number of predictive models have been developed to predict the permeability of various textile structures by means of meso-modelling [49-52]. The geometry of a representative volume of the fibre reinforcement is used as an input to detailed flow models, of which the relation between the flow field and the pressure difference determines the averaged permeability tensor. An international benchmark exercise has started from the Esaform community to measure and simulate the permeability of a set of fabrics and of reference material with a representative well defined meso geometry.

## 6.3  The future

Process control is the driving force of the ongoing research work on resin injection. The tooling should be designed quickly and right-first-time, resulting in process optimised product quality. The predictive methods under development are required for design tools in which the product and process can be optimised quickly and virtually, hence with minimum hardware costs. This type of processing information can further be elaborated to develop control strategies, for inline process control, steering gates and vents based on sensor information [53].

In this particular area there will be ample attention for the statistics: how do the distributions of the fabrics' meso-geometry affect the overall permeability values? The benchmark activities will hopefully tell us more about the proper measurement procedures and the statistics in the actual permeability.

The transient aspects of infusion will be further elaborated in the years to come. Think of wetting phenomena, with effects on fibre rearrangement and on void generation. Extending the work on infusion to consolidation processes, e.g. during prepregging, opens a wide field with a challenging interplay between permeation, fabric deformation and the temperature distribution in a transient multiphase system. The dynamics void formation and mobility during infusion must be understood to predict the resulting composite material quality. Multiphase analyses will play an important role also in this respect.

## 7    Benchmark effort on material testing and forming of woven composites fabric

Woven-fabric reinforced composites (hereafter referred to as woven composites) have attracted a significant amount of attention from both industry and academia, due to their high specific strength and stiffness as well as their supreme formability characteristics. However, applications of these materials have been hampered by a lack of low-cost fabrication methods, as well as robust simulation methods. Designing low cost manufacturing processes requires accurate material modelling and process simulation tools. Recognizing these requirements, a group of international researchers gathered at the University of Massachusetts Lowell for the NSF Workshop on Composite Sheet Forming in September 2001. The main objectives of that workshop were to better understand the state-of-the-art and existing challenges in both materials characterization and numerical methods required for robust simulations of forming processes. One direct outcome of the workshop and the effort to move towards standardization of material characterization methods was a web-based forum exclusively for research on forming of woven composites, established in September 2003, at http://nwbenchmark.gtwebsolutions.com/. Meanwhile, a benchmark activity was starting to form.

Material property characterization and material forming characterization were two main areas related to material testing identified at the 2001 NSF Composite Sheet Forming workshop. Standard material testing methods are necessary for researchers to understand the formability of the material, the effect of process variables on formability, and to provide input data and validation data for numerical simulations. Thus, the researchers embarked on a benchmarking project to study, understand and report the results of material testing efforts currently in use around the world for woven composites to make recommendations for best practices.

- Material property characterization: Three different commingled fiberglass-polypropylene woven-composite materials were used for this research. The materials were donated by Vetrotex Saint-Gobain in May 2003 and were distributed in July 2003 to the following research groups: Hong Kong University of Science and Technology (HKUST) in Hong Kong, Katholieke Universiteit Leuven (KUL) in Belgium, Laboratoire de Mécanique des Systèmes et des Procédés (LMSP) in France, Northwestern University (NU) in the U.S.A., University of Massachusetts Lowell (UML) in the U.S.A., University of Twente (UT) in the Netherlands, and University of Nottingham in U.K.

- Material forming characterization: Martyn Wakeman of EPFL and Patrick Blanchard of Ford have led the effort on stamping tests in 2004 and 2005, utilizing a double-dome geometry designed for model validation through measurement of local geometric deformation and local mechanical characterization.

- Friction characterization: University of Massachusetts, Lowell and Hong Kong University of Science and Technology (HKUST) have performed yarn-to-yarn friction tests and fabric-to-tool friction tests.

Since 2002, the group has taken ESAFORM as a venue, in addition to the benchmark website, to exchange research findings related to composite material characterization and forming characterization. The first group paper on the material characterization was published in the Proceedings of ESAFORM 2004 [54]. Continuous efforts have been made in finishing the second phase of this benchmark activity and in publishing the results in a special journal issue.


# 8   Conclusion


Research concerning composites forming is an active field that is supported by the development of composites in various industrial applications and in particular aeronautics. Good progress has been achieved during these ten last years as described in this chapter. Nevertheless, the composites forming processes are numerous, complex and often new. As a consequence, considerable experimental, modelling and simulation efforts are required in this field

The Composites Forming group is very active within the ESAFORM association. Practically all international research groups working in this field participate in the "Composites Forming" mini-symposium thanks to the focus of this mini-symposium on their field of research. Hopefully, the ten next years of ESAFORM will be at least as profitable as the past ten years in the field of composites forming.


# References


1. McGuiness G B and O'Bradaigh C M, Development of rheological models for forming flows and picture frame testing of fabric reinforced thermoplastic sheets, Journal of Non-Newtonian Fluid Mechanics, 73 (1997) 1-28
2. Long A C, Rudd C D, Blagdon M and Johnson M S, 'Experimental analysis of fabric deformation mechanisms during preform manufacture', Proc. 11th International Conference on Composite Materials (ICCM-11), Gold Coast, Australia, (1997) 238-248
3. Prodromou A G and Chen J, 'On the relationship between shear angle and wrinkling of textile composite preforms', Composites Part A, 28A (1997) 491-503

4. Harrison P, Clifford M J and Long A C, 'Shear characterisation of woven textile composites: a comparison between picture frame and bias extension experiments' Composites Science and Technology, 64 (10-11) (2004) 1453-1465

5. Wang J, Page R and Paton R, 'Experimental investigation of the draping properties of reinforcement fabrics', Composites Science & Technology, 58 (1998) 229-237

6. Harrison P, Clifford M J, Long A C and Rudd C D, 'A constituent based predictive approach to modelling the rheology of viscous textile composites', Composites Part A  37-7-8 (2005) 915-931

7. Zouari B, Dumont F, Daniel J L, Boisse P, 'Analyses of woven fabric shearing by optical method and implementation in a finite element program', Proc. 6$^{th}$ Int. ESAFORM Conf., Salerno, (2003), 875-878

8. Lin H, Long A C, Clifford M J, Wang J, Harrison P, 'Predictive FE modeling of prepreg forming to determine optimum processing conditions', Proc. 10th Int. ESAFORM Conf., Zaragoza, (2007)

9. Creech G, Pickett A K, Greve L, 'Finite element modelling of non-crimp fabrics for draping simulation', Proc. 6th Int. ESAFORM Conf., Salerno, (2003) 863-866

10. Boisse P, Borr M, Buet K and Cherouat A, 'Finite element simulations of textile composite forming including the biaxial fabric behaviour', Composites Part B, 28-4 (1997) 453-464

11. Boisse P, Gasser A and Hivet G, 'Analyses of fabric tensile behaviour: Determination of the biaxial tension-strain surfaces and their use in forming simulations', Composites Part A, 32-10 (2001) 1395-1414

12. Scherer R and Friedrich K, 'Inter- and intraply-slip flow processes during thermoforming of CF/PP-laminates', Composites Manufacturing, 2-2 (1999) 92-96

13. Murtagh A M, Characterisation of shearing and frictional behaviour in sheetforming of thermoplastic composites, PhD Thesis, University of Limerick, 1995

14. Wilks C E, Processing technologies for woven glass/polypropylene composites, PhD Thesis, University of Nottingham, 2000

15. Gorczyca J L, Sherwood J A, Chen J, 'Friction at the tool-fabric interface during the thermostamping of woven commingled glass-polypropylene composite fabrics', Proc. 7th Int. ESAFORM Conf., Trondheim, (2004) 337-340

16. S.W. Hsiao and N. Kikuchi, "Numerical analysis and optimal design of composite thermoforming process", Comp. Meth. in Appl Mech. Engrg. 177:1-34 (1999)

17. T.G. Rogers, "Rheological characterisation of anisotropic materials", Composites, 20 (1989) 21-27

18. A.J.M. Spencer, "Theory of fabric-reinforced viscous fluids", Composites Part A, 31 (2000) 1311-1321

19. S.V. Lomov, Tz. Stoilova and I. Verpoest, "Shear of woven fabrics: theoretical model, numerical experiments and full field strain measurements", Proc. 7th Int. Conf. on Material Forming (ESAFORM 2004), Trondheim, Norway, (2004) 345-348

20. G. Hivet and P. Boisse, "Analytical mechanical model for biaxial tensile behaviour during forming of woven composite reinforcements", Proc. 9th Int. Conf. on Material Forming (ESAFORM 2006), Glasgow, United Kingdom, (2006) 787-790

21. P. Harrison, A.C. Long, M.J. Clifford and C.D. Rudd, "Constitutive modelling of shear deformation for impregnated textile composites", Proc. 6th Int. Conf. on Material Forming (ESAFORM 2003), Salerno, Italy, (2004) 847-850

22. X.Q. Peng, J. Cao, J. Chen, "Stamping simulation of woven composites using a non-orthogonal constitutive model", Proc. 6th Int. Conf. on Material Forming (ESAFORM 2003), Salerno, Italy, (2004) 827-830

23. B. Hagège, P. Boisse and J.L. Billoët, "Analysis and simulation of the constitutive behaviour of fibrous reinforcements", Proc. 7th Int. Conf. on Material Forming (ESAFORM 2004), Trondheim, Norway, (2004), 317-320

24. R.H.W. ten Thije, R.Akkerman and J. Huétink, "Large Deformation Simulation of Anisotropic Material", Proc. 9th Int. Conf. on Material Forming (ESAFORM 2006), Glasgow, United Kingdom, (2006), 803-806.

25. Van Der Ween F., Algorithms for draping fabrics on doubly curved surfaces, International Journal of Numerical Method in Engineering, 31, (1991) 1414-1426

26. L. Dong, C. Lekakou and M. G. Bader, Processing of Composites: Simulations of the Draping of Fabrics with Updated Material Behaviour Law, J. Comp.. Mat. 35 (2001) 138-163

27. X. Peng, J. Cao, A continuum mechanics-based non-orthogonal constitutive model for woven composite fabrics, Composites Part A, 36 (2005) 859–874

28. Hagège B., Boisse P.  and Billoët J.-L, 'Finite element analyses of knitted composite reinforcement at large strain', European J. of Comput. Mechanics, 14-6-7 (2005) 767-776

29. P. Boisse, A. Gasser, B. Hagege, J.L. Billoet, Analysis of the mechanical behaviour of woven fibrous material using virtual tests at the unit cell level,  Int. J. Mater. Sci. 40 (2005) 5955-5962

30. Ben Boubaker, B., Haussy, B. and Ganghoffer J.F.(2005), 'Discrete models of fabrics accounting for yarn interactions', European J. of Comput. Mechanics, 14 (6-7) 653-676

31. Skordos A., Monroy Aceves C., X, Sutcliffe M., (2005), 'Development of a simplified finite element model for draping and wrinkling of woven material' , Proceedings of the 8th Int.Conf. ESAFORM on Material Forming, Cluj-Napoca (Romania)

32. Pickett A. K., Creech G. and de Luca P., 'Simplified and Advanced Simulation Methods for Prediction of Fabric Draping', European J. of Comput. Mechanics, 14-6-7 (2005) 677-691

33. D. Durville, Numerical simulation of entangled materials properties, Journal of Materials Science, 40 (2005) 5941-5948

34. P. Boisse, K. Buet, A. Gasser, J. Launay Meso-macro mechanical behaviour of textile reinforcements of thin composites, Composites Science and Technology, 61-3, (2001) 395-401,.

35. B.Zouari, J. L.Daniel , P. Boisse, A woven reinforcement forming simulation method Influence of the shear stiffness, Computers and Structures, 84-5-6, (2006) 351–363

36. R. Akkerman, E.A.D. Lamers, S. Wijskamp, An Integral Process Model for High Precision Composite Forming, European Journal of Computational Mechanics, 15-4 (2006) 359-378

37. Lomov, S.V., I. Verpoest, and F. Robitaille, Manufacturing and internal geometry of textiles, in Design and manufacture of textile composites, A. Long, Editor, Woodhead Publishing Ltd. (2005) 1-60

38. Lomov, S.V., Virtual testing to establish material formability, in Composite Forming Technologies, A.C. Long, Editor, Woodhead Publishing Ltd. (2007), 1-60

39. Harrison, P., M.J. Clifford, A. Long, and C.D. Rudd A constituent-based predictive approach to modelling the rheology of viscous textile composites. Composites part A, 35 (2004) 915-931

40. Verpoest, I. and S.V. Lomov Virtual textile composites software Wisetex: integration with micro-mechanical, permeability and structural analysis. Composites Science and Technology, 65-15-16 (2005) 2563-2574

41. Boisse, P., B. Zouari, and A. Gasser A mesoscopic approach for the simulation of woven fibre composite forming. Composites Science and Technology, 65 (2005) 429-436

42. Lomov, S.V., A. Willems, I. Verpoest, Y. Zhu, M. Barburski, and T. Stoilova Picture frame test of woven fabrics with a full-field strain registration. Textile Research Journal, 76-3 (2006) 243-252

43. Boisse, P., B. Zouari, and J.L. Daniel Importance of in-plane shear rigidity in finite element analyses of woven fabric composite preforming. Composites A, 37-12, (2006), 2201-2212

44. Lomov, S.V. and I. Verpoest Model of shear of woven fabric and parametric description of shear resistance of glass woven reinforcements. Composites Science and Technology, 66 (2006) 919-933

45. Crookston, J.J., M.N. Sherburn, L.G. Zhao, J. Ooi, A.C. Long, and I.A. Jones, Finite element analysis of textile composite unit cells using conventional and novel techniques, in Proceedings ICCM-15 Durban, CD edition (2005)

46. C D Rudd, A C Long, K N Kendall and C Mangin, Liquid moulding technologies: Resin transfer moulding, structural reaction injection moulding and related processing techniques, Woodhead Publishing Ltd, (1997)

47. K. Hoes, H. Sol, D. Dinescu, S.V. Lomov and R. Parnas, "Study of nesting induced scatter of permeability values in layered reinforcement fabrics" Composites Part A, 35 (2004) 1407-1418

48. R. Loendersloot and R. Akkerman, "A permeability prediction for non-crimp fabrics" Proc. 9th Int. Conf. on Material Forming (ESAFORM-9), Glasgow, United Kingdom, (2006), 799-802.

49. B. Laine, G. Hivet, P. Boisse, F. Boust and S.V. Lomov, "Permeability of the Woven Fabrics: A Parametric Study" Proc. 8th Int. Conf. on Material Forming (ESAFORM-8), Cluj-Napoca, Romania, (2005) 995-998

50. I. Verpoest and S.V. Lomov, "Virtual textile composites software Wisetex: integration with micro-mechanical, permeability and structural analysis" Comp. Sci. and Tech., 65 (2005) 2563-2574

51. C C Wong, A C Long, M Sherburn, F Robitaille, P Harrison and C D Rudd, "Comparisons of novel and efficient approaches for permeability prediction based on the fabric architecture", Composites Part A 37 (2006) 847-857

52. B. Verleye, D. Roose, S.V. Lomov, I. Verpoest, G. Morren and H. Sol, "Computation of Permeability of Textile Reinforcements", Proc. 9th Int. Conf. on Material Forming (ESAFORM-9), Glasgow, United Kingdom, (2006) 735-738.

53. D.K. Modi, M.S. Johnson, A.C. Long and C.D. Rudd "Active control of the vacuum infusion process", Proc. 8th Int. Conf. on Flow Processes in Composite Materials (FPCM-8), Douai, France, (2006) 355-362

54. J. Cao, H.S. Cheng, T.X. Yu, B. Zhu, X.M. Tao, S.V. Lomov, Tz. Stoilova, I.Verpoest, P. Boisse, J. Launay, H. Hivet, L. Liu, J. Chen, E.F. de Graaf and R. Akkerman, "A Cooperative Benchmark Effort on Testing of Woven Composites" in Proceedings of ESAFORM 2004, edited by S. Stören, (2004) 305-309

# Current Status of Semi-Solid Processing of Metallic Materials

Helen Atkinson

Department of Engineering, University of Leicester, University Rd., Leicester, LE1 7RH, UK., hva2@le.ac.uk

**Abstract.** Semi-solid processing involves forming metallic alloys between the solidus and the liquidus. The microstructure must consist of solid spheroids in the liquid matrix, rather than dendrites, for the process to operate. The material then behaves thixotropically i.e. it flows when it is sheared but thickens again when allowed to stand. The behaviour was first discovered by Flemings and co-workers in the 1970s and is utilized in a family of processes, some now applied commercially. Here the current status of semi-solid processing, both technologically and from a research point of view, will be reviewed.

**Keywords:** semi-solid processing; thixoforming; rheoforming; thixocasting; thixoforging.

## 1 Introduction

In the early 1970s, a PhD student at MIT in the US was studying hot tearing of casting alloys by using a rheometer to measure the viscosity as he cooled into the semi-solid state. He found that if he was stirring the material continuously during cooling it showed less resistance to shear than if the material was cooled into the semi-solid state and then stirred [1]. The student and his supervisor identified that the microstructure of the continuously stirred materials was spheroidal (i.e. consisted of spheroids of solid in the liquid matrix) whereas the material which was cooled into the semi-solid state without stirring was dendritic (figure 1). The material with the spheroidal microstructure in the semi-solid state was *thixotropic* i.e. when it is sheared it thins and flows, when it is allowed to stand it thickens again [2]. This property can be exploited in semi-solid processing where the material is forced to fill a die. In comparison with die casting, where the flow is turbulent, here the flow is smooth and laminar and hence the resulting mechanical properties are enhanced. Consequently, parts can be made with thinner sections and hence lighter weight. As the material does not have to cool all the way from the liquid state there is less solidification shrinkage and therefore parts can be made closer to net shape. The lower temperatures than die casting mean less die attrition. In addition, parts can be heat treated and the surface finish is improved. In comparison with forging, complex shapes can be made in one shot, including undercuts, and less force is needed, hence the machinery can be on a smaller scale.

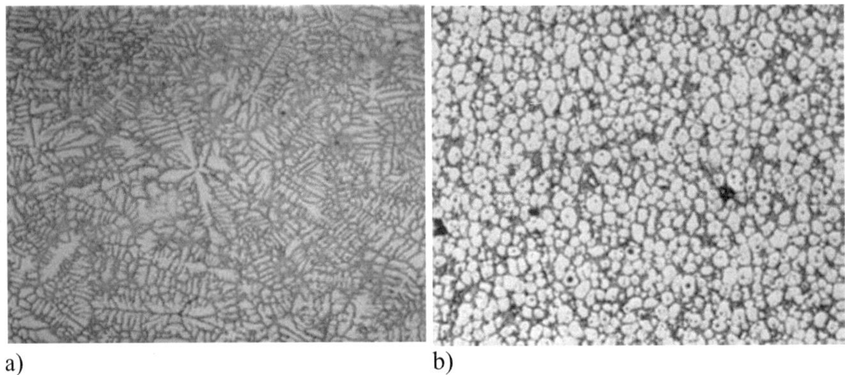

**Fig. 1.** a) Dendritic cast microstructure and b) spheroidal microstructure.

The discovery of the thixotropic behaviour of semi-solid metallic alloys has led to the development of a family of processing routes. Initially, a spheroidal microstructure has to be obtained. Sometimes this is a prior step (as in *thixoforming*, *thixoforging* and *thixocasting*) and sometimes it is integral to the forming process itself (as in *rheocasting*, *thixomolding* and *rheodiecasting*).

One route to a spheroidal microstructure is stirring. There are, however, a range of potential feedstock production routes and these will be discussed further in section 2. After stirring, the material can be solidified and cut into billet. It is then reheated. If the reheated billet is placed into the shot chamber of a die casting machine and forced into the die, this is termed *thixocasting* and usually the fraction of *liquid* is around 60% or more. If the semi-solid billet is placed between closed dies then it is *thixoforging* and the fraction *solid* will be quite high (around 60% or more). *Thixoforming*, as a term, is sometimes used to cover both thixocasting and thixoforging, or to cover a hybrid process where the billet is placed on a ram (rather than in a shot sleeve) and forced into a die. In *rheocasting*, (and in the related routes *rheodiecasting* and *thixomolding*®), the feedstock is not solidified after stirring but rather taken straight from the semi-solid state into the die casting machine.

In section 2, feedstock production routes will be discussed in more detail and in section 3 the semi-solid processing routes. Section 4 will cover industrial applications, section 5 rheology of semi-solid material, section 6 modelling and section 7 the current status with respect to which alloys can be semi-solid processed. The field has been reviewed in [3-8].

## 2   Routes to Spheroidal Feedstock

Routes to spheroidal feedstock are based around either the liquid state or the solid state. From the liquid state, spheroids can form by dendrite fragmentation with the arms breaking off at their roots due to shear forces, or the dendrite arms melting at their roots, or dendrite arm bending causing dislocation generation followed by the formation of grain boundaries and grain boundary wetting [4]. Alternatively, the conditions for dendrite formation may not be met. For example, constitutive supercooling will not occur if liquid is stirred to homogenise solute concentrations or if the material is held isothermally. The solid state routes to spheroidal feedstock are essentially based around the fact that deformed material will recrystallise as it is heated and the liquid will tend to form at the recrystallised boundaries [5].

## 2.1  Liquid Metal Routes

Liquid metal routes can be divided into those which involve agitation and those which do not. Semi-solid material can be vigorously stirred with a mechanical stirrer but this tends to lead to unacceptable erosion of the stirrer and contamination. In the shear cooling roll technique [9], liquid metal is poured between a rotating roll and a stationary cooling shoe. Sprayforming is the atomisation of melt into a spray followed by collection of droplets on a substrate. This gives a very fine spheroidal microstructure when remelted and allows alloys which cannot be made by other routes to be fabricated [10]. However, it is relatively expensive. The 'new MIT process' is a hybrid of stirring and near liquidus casting [11]. The melt is held just above the liquidus and a cooling finger, which also acts as a stirrer, is inserted. The melt is cooled to just below the liquidus and the finger withdrawn. This generates a cloud of nuclei. This process is very interesting from a commercial point of view. The most important agitation route to date from a commercial point of view though has been magnetohydrodynamic stirring (MHD) [3]. High local shear is generated by rotating electromagnetic fields (thereby avoiding contamination, gas entrapment and stirrer erosion), with stirring deep in the sump of the liquid. The alloy is filtered and degassed before treatment and a fine spheroidal size of about 30µm results. There is some lack of uniformity and spheroids may retain some 'rosette' character, a factor which influences flow. The majority of material which has been thixoformed commercially in the last fifteen years has been produced by the MHD route.

The 'new rheocasting (NRC) process' patented by UBE [12,13] is an example of a route which does not involve agitation. Molten metal at near-liquidus temperature is poured into a tilted crucible. Grain nucleation occurs on the side of the crucible and the grain size is fine because the temperature is near-liquidus. To date this has been the most important of the rheocasting-type routes, for aluminium alloys, from a commercial point of view.

Other non-agitation routes include the Direct Thermal Method (DTM) [14] where liquid alloy with a low superheat is poured into a cylindrical metallic mould of very low thermal mass but high thermal conductivity. Heat matching between the alloy and the mould results in a pseudo-isothermal hold within the solidification range. This is a low cost method ideal for the laboratory environment but with a strong limitation on the size of the billet that can be produced.

In the Cooling Slope method [15], liquid metal is poured down a cooled slope and collects in a mould. Nucleation on the slope ensures the spheroid size is fine. This shows considerable potential for combination with rolling but there are issues of gas entrapment and oxide formation on the slope.

A number of manufacturers have aimed to use grain refinement to suppress dendrite growth eg. titanium diboride in aluminium alloys. It has proved difficult to obtain uniform grain sizes less than 100µm and the volume of liquid entrapped in spheroids tends to be high (entrapped liquid does not contribute to flow) [16].

WPI in the US have recently developed a new route [6] involving mixing two hypoeutectic alloys and this is gaining considerable commercial interest.

## 2.2  Solid State Routes

The main solid state routes are Strain induced Melt Activated (SIMA) [17] and (RAP) [18], both of which have been used commercially. Essentially, the material is deformed, for example by extrusion, and then reheated into the semi-solid state. Recrystallisation occurs and the liquid forms around the grains which are fine and equiaxed. The distinction between SIMA and RAP is that the former involves hot working (above the recrystallisation temperature) and the latter warm working. SIMA therefore requires an intermediate additional cold work step. RAP is, however, more limited in the billet size because it is difficult to introduce deformation uniformly across the section (and for successful thixoforming a uniform spheroidal microstructure is required). The advantages of these routes are that many alloys are supplied in the extruded (or rolled) state anyway and the spheroids tend to be more rounded, giving better flow than 'rosette-type' structures sometimes found, for example, with the MHD route.

# 3.  Semi-Solid Processing Routes

## 3.1  Rheocasting

In rheocasting, alloy is cooled into the semisolid state and injected into a die without intermediate solidification. The non-dendritic microstructure can be obtained (see section 2) by: mechanical stirring; stimulated nucleation of solid particles (as in the NRC process); electromagnetic stirring in the shot sleeve (the Hitachi process [19]); or by sub-liquidus casting (the THT process [20]). The NRC process is shown in figure 2.

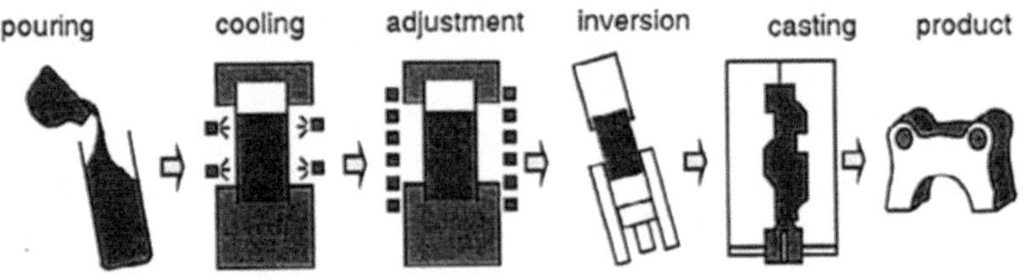


**Fig. 2.** The UBE new rheocasting (NRC) process [12]

The inversion of the billet causes the oxide skin to run into the biscuit and runner. With the NRC process, there is no need for specially treated thixoformable feedstock and scrap can be readily recycled within the plant. However, there is a lack of flexibility; the crucible size and heating/cooling arrangements are specific to that volume of metal.

In sub-liquidus casting [20], no processing equipment other than a die casting machine with a large shot diameter and a short stroke is required. The product emerges free of gates so no trimming is required. The melt is chemical grain refined.

## 3.2. Rheomoulding

Rheomoulding is allied to polymer injection moulding. It employs either a single screw [21] or a twin screw [22,23]. Liquid metal is fed into a barrel where it is cooled while being mechanically stirring by a rotating screw. The semisolid is then injected into a die cavity. This allows the continuous production of large quantities of components and no specially produced feedstock material is needed. Workers at Brunel University are in commercial trials of what they call the rheodiecasting (RDC) process [24] (Fig. 3). They have also found a solution to the challenge of what material to use for the screw and the barrel to avoid interaction with aluminium alloys.

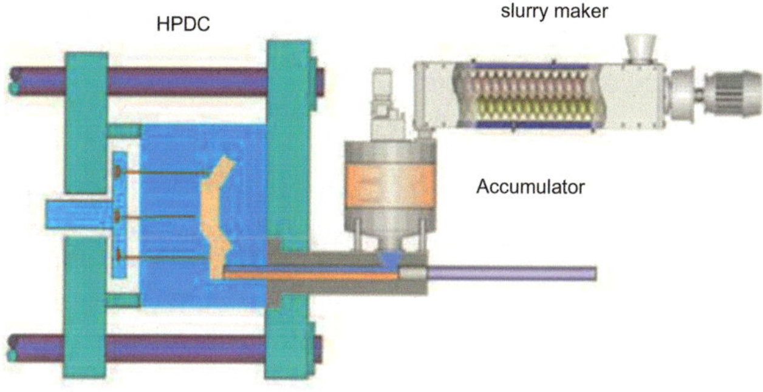


**Fig. 3.** Rheodiecasting

## 3.3 Thixomoulding

Thixomoulding is licensed by a firm called Thixomat. Many firms in Japan, the US and the Far East use the process to produce components from magnesium alloys. Magnesium alloy pellets are fed into a continuously rotating screw. The process is therefore similar to rheodiecasting but the starting material is solid pellets rather than liquid. The energy generated by shearing is enough to heat the pellets into the semisolid state. The screw action produces the spheroidal microstructure and the material is fed into a die.

## 3.4 Thixoprocesses

In thixocasting, the material is solid initially and the alloy has been treated in such a way that when it is reheated into the semisolid state a spheroidal microstructure is obtained. The liquid content is relatively high (>50vol.%). Reheating can be in a radiant furnace or an induction heating furnace. The high liquid fractions mean that the billet tends to collapse and is therefore tipped into the shot sleeve from a container. This process has been used commercially for making fuel rails for cars.

Thixoforging [25] involves heating the feedstock into the semisolid state and placing it between die halves. The parts of the die are then brought together. Material is used efficiently because of the lack of runners, gates and press discard. The liquid volume fraction is generally less than 50vol.% so that the billet can be handled. The process has not been used commercially, perhaps because of the difficulty is obtaining repeatability.

In thixoforming [26], the billet is heated as a vertical cylinder (in contrast with thixocasting) and then injected into the die, either horizontally or vertically. The liquid content is between 30 and 50 vol.%. This process has been used in industry to produce a range of automotive components. Billets are heated on a carousel with induction. Process control is demanding but cycle times are very comparable with those in die casting, and indeed may be faster because full solidification is not required.

## 4  Industrial Practice and Applications

### 4.1  Process Issues

For the rheocasting routes, heating of the feedstock billet is not required. However, for the thixoprocesses, the general method is to use induction heating in a carousel arrangement so that a billet is always ready for processing. The induction heating sequence may be step-wise in order to obtain as uniform a temperature profile across the billet as possible (given that induction heating works through the skin effect). Process control is demanding because the spheroidal microstructure must be as uniform as possible across the cross-section with a consistent liquid/solid fraction. The viscous slug must retain its shape until the shear force is applied. Experience shows that it is difficult to obtain satisfactory thixoprocessing unless the equipment has real-time control and closed loop control. For example, feedback must occur so that, as the ram makes its stroke, the velocity is constantly monitored and controlled to given values. Otherwise, as the ram forces material into the die and encounters increasing resistance, it will slow down. In practice, successful semi-solid processing is based around optimised profiles for pressure and velocity versus time.

### 4.2  Examples of Commercial Components

Components made by thixoforming because of the near net shape requirement include: computer heat sinks; industrial bolt connectors; racing bike swing arms; computer disk drive motor bases and mountain bike suspension parts. For other parts, it is the capability to achieve enhanced mechanical properties which is significant eg. motorcycle chassis frame arms and mountain bike forks. In some cases, thixoforming can produce a component which is pressure tight where die casting cannot, for example, intricately shaped gas control valves. Other examples of leak-tight, pressurized applications include: master brake cylinders; ABS system valves; missile connector supports and airbag canisters. All these are made from aluminium alloys, usually A356 or A357 in the heat treated condition.

Many components for mobile phones, laptops and automotive applications are made from magnesium alloys by thixomoulding.

## 5 Experimental Determination of the Parameters for Semi-Solid Processing

### 5.1 Background Rheology

In a Newtonian fluid, the shear stress, is proportional to the shear rate, and the constant of proportionality is the viscosity $\eta$. Thixotropic fluids are non-Newtonian i.e. the shear stress is not proportional to the shear rate. The viscosity is then termed the apparent viscosity and is dependent on shear rate, pressure, temperature and time. Some non-linear fluids also show viscoelasticity i.e. they store some of the mechanical energy as elastic energy. Thixotropic materials do not store energy elastically and show no elastic recovery when the stress is removed.

There is dispute over whether thixotropic semi-solid alloys display yield and whether they should be modelled as such. Barnes [27] concluded that the presence of a yield stress as reported by some workers for thixotropic materials (but not semisolid alloys) is probably due to the limitations of their experimental apparatus in not being able to measure shear stresses at very low shear rates. Koke and Modigell [28] have used a shear stress controlled rheometer to measure yield stress directly on Sn15%Pb. They distinguish between a static yield stress where the fluid is at rest prior to the application of a shear stress, and a dynamic yield stress where the fluid is being continuously sheared. The yield stress increases with rest time prior to deformation because of the increasing degree of agglomeration. In terms of modeling semi-solid alloy die fill, the use of a yield stress may be appropriate because a vertical billet does not collapse under its own weight unless the liquid fraction is too high. In addition, in rapid compression experiments (to be described later) an initial peak in the load versus displacement curve is detected. Contrary to this though is the fact that at the 'thixoforming temperature' the initial peak is so small as to be undetectable.

For a thixotropic material at rest, when a step increase in shear rate is imposed, the shear stress will peak and then gradually decrease until it reaches an equilibrium value for the shear rate over time (Fig. 4). The higher the shear rate after the step, the lower the equilibrium viscosity. The peak viscosity encountered will increase with increasing rest time before it recovers back to the equilibrium viscosity of the shear rate specified.

In semi-solid metallic alloys (as opposed to other thixotropic materials such as clays, mousses, emulsions and flocs), the thixotropic behaviour is associated with agglomeration and disagglomeration. When the material is allowed to stand, or stirred at a low shear rate, spheroids link together to form agglomerates (Fig. 5). With a step change up in shear rate, the agglomerates are broken down and the viscosity decreases towards a new equilibrium level.

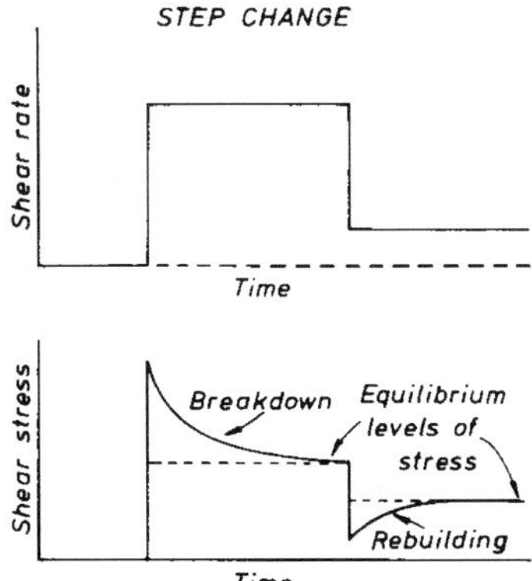



**Fig. 4.** Behaviour of a thixotropic material with a step-change in shear rate.

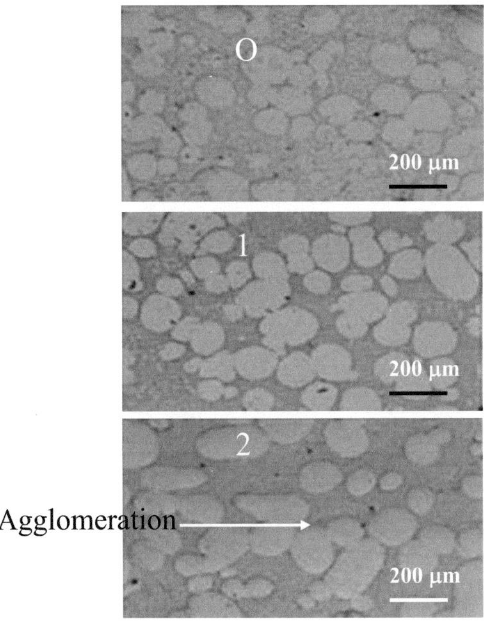



**Fig. 5.** Microstructures of Sn 15%Pb alloy (fraction solid 0.36) at 0, 1 and 2 hour rest times (after stirring at 30 s-1) [29].

The higher the shear rate (faster stirring) and the lower the solid fraction, then the lower the viscosity. If the shear rate is changed then over a period of time the viscosity will reach a new equilibrium value (steady state). The viscosity in the steady state depends on the balance between agglomeration and disagglomeration and also on particle morphology; pure spheres give the lowest viscosity. Liquid entrapped within the spheres does not contribute to shear flow.

## 5.2  Transient Versus Steady State Behaviour

This has previously been reviewed [30]. In semi-solid processing, the injection into the die takes less than one second. If the material is initially at rest (as in thixoforming), in that fraction of a second the viscosity is changing by orders of magnitude from $10^6$ Pas to $10^3$ Pas (equivalent to that of heavy machine oil). (The viscosity change will be less in rheocasting type processes where there has been insufficient time in a rest state for the agglomeration to occur to a great extent.) For modelling, we need to use computational fluid dynamics, but the issue is essentially where to derive the material parameters from. Several workers have carried out rheological experiments to obtain data on the steady state but there is little information on transients.

After a shear rate increase, there is a characteristic time for the slurry to achieve the steady state. Quaak et al. [31] deduced that in fact there are at least two processes occurring and therefore the breakdown should be represented with a double exponential. At the instant after a jump up in shear rate, they argue that the structure is the same ('isostructure'). Very quickly deagglomeration occurs and then over a slower time scale the fragments spheroidise and coarsen by diffusion. It is the very fast process which is most relevant to transient behaviour in thixoforming and hence we must design experiments to investigate that.

**Rheometry.** A rheometer can be used to carry out rapid changes in shear rate. The advantage of a rheometer is that the shear rate in the gap between the bob and the cup can essentially be treated as uniform (but note some concerns which are now emerging [32]). A typical experiment would involve stirring the alloy (usually Sn15%Pb as a model because few rheometers will operate at semisolid aluminium temperatures) while cooling from above the liquidus point to the required semisolid temperature (to within 1°C) to obtain a non-dendritic microstructure. Stirring continues until an apparent steady state is obtained with the specimen protected from oxidation by passing inert gas over the material. Shear rate jumps for several different rest times and initial and final shear rates can be obtained. Typical results are shown in Fig. 6 [29].

Breakdown times can be obtained by fitting an exponential to the data obtained in the second after 90% of the specified final shear rate has been achieved [29]. The longer the rest time prior to the shear rate jump, the lower (ie. slower) the breakdown time due to increased solid particle sizes and degree of agglomeration. These increases would impede the movement of particles when the shear stress is imposed. The peak viscosity increases with rest time because of the greater degree of agglomeration. For Sn15%Pb, the breakdown time for the fast disagglomeration process is about 0.15s. The ease with which the particles flow past each other depends on the particle size, the fraction liquid and the degree of agglomeration. The breakdown time is independent of the initial shear rate but controlled by the final specified shear rate. Breakdown is faster for higher final shear rates and faster than recovery.

The challenge now for rheometry is to obtain data for aluminium alloys despite the difficulties with operating rheometers at these higher temperatures.

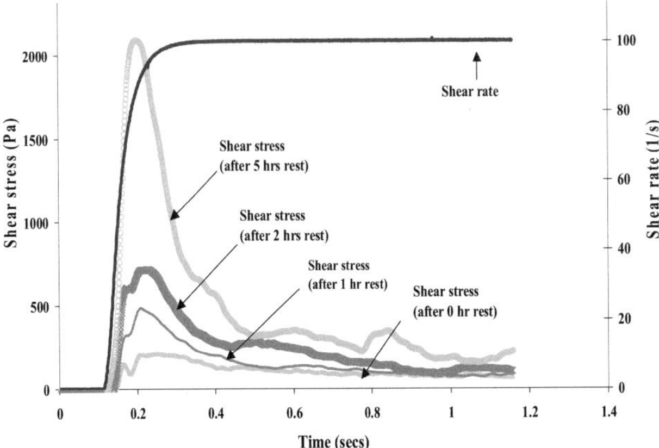


**Rapid Compression Experiments.** There are few rheometers that can operate above about 50% solid (because the torque that is required is too high). However, thixoprocessing often takes place with around 60%. One possibility is then to use the thixoformer itself (or a hydraulic press) to carry out rapid compression experiments or to use a drop forge viscometer [33,34]. The shear rates and temperatures are then akin to those in the process itself. However, in contrast with a rheometer, the shear rate varies from point to point within the volume of the material. There are two ways of approaching this. One is to assume an average shear rate applies across the material volume. The other is to model the compression process with Computational Fluid Dynamics (CFD), which, in itself, requires the material parameters one is aiming to measure. Those material parameters then have to be obtained by an iterative process, fitting the CFD results to the experiment.

Fig. 7 shows a typical signal response to rapid compression [34]. The peak load is thought to represent the resistance to flow of the skeletal structure connecting agglomerates. There is rapid breakdown under shear (within 10ms). The peak decreases with increase in temperature (and hence fraction liquid). It increases with increasing soaking time prior to compression as the skeletal structure is built up.

The load versus displacement response can be obtained by using a load cell arrangement [34]. Recent experiments have shown that the load can also be obtained from data from the ram operation without the need for a load cell and that the results are equivalent [35]. Typical results obtained using the averaging approach (and assuming Newtonian behaviour in the analysis; a shortcoming) are shown in Fig. 8. Shear thinning occurs with an increase in shear rate. With increase in temperature, the viscosity decreases. The CFD type of approach has not been fully developed and is discussed further below because it is inherently a modelling approach.

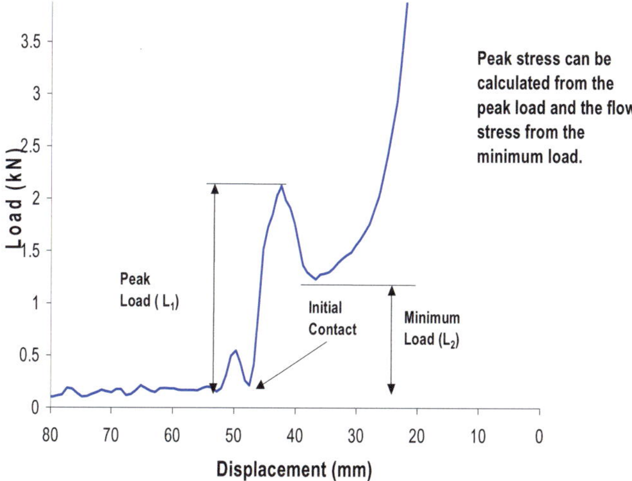



**Fig. 7.** A typical signal response to rapid compression [34].

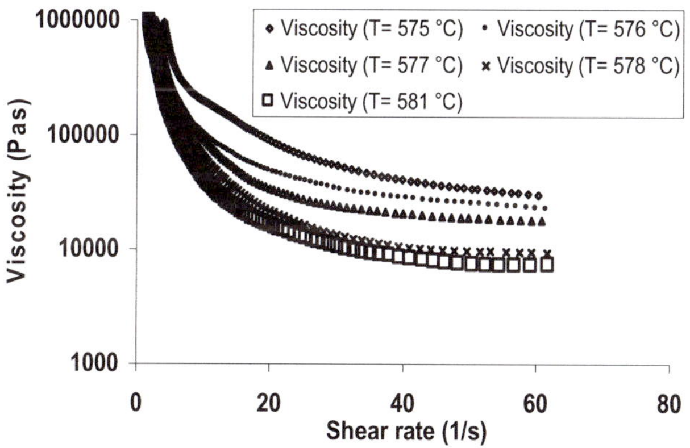



**Fig. 8.** Effect of temperature on viscosity under rapid compression of Alusuisse A356 [34]

# 6 Modelling

There have been a variety of approaches to modelling [reviewed in 8 and 36]. Illustrative highlights are given here. CFD approaches include finite difference, both one and two phase, and finite element, again both one and two phase. Micro-macro modelling is also under investigation [37].

### 6.1  Finite Difference Modelling

Modigell and Koke [38] have used a one phase finite difference model (based on FLOW3D from FLOWSCIENCE Inc) to examine flow around an obstacle. Fig. 9 illustrates the startling difference in behaviour between a Newtonian fluid and a thixotropic one.

It is important to check that the model can predict not only the flow front but also the rheometer results [36].

A number of workers have used the MAGMAsoft package with the thixotropic module (finite difference, one phase) and obtained reasonable agreement between interrupted filling tests and the simulation [e.g.39].

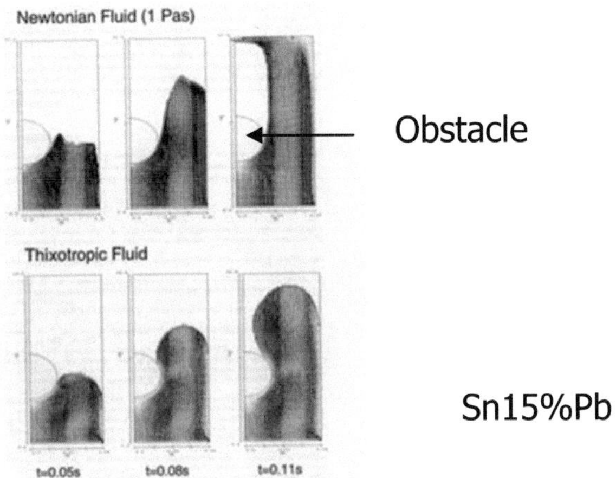


**Fig. 9.** Results from Modigell and Koke using a finite difference, one phase model [38].

### 6.2  Finite Element Modelling

Alexandrou and co-workers [e.g.40] have carried out important work with a finite element one phase model, with a continuous Bingham law to avoid the discontinuity at the yield stress. The work has explored the relative importance of the inertial, yield stress and viscous flow effects. They have been able to predict the regimes under which various flow patterns occur (Fig. 10) and in particular the conditions where defects such as the 'toothpaste instability' are found.

Orgeas et al. [41] use a Power Law Cut-Off model in the PROCAST package (Fig. 11). Shear thinning only occurs if a cut-off value is exceeded. This can be set at different values in different parts of the component. The behaviour is, therefore, ratchet-type i.e. an increase of shear rate beyond the largest shear rate experienced so far will lead to decrease in viscosity. Otherwise the viscosity is unchanged. Orgeas et al. also observe that, under certain conditions, flow leads to the concentration of liquid in veins in the structure.

Two phase models can predict phase separation. However, the determination of the rheological parameters required is not straightforward. Modigell and co-workers [42] have used a finite element two phase model to predict regimes of behaviour and validated the results with Sn15%Pb by filming die filling [43].

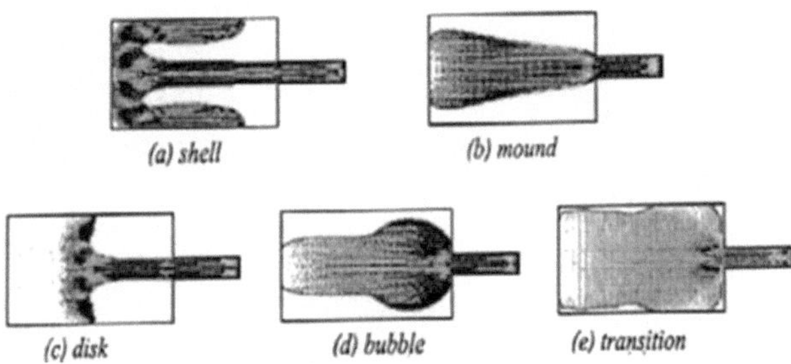


**Fig. 10.** Results from PAMCASTSIMULOR showing flow patterns under a variety of flow conditions [40].

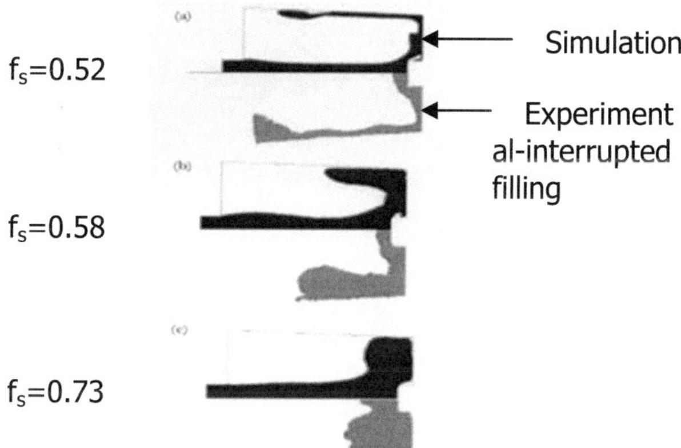


**Fig. 11.** Results from Orgeas et al. [41] comparing simulation and experiment with a Power Law Cut-Off Model

## 6.3  Experimental Validation

Apparatus for filming die filling has been reported [43,44]. One of the challenges which has been overcome [44] is the need for an effective arrangement with the window, which is also inexpensive, for use with aluminium alloys, where the high temperatures are problematic. There are advantages to observing the flow *in situ.* Interrupted filling tests have disadvantages as the material can continue to travel, because of inertia, after the ram is stopped. Results are shown in Fig. 12. In general, with a broader obstacle, the flow fronts meet when the die is more full.

Overall, for modelling, there are few examples of direct comparison between models, few examples of quantitative comparison between flow fronts and predictions, and few examples of in situ observation of die filling.

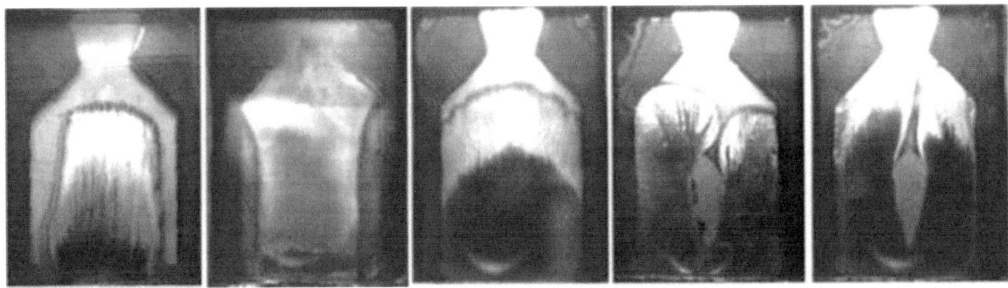

**Fig. 12.** Shots from filmed die filling with Al A357;( left to right) 576°C, 0.25 ms$^{-1}$, parallel entrance; 576°C 1ms$^{-1}$, parallel entrance; 1ms$^{-1}$splayed entrance, 0.25 ms-1 splayed entrance with obstacle and the same at 1ms$^{-1}$, all at 577°C [44].


# 7  Alloys for Semi-Solid Processing

The range of alloys available for semi-solid processing and the research to widen the range of applicability has been reviewed [45,46]. Virtually all the existing commercial activity is with aluminium casting alloys, particularly A356 and A357, and magnesium alloys (in the case of thixomolding). There is extensive interest in semi-solid processing higher strength aluminium alloys [45], which are normally wrought, and higher temperature materials such as copper based alloys, cast irons and steels. In addition, there is increasing interest in composites.

The problem with the normally wrought aluminium alloys is the ductility; the yield strengths are close to target. The difficulty here is separating out the processing defects which are inherent to the composition of the alloy and those which are due to laboratory scale methods of making spheroidal microstructure material. For example, some results have been obtained with material from a cooling slope and there are questions as to whether oxide has been incorporated as the material has flowed down the slope. However, wrought composition alloys are inherently vulnerable to hot tearing, and, at least in part, the ductility problems are associated with that. This is because the freezing range tends to be relatively wide with a long 'tail' during the final stages of solidification. There has been extensive discussion in the literature about the design of high strength alloys specifically for semi-solid processing [47-49], with consideration of the freezing range and the rate of liquid formation within the semi-solid processing range. Thermodynamic prediction is potentially a useful tool in this. There is no commercial production of high strength alloys yet although some intermediate level alloys such as A390 are in use.

There has been some exploration of copper alloys, particularly in the context of motor squirrel cages (where the alloying element can, in fact, be oxygen, to obtain a freezing range, whilst maintaining a relatively high electrical conductivity) [50]. However, currently there is no commercial application.

Considerable attention is being focussed on steels [51,52], and has included the development of steel compositions specifically designed for thixoforming. The highly alloyed steels (such as tool steels) are the most amenable and show good results for short runs [53]. However, to some extent, the lack of availability of commercial equipment suitable for such production (eg. with the facility to avoid oxidation during the high temperature heating) has limited the exploration of long run production issues. Tool materials need development.

Other high temperature materials under investigation include ductile irons [54], stellites [55] and high performance steels [53]. Metal Matrix Composites (MMCs) are gaining increasing interest.

# 8  Concluding Remarks

The current status of semi-solid processing has been reviewed. The discovery of thixotropic behaviour in semi-solid alloys with spheroidal rather than dendritic microstructures has been described. The terminology has been introduced and the origin of the behaviour identified. Routes to feedstock material, and the continuing fertility of ideas in this area, have been described. The distinctions between process routes have been highlighted and again, the continuing developments in improving routes so that the economics become increasingly competitive, are of note. Some examples of industrial applications have been given. Background rheology has been summarised so that efforts to obtain experimental parameters for modelling can be understood. The variety of models has been discussed and future challenges identified. Finally, the range of alloys which are currently semi-solid processed commercially and which have been investigated in R&D units is described. There are challenges in widening the range of applicability but significant progress has been made. The technology is gaining increasing industrial application.

**Acknowledgements**
The author would like to acknowledge her co-workers over 15 years of working in this area, particularly A. R. A. McLelland, N. G. Henderson, H E Pitts, K E Burke, S A Chayong, W Jirattiticharoean, E Cardosa Legoretta, P J Ward, P Kapranos, C M Sellars and D H Kirkwood. In addition, a range of industrial collaborators, the EU and the UK DTI and EPSRC funding bodies are acknowledged.

**References**
1. Spencer, D.P., Mehrabian, R., Flemings, M.C: Metall. Trans. 3 (1972) 1925-32.
2. Joly, P.A., Mehrabian, R.: J. Mater. Sci. 1 (1976) 1393-1418.
3. Kenney, M.P., Courtois, J.A., Evans, R.D., Farrior, G.M., Kyonka, C.P., Koch, A.A., Young, K.P.: Metals Handbook, 9$^{th}$ edition, 15 (1988) 327-38, ASM International, Metals Park, Ohio, USA.
4. Flemings, M.C.: Metall. Trans. A, 22 (1991) 957-81
5. Kirkwood, D.H.: Inter. Mater. Rev. 39 (1994) 173-89.
6. Figueredo A. de (editor): Science and Technology of Semi-Solid Metal Processing, publ. North American Die Casting Association, Rosemont, Illinois, US, 2001.
7. Fan, Z.: Inter. Mater. Rev. 47 (2002) 49-85.
8. Atkinson, H.V.: Prog. Mater. Sci.50(3) (2005) 341-412.
9. Kiuchi, M, Sugiyama, S. In: Proc 3rd Int Conf on Semi-Solid Processing of Alloys and Composites, Tokyo, Japan, 1994, edited by Kiuchi, M., published by Tokyo Institute of Industrial Science, 1994. p. 245-57.

10. Ward, P.J., Atkinson, H.V., Anderson, P.R.G., Elias, L.G., Garcia, B., Kahlen, L., Rodriguez-Ibabe, J.M.: Acta Metallurgica et Materialia, 44 (1996) 1717-27.

11. Martinez, R.A.: A new technique for the formation of semi-solid structures. MSc Thesis, MIT, Cambridge, MA, June 2001.

12. European patent 0745694A1, 'Method and apparatus of shaping semisolid metals', UBE Industries Ltd, 1996.

13. Hall, K., Kaufmann, H., Mundl, A.: In: Proc 6th Int Conf Semi Solid Processing of Alloys and Composites, Turin, Italy, Sept. 2000, edited by Chiarmetta GL, Rosso M, Published by Edimet Spa, Italy, 2000. p. 23-8.

14. Browne, D.J., Hussey, M.J., Carr, A.J., Brabazon, D.: Int. J. Cast Metals Research, 16 (2003) 418-26.

15. Haga, T., Kouda, T., Motoyama, H., Inoue, N., Suzuki, S. In: Proc ICAA7, Aluminium Alloys: Their Physical and Mechanical Properties, 1 (1998) 327-32.

16. Wan, G., Witulski, T., Hirt, G. In: Proc Int Conf on "Aluminium Alloys: New Process Technologies", Ravenna, Italy, 1993, Associazona di Mettalurgia, p. 129-41.

17. Young, K.P., Kyonka, C.P., Courtois, J.A: Fine grained metal composition, US Patent No. 4,415,374, 30th March 1982.

18. Kirkwood, D.H., Sellars, C.M., Elias Boyed, L.G: Thixotropic Materials, European patent 0 305 375 B1, 28th October 1992.

19. Kaneuchi, T., Shibata, R., Ozawa, M. In: Proc 7th Int Conf on Advanced Semi-Solid Processing of Alloys and Composites, Tsukuba, Japan, Sept 2002, edited by Tsutsui, Y., Kiuchi, M., Ichikawa, K., published by National Institute of Advanced Industrial Science and Technology, Japan, 2002. p.145-150.

20. Jorstad, J., Thieman, M., Rogner, R.: In: Proc.8th Int ESAFORM Conf., 27-29 April 2005, Cluj-Napoca, Romania, Editor Banabic, D., Published by the Publishing House of the Romanian Academy, ISBN:973-27-1173-6, pp.1115-8.

21. Peng, H., Hsu, W.M. In: Proc 6th Int Conf. Semi Solid Processing of Alloys and Composites, Turin, Italy, Sept. 2000, edited by Chiarmetta, G.L., Rosso, M., Published by Edimet Spa, Italy, 2000. p. 313-7.

22. Fan, Z., Bevis, M.J., Ji, S.: 'Method and apparatus for producing semisolid metal slurries and shaped components'. Patent PCT/WO/O1/21343A1 1999.

23. Ji, S., Fan, Z., Bevis, M.J: Mater. Sci. Eng. A299 (2001) 210-17.

24. Ji, S., Qian, M., Fan, Z: Metall and Mater. Trans., A, 37A (2006) 779-787.

25. Kopp, R., Neudenberger, D., Winning, G: J. Mater. Process. Technol., 111 (2001) 48-52.

26. Chiarmetta, G. In: Proc 6th Int Conf Semi Solid Processing of Alloys and Composites, Turin, Italy, Sept. 2000, edited by Chiarmetta, G.L., Rosso, M., Published by Edimet Spa, Italy, 2000. p. 204-7.

27. Barnes, H.A: J. Non Newtonian Fluid Mech., 70 (1997) 1-33.

28. Koke, J., Modigell, M: J. Non Newtonian Fluid Mech., 112 (2003) 141-60.

29. Liu, T.Y., Atkinson, H.V., Ward, P.J., Kirkwood, D.H: Metall. and Mater. Trans. A., 34A (2003) 409-17.

30. Atkinson H.V., Liu, T.Y., Ward, P.J., Kirkwood, D.H., Kapranos, P.: Proc. 6th int. ESAFORM Conf., Salerno, Italy, April 2003, Editor Brucato, V., Publ. Nuova Ipsa, ISBN 88-7676-211-6, pp. 315-318.

31. Quaak, C.J., Katgerman, L., Kool, W.H. In: Proc 4th Int Conf Semi Solid Processing of Alloys and Composites, Sheffield, UK., 19-21 June 1996, edit. Kirkwood, D.H., Kapranos, P. Publ. Dept. Engineering Materials, University of Sheffield, UK, 1996, p. 35- 9.

32. Alexandrou, A., Georgiou, G., Tonmukayakul, N., Apelian, D.: Solid State Phenomena, 116-117 (2006) 429-432.

33. Yurko, J.A., Flemings, M.C: Metall. Mater. Trans. A., 27 (2002) 2737-46.

34. Liu, T.Y., Atkinson, H.V., Kapranos, P., Kirkwood, D.H., Hogg, S.C: Metall. and Mater. Trans. A. 34A (2003) 1545-54.

35. Omar, M.Z., Atkinson, H.V., Palmiere, E.J., Howe, A.A., Kapranos, P: Solid State Phenomena, 116-117 (2006) 677-680.

36. Atkinson, H.V: Proc. 7$^{th}$ Int. ESAFORM Conf, Trondheim, Norway, April 28030, 2004, Edited Stőren, S., publ. Norwegian Institute of Science and Technology, Trondheim, Norway, 2004, ISBN 82-92499-02-04, p. 405-408.

37. Cezard, P., Favicr, V., Bogit, R., Balan, T., Berveiller, M: Computational Materials Science 32 (2005) 323-328.

38. Modigell, M., Koke, J: Mechanics of Time Dependent Materials, 3(1999)15-30.

39. Kim, N.S., Kang, C.G: J. of Mater. Processing Technology, 103 (2000) 237-46.

40. Alexandrou, A.N., Duc, E., Entov, V: J. Non-Newtonian Fluid Mech. 96 (2001) 383-403.

41. Orgéas, L., Gabathuler, J.-P., Imwinkelried, Th., Paradies, Ch., Rappaz, M: Modelling Simul. Mater. Sci. Eng., 11 (2003) 553-74.

42. Modigell, M., Kopp, R., Sahm, P.R., Neuschutz, D., Petera, J: In: Proc 7$^{th}$ Int Conf on Advanced Semi-Solid Processing of Alloys and Composites, Tsukuba, Japan, Sept 2002, edit. Tsutsui, Y., Kiuchi, M., Ichikawa, K., publ. National Institute of Advanced Industrial Science and Technology, Japan, 2002, p.77-82.

43. Petera, J., Modigell, M., Hufschmidt, M: In: Proc 5$^{th}$ ESAFORM Conf on Material Forming, Ed, Pietrzyk, M., Mitura, Z., Kaczmar, J., Kraków, Poland, April 14-17 2002, Publ. Akademia Górniczo-Hutnicza, p.675-8.

44. Atkinson, H.V., Ward, P.J: Proc 8$^{th}$ Int. ESAFORM Conf., Cluj-Napoca, Romania, 27-29 April 2005, Edit. Banabic, D., Publ. Publishing House of the Romanian Academy, 2005, ISBN: 973-27-1173-6, p. 1107-1110.

45. Atkinson, H.V: Proc. 5$^{th}$ Int. ESAFORM Conf., Cracow, 14-17 April 2002, Edit. Pietrzyk, M., Mitura, Z., Kaczmar, J., Publ. Akapit, Cracow, Poland, pp. 655-658.

46. Atkinson, H.V., Liu, D: Proc. 7$^{th}$ Int. Conf. On 'Advanced Semi-Solid Processing of Alloys and Composites', Tsukuba, Japan, Sept. 2002, Edited by Tsutsui, Y., Kiuchi, M., Ichikawa, K., Publ. National Institute of Advanced Industrial Science and Technology, pp. 51-56.

47. Kazakov, A.A: Advanced Materials & Processes, 3 (2000) 31-4.

48. Maciel Camacho, A., Atkinson, H.V., Kapranos, P., Argent, B.B: Acta Mater., 51 (2003) 2331-2343.

49. Liu, D., Atkinson, H.V., Jones, H : Acta Mater., 53 (2005) 3807-3819.

50. Ward, P.J., Atkinson, H.V: Proc. 2$^{nd}$ Int. ESAFORM Conf., Guimaraes, Portugal, 13-17 April 1999, Ed. Covas, J.A., ISBN 972-98103-0-3, pp.213-216.

51. Fraipont, C., Lecomte-Beckers, J: Proc. 10$^{th}$ Int. ESAFORM Conf., Zaragosa, 18-20 April 2007, To be published.

52. Rassili, A., Robelet, M., Fischer, D: Proc. 9th Int. ESAFORM Conf., Glasgow, UK., 26-28 April 2006, Edited Juster, N., Rosochowski, A., ISBN83-89541-66-1, pp.819-822.

53. Omar, M.Z., Atkinson, H.V., Palmiere, E.J., Howe, A.A., Kapranos, P: Proc. 9th Int. ESAFORM Conf., Glasgow, UK., 26-28 April 2006, Edited Juster, N., Rosochowski, A., ISBN83-89541-66-1, pp.847-850.

54. Goka, M., Kato, T., Kiuchi, M: Solid State Phenomena, 116-117 (2006) 9-15.

55. Hogg, S.C., Atkinson, H.V., Kapranos, P: Proc. 10th Int. ESAFORM Conf., Zaragosa, 18-20 April 2007, To be published.

# Microforming and Nanomaterials

Ulf Engel[1], Andrzej Rosochowski[2], Stefan Geißdörfer[1] and Lech Olejnik[3]

[1] Chair of Manufacturing Technology, University of Erlangen-Nuremberg,
Egerlandstrasse 11, 91058 Erlangen, Germany
[2] Department of Design, Manufacture and Engineering Management, University of
Strathclyde, 75 Montrose Street, Glasgow G1 1XJ, United Kingdom
[3] Institute of Materials Processing, Warsaw University of Technology, 85 Narbutta Street,
02-524 Warsaw, Poland
engel@lft.uni-erlangen.de, a.rosochowski@strath.ac.uk,
geissdoerfer@lft.uni-erlangen.de, l.olejnik@wip.pw.edu.pl

**Abstract.** Micro technology and nano materials are gaining increasing interest in the metal forming community. This can be explained by a large number of new applications and products pushed to market in the past, yielding smaller geometrical dimensions of the final products and thus demanding for smallest components to be manufactured in large quantities. Up to now, most of these parts are being manufactured by machining technology well suited for the production of small series. The analysis of the current market situation shows, that the steadily increasing trend towards smallest products is continuing in the future and thus requesting for new manufacturing technologies for large quantity production. Forming technology seems to be well suited due to its high production output and high accuracy. The aim of this chapter is to give an overview on current research activities related to microforming and nanomaterials technology showing the capabilities and problems. The analysis of the current state will give a hint on existing gaps in knowledge and will finally describe the future demands.

## 1. The Past

Micro system technology, one of the future key technologies, is gaining increasing interest in industrial and academics community. This can be easily explained by the ongoing trend towards smaller product dimensions, in particular in the field of electronics production, but also in other fields like micro system technology or in the medical sector. This trend is also confirmed by recent market analysis of the micro systems worldwide turnover, increasing from 50 bn. US $ in 2003 to more than 200 bn. US $ estimated for 2010 [1]. Latest products like mobile phones with integrated digital camera and optical zoom or keyhole surgical instruments are composed of smallest micromechanical components, such as connector pins, smallest screws or contact springs. The characteristic of all of these parts is that at least two geometric dimensions are at sub-millimetre range. Depending on the required functionality of the parts and the production numbers, different manufacturing technologies are available like machining, moulding and forming. In case of smallest metallic parts, up to now the preponderant numbers of these parts are produced using machining technologies like turning, grinding or milling. For small batch production these technologies may be justified, if large quantities are requested, forming technology is more adequate due to its high production rate and remarkable accuracy. However, investigations on microforming processes have shown significant differences in the forming behaviour of smallest parts compared to conventional length scale forming, preventing forming technology from being used at micro scale.

The first steps in direction to microforming applications were done more than one decade ago. At this time, in the 1990's, only few miniaturized products have been required by industry predominantly being produced conventionally by machining operations. Despite the few applications at this time, the first steps in direction to the nowadays still increasing trend towards miniaturized products have been done. One of the first documents released to the public can be found in the 1989 Annual Report of the JSTP given by T. Maeda who proposed the development of a superprecision micro press machine [2]. This was the starting shot for a wider community to become aware of micro technology, taking notice that there was nearly no basic knowledge in this area of research. Thus, a few institutes decided to initiate basic research firstly focused on the process itself. This pioneering work has been appreciated in [3,4]. Very soon it became clear that there were many specifics at micro scale to be considered if products were to be manufactured at industrial scale. Several fields have been identified where the know-how of conventional forming processes cannot simply be transferred to micro applications. For the smallest specimen dimensions, it has been shown that the flow behaviour is significantly influenced by the microstructure and the specimens' surface topography [3]. The role of the microstructure in microforming suggests that the reduction of grain size may help resolve many of the product downsizing problems. Traditionally, grain refinement in metals was a matter of metallurgical intervention which, by experimenting with chemical composition and thermo-mechanical treatment of the materials, managed to reduce the grain size to a few micrometers. However, the number of such materials was limited as they were mainly made for superplastic forming. This situation changed when a new method of refining grain structure of bulk metals, based on severe plastic deformation (SPD), emerged. Research on SPD started in the 1970s. However, a more substantial interest in metals with refined grain structure has been observed for the last 10 years, that is in the same period when microfoming came up. The strong alliance of both the microforming technology and nanomaterials became increasingly visible at international conferences on either "nanoSPD" or "micromanufacture" attended by people of both communities. For example, conferences on nanomaterials have been held in Russia (1999), Japan (2001), Austria (2002), Australia (2003) Japan (2005) and China (2005). It is interesting that both disciplines are increasingly mixed through common minisymposia within other conferences such as the ICTP (2003), the annual ESAFORM (2005, 2006) and 4M (2005, 2006). The American equivalents are biennial symposia like TMS and ICOMM. There are also other smaller conferences, symposia and workshops organised every year. No wonder that the number of related publications in the last 15 years has grown rapidly (e.g. 2300 on SPD [5]).For a reader willing to get the idea about these disciplines rather quickly, there are an increasing number of overviews available. A good starting point are [3,6,7] for microforming and [8,9] for SPD.

Additionally to the microstructure controlling the forming behaviour, it has been shown that accuracy and repeatability of the micro production process are also influenced by the handling, tool and machine systems. As this was recognized, many activities were started in the late 1990's world widely to find solutions for the problematic effects observed. Recently, these efforts have been intensified all over the world as seen in the networks [10], integrated projects [11], and studies on the state of the art [12].

## 2. The present

As the first step to a better understanding of these effects, it is useful to consider the whole micro forming system divided into four main groups [3]:
- material
- process
- tools
- machines/equipment

Of course these four groups cannot be considered to be independent as the material behaviour of scaled specimen is influencing the process results like accuracy and reliability but in all these areas specific effects of miniaturization are observed. The material characteristics (e.g. flow stress, anisotropy, forming limit) are strongly depending on the specimen size and the microstructure. The process, influenced by the material as noted before, shows additional effects concerning the forming forces, tribology, springback and the scatter of the results. With respect to tooling systems, the main problem is the manufacturing with high precision and narrow tolerances. If accuracy comparable to conventional tool manufacturing is to be achieved, shape accuracy and surface roughness in nanometre scale is demanded. To overcome these difficulties, new innovative manufacturing methods have to be developed. Furthermore, the production rate is also a challenge for micro machines. In the case of serial production, if several hundreds of parts have to be produced per minute, new challenges for the machine and the handling systems appear. The required punch stroke of only a few hundred microns, the requested accuracy of the punch centring is in the range of a few microns and the corresponding accuracy of the handling system show clearly that new machine and handling concepts are needed to solve the problems occurring at the microscale.

### 2.1 Basic Research on Microforming

In the following sections, the size effects, associated with each of the four main groups, will be discussed in detail, based on the results of research activities.

### 2.1.1 Material behaviour

Since the most relevant parameters describing the material behaviour in forming processes are the flow stress and the flow curve, standard tests geometrically scaled according to the theory of similarity [13] have been performed to obtain these

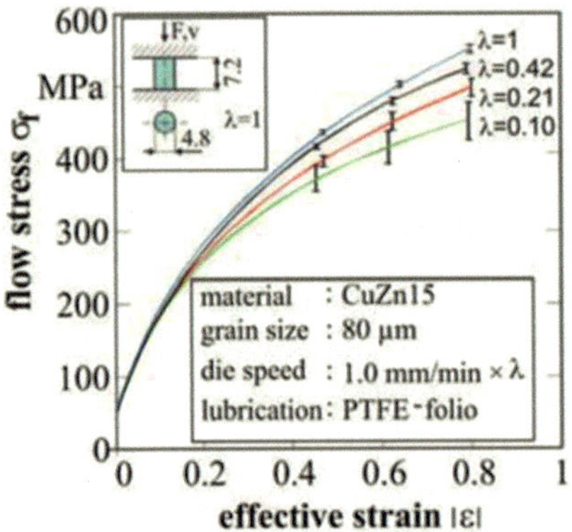



**Fig. 1.** Size dependency of the material properties (LFT)

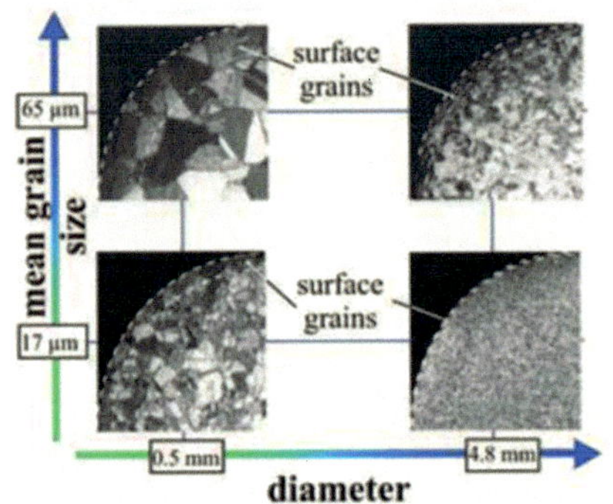



**Fig. 2.** Increasing share of surface grains (LFT)

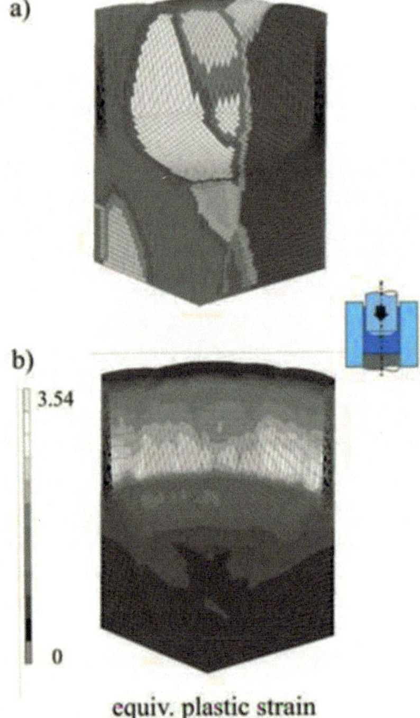



**Fig. 3.** FE-simulation of a can-backward extrusion process using the mesoscopic model a) synthetic material structure b) equivalent plastic strain (LFT)

parameters valid at a micro scale. When carrying out tensile tests using CuZn15, CuNi18Zn20 [14], copper [15] and aluminium [16] as well as upsetting tests using copper, CuZn15 and CuSn6 [17, 18, 19], two significant effects have been detected: a reduction of the flow stress when reducing the specimen size and thus changing the ratio between surface to volume as well as an increasing scatter (Fig. 1).

The first approach to describe the decreasing yield stress has been based on [20,21] introducing the so-called surface layer model. It uses the assumption, that grains positioned on a free surface have fewer constraints to fulfil, compared to grains which are surrounded by the material and thus the local flow behaviour of the surface grains must be different. As the dislocations induced by a deformation process are able to pile up at grain boundaries but not at a free surface, lower hardening occurs in the vicinity of free surface. When the specimen size decreases, the share of surface grains increase yielding to a reduced overall flow curve of the specimen (Fig. 2).

Based on the surface layer model, an extended approach named mesoscopic model to describe the size-effects by means of finite element simulation has been developed [22,23,24,25]. By means of the Monte-Carlo-Potts model [26], a synthetic material structure is generated representing the real material structure. For each of the synthetic grains, material properties are calculated using the theories of Ashby [27], Hall-Petch and Meyers and Ashworth [28]. As the material structure is included in FE-simulation, the process scatter as well as the influence of the material structure on the process results can be predicted (Fig. 3).

The lower flow stress relative to the product dimensions, leads to a lower process force, when the deformed area has free surfaces like in the air bending process. This type of sheet metal forming process has been investigated by several researchers [15, 29,30,31]. In the case of metal forming processes without free surfaces like blanking or extrusion, there is no detectable decrease in the process force. The force even increases, caused, among other things, by increased friction, as will be shown later. Additional material effects have been observed and are described in [,32]: The normal mean anisotropy of sheets decreases with miniaturization. As a result, the forming characteristics are getting worse with miniaturization, because the thickness reduction, which is an undesired phenomenon in deep drawing processes, increases. Furthermore, the uniform elongation (the strain, when global necking starts in tensile tests) decreases with miniaturization, and also the necking elongation, which is the elongation between uniform and fracture elongation. A further effect, which has been observed in many experiments, is the increase of scatter with decreasing specimen size. The explanation of this effect is given by the increasing influence of the orientation and size of each single grain which leads to an inhomogeneous forming behaviour of the material.

Finally, an apparent paradox, observed by Kals [] and Raulea [] should be mentioned. Both of them independently carried out bending tests using specimens with different grain size d, sheet thickness $s_0$ resulting in different ratios of sheet thickness to grain size r. As pointed out before, the yield strength and the process force decrease with miniaturization, i.e. when $s_0$ or r decreases. However, the process forces increase again with a certain value of r. This means that, when only very few grains are located over the thickness of the sheet, the process force increases with the increasing grain size. This contradicts the theory of metal forming for conventionally sized components, where the flow stress decreases with increasing grain size (Hall-Petch-relation) [33]. This effect has been

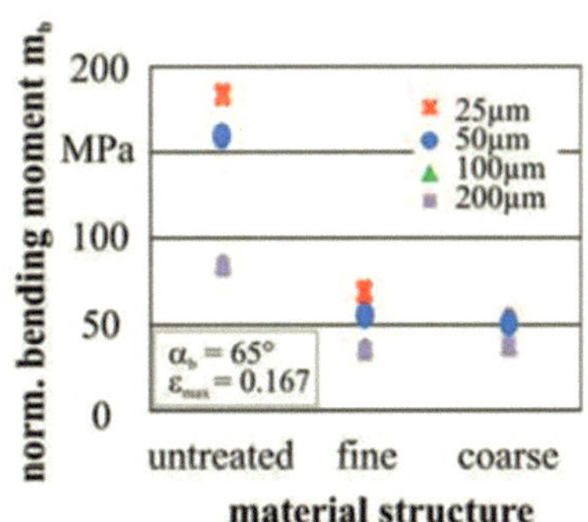


**Fig. 4.** Bending moment in dependency of the sheet thickness and the material structure (LFT)

confirmed in [34] performing scaled bending experiments on metallic foils. It has been shown, that depending on the thickness of the sheet (scaled from 200 microns down to 25 microns) and on the material structure, the process force, bending moment and thus the springback angle increase again when scaling down. This behaviour can be explained by the previously described theory of strain gradient plasticity [,35,36,37], which considers the increasing strain gradient in direction perpendicular to the foil when reducing the foil thickness [34].

When decreasing the foil thickness, two contrary effects appear: the effect of a flow stress reduction due to an increasing share of surface grains and the effect of flow stress increment caused by the increasing density of geometrically necessary dislocations. When the foil thickness is getting smaller, the latter effect gains a superior influence, which results in both a higher normalized bending moment (see Fig. 4) and a higher springback angle.

Visioplasticity and optical strain measurement have been used in order to study the deformation behaviour of homogeneous and inhomogeneous materials and to verify the above described model. In this technique, a sequence of pictures of a forming process is taken by a CCD camera. The strains are calculated by identifying characteristic spots on the surface and following their paths. To ensure reliable strain measurements, the contrast of the considered surface has to be high. There are some possibilities to create a better contrast, for example to treat the surface with powder or to apply a micro grid as it is described in [38]. Raulea [39] investigated the blanking processes and showed the effect of the geometry of the sheared edge on the orientation of grains in the plastic region. Since in micro forming processes often only a few grains are deformed, he carried out his experiments with recrystallized, coarsely grained aluminium sheets with only one grain over the sheet thickness. Sheets made of CuZn15 with a thickness of 0.5 mm were bent [40,41]. In the case of hard material with fine grains (grain size approx. 10 μm), the strain distribution was typical for the bending process (Fig. 5a) with an elongated material on the outer surface, a compressed material on the inner surface and a neutral plane in the centre. This layered structure was disturbed when the material was coarsely grained (Fig. 5b).

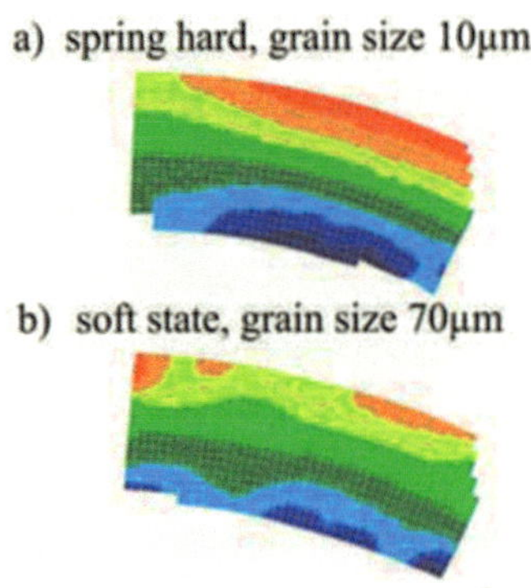


**Fig. 5.** strain distribution a) fine grained material b) coarse grained material (LFT)

In this case, the grain size was about 70 μm. This means, that different grain sizes and orientations, cause different resistance against deformation, have an increasing effect on the strain distribution. In this case, the idea of homogeneous strain, as described by Taylor [42], is no longer valid. Thus the paradox of increasing forces caused by increasing relative grain sizes in micro forming, can be explained by the deformation behaviour of the material. In the case of a polycrystalline material, the strain is homogeneously distributed while in the case of few grains, only the grains with favourable orientation are deformed. If the number of grains is even smaller, each single grain will be deformed according to the shape of the tool, regardless of possibly unfavourable orientation.

Additional investigations have confirmed the size-effects of the material by applying micro-hardness measurements on the upset specimens [43], revealing a significant dependency of the hardness value on the specimen size.

## 2.1.2 Process

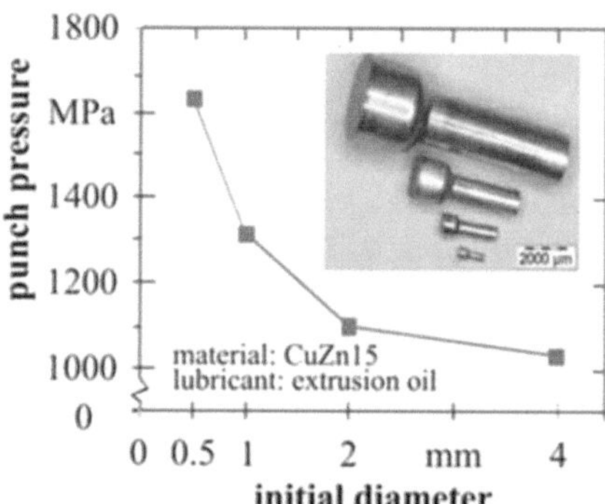

Fig. 6. Full forward extrusion process, increase of the related forming force (LFT)

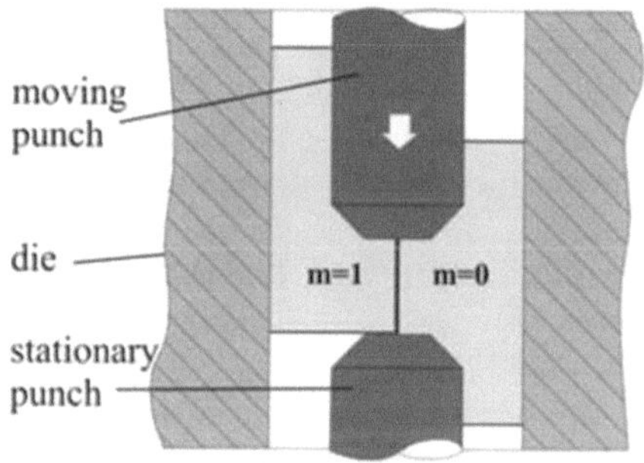



Fig. 7. Setup of the double cup extrusion test (LFT)

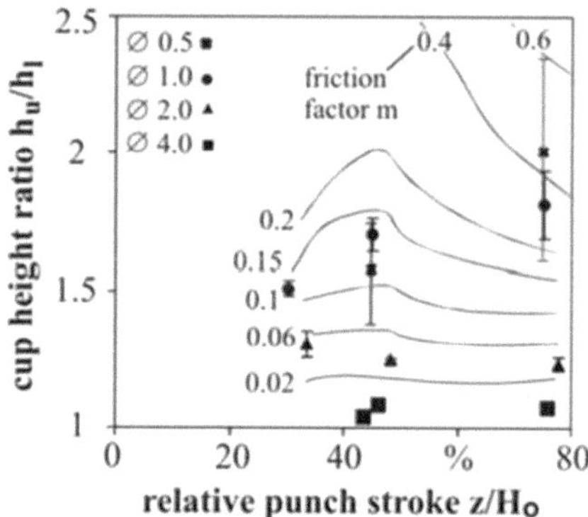



Fig. 8. Results of the DCE-test, curves of constant friction deter-mined by numerical identification using FE-simulation (LFT)

In addition to the above mentioned size effects due to material behaviour, the process parameters and layout, the specimens' geometrical dimensions and accuracy are influencing the process results. To analyze their impact on the micro forming results, several forming processes have been tested. In the case of a forward extrusion process, scaled from an initial specimen diameter of 4 mm down to 0.5 mm, an increase in the relative punch force was observed as shown in Fig. 6. A possible explanation for this effect can be increasing friction caused by the decreasing specimen dimensions. To verify this hypothesis, further tests on the scale dependency of friction conditions have been done by Messner using the ring compression test [44]. The analysis of the experiments has confirmed these findings but has also shown that further experiments are necessary to get more detailed information on the correlation between specimen size and friction. This work has been done by [45] using the double cup extrusion (DCE) test which was first proposed by [46] and applied by [47]. In this test, well suited to represent the large strain occurring at bulk metal forming processes, a cylindrical billet is positioned between a stationary and a moving punch. Theoretically, in the case of zero friction (m = 0), both cups are supposed to have the same height; the higher the friction gets, the more is the height of the upper cup (Fig. 7). Thus, the change in the height ratio between the upper and the lower cup is quite a sensitive measure to characterize the change in the friction conditions. If the absolute values of the friction factors are requested, the method of numerical identification can be used (cf. Fig. 8).

The experimental investigations have been performed by [], where the

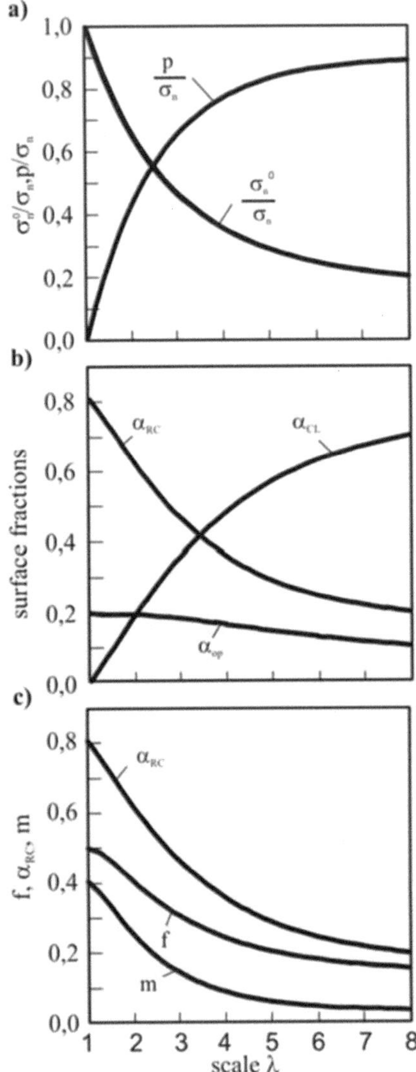



**Fig. 9.** Effect of scaling factor $\lambda$ on p and $\sigma_{n0}$ normalized to $\sigma_n$ (a), on $\alpha_{RC}$, $\alpha_{CL}$ and $\alpha_{OP}$ (b), and on m, f and $\alpha_{RC}$ (c) (LFT)

specimens have been scaled down, remaining geometrically similar, from a diameter of 4 to 0.5 mm with a ratio of diameter to height $D_0 / H_0 = 1$. As it can be seen in Fig. 8, friction increases with the decreasing specimen size from a friction factor of about m = 0.02 for the largest specimen up to m = 0.4 for the smallest specimen. An attempt to describe the frictional behaviour on a topographical level is given by a mechanical-rheological model [], where the relation between real contact, open and closed lubricant pockets and their impact on friction are considered. If a forming load is applied to a specimen surface, the roughness related asperities start to deform plastically. From this point on, the lubricant is either trapped and pressurized within the closed lubricant areas $\alpha_{CL}$ or squeezed out if a connection to the edge of the surface exists.

The forming load can be transmitted onto the workpiece either by the pressurized lubricant or the flattened asperities. Due to the scale-invariant production process of specimens and thus the scale-invariant surface topography, the region where open lubricant pockets remain is constant when scaling down geometry. Simultaneously, the area of closed lubricant pockets is reduced and thus the real contact area $\alpha_{RC}$ is increased. This leads to an increase in the friction factor.

Based on the mechanical-rheological, model further investigations have been performed in order to describe the size-dependent friction factor analytically [48]. Using the Wanheim/Bay's friction law [49] and the geometrical boundary conditions, it can be shown that the friction factor changes, caused by the different surface topography, are in a good agreement with experimentally obtained results (cf. Fig. 9 a-c).

Besides these basic considerations there are also efforts to counteract the tribological size effect. Exemplarily, two measures successfully proved to reduce

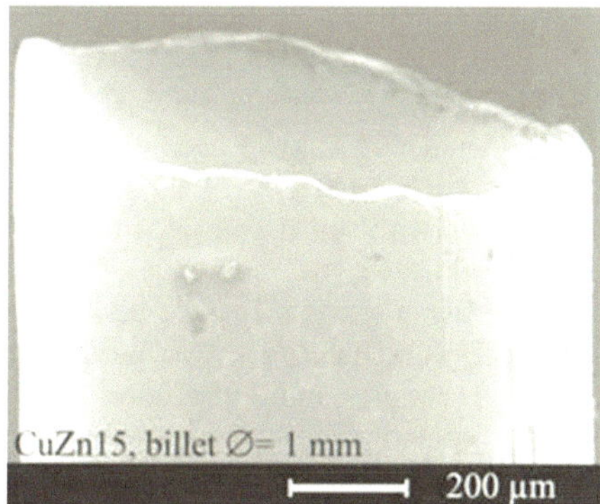


**Fig. 10.** Effect of specimen size and microstructure on the shape of the extruded parts. (LFT)

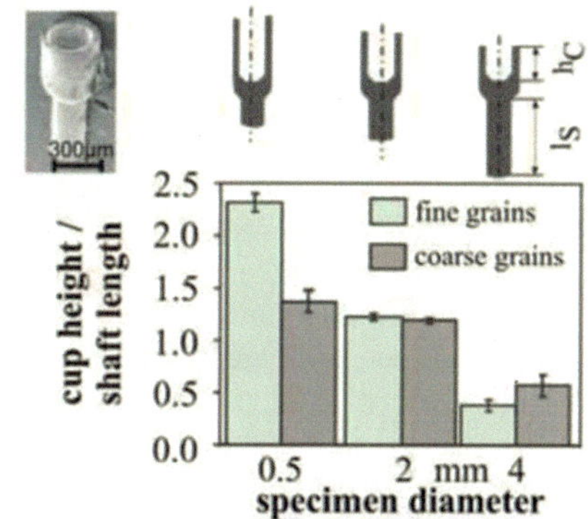


**Fig. 11.** Combined full forward cup backward extrusion: effect of microstructure and friction on the shape of the extruded parts. (LFT)

friction should be mentioned, by simultaneously inducing vibration [50,51], and by applying high strength, low friction coatings [52].

The investigations of micro extrusion processes with high aspect ratios and large strains showed a significant dependency of the forming results on the material structure [53,]. For the case of a backward can extrusion process, the cup geometry was chosen with a cup wall thickness of about 8 microns, the SEM picture revealed a strong influence of the material structure on the shape evolution process as illustrated in Fig. 10 by an uneven cup height.

Further investigations, using micro hardness measurements to evaluate local material flow, also confirmed the above described results. In the case of grains being larger than the feature size, the grains are forced to flow into such a feature and thus the flow dependency on the grain size and orientation causes the uneven cup height. Accordingly this effect is less distinct in the case of fine grained material.

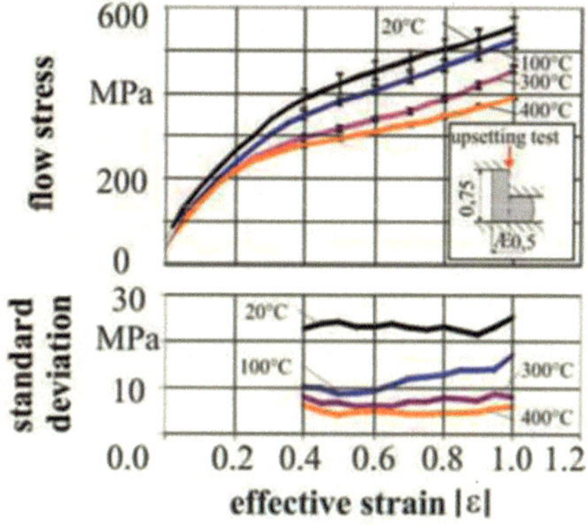


**Fig. 12.** Dependency of the flow curve from temperature using CuZn15 (LFT)

The increase of the friction factor when scaling down leads to an increase of the ratio between cup height and shaft length for both the coarse grained and the fine grained material. The minor increase in the case of coarse grained material can be explained by the fact that the grain size is in the same range as the feature size. Thus, the material behaviour cannot be considered to be polycrystalline; it is easier for the material to flow in the shaft than into the cup.

An important microscale size-effect is the large scatter of the process results in cold forging, which prevents the application at serial manufacturing processes. As it is known from forming at a conventional length scale, increasing forming temperature leads to the activation of additional slip systems in the material and thus to a decreasing flow stress and better formability.

To quantify the influence of the elevated temperature on the flow stress at microscale, upsetting tests at 20°C, 100°C, 200°C and 300°C have been performed [54,55]. The materials used were CuZn15 and steel X4CrNi1810. As it can be seen in Fig. 12, increasing process temperature results in a significant decrease of the flow curve as well as a decrease of the process scatter.

Further investigations on the backward can extrusion process [] have confirmed these findings. A detailed analysis of the material work hardening was performed using a micro hardness measurement system. It enables high resolution strain measurements indirectly on a deformed specimen. Since, the measurement load is set to 20 mN, the minimum distance between two measurement points is 40 microns. The analysis of the local hardening shows, that in contrast to the forming process performed at room temperature, the plastically deformed area at elevated temperature is smaller and more concentrated around the punch. Thus, it can be stated that the forming behaviour of the material at an elevated temperature is closer to the one observed at the macro scale.

Studies on coining aluminium sheets have been performed by Ike and Plancak [56], who compressed half hard, commercially-pure aluminium disks 2 mm in thickness and 30 mm in diameter, using dies with holes with diameters from 0.05 to 1.6 mm.

The coining results were evaluated in terms of the aspect ratio, the radial position of the hole and the die and the forming load. It has been shown that the beginning of plastic flow is independent from the radial position of the hole, but the height of the pins is clearly linked to the position of the holes and their diameter.

Also at LWP Saarbrücken, Germany [57], coining of small cavities with a thickness of about 400 microns has been investigated. The results show, that in the case of micro-coining the influence of the tool geometry and the tool deflection on the forming results must be considered. This is due to rather high normal stresses (minimum three times of the flow stress of the coined material) having a large influence on the tool system through tool deformation and damage.

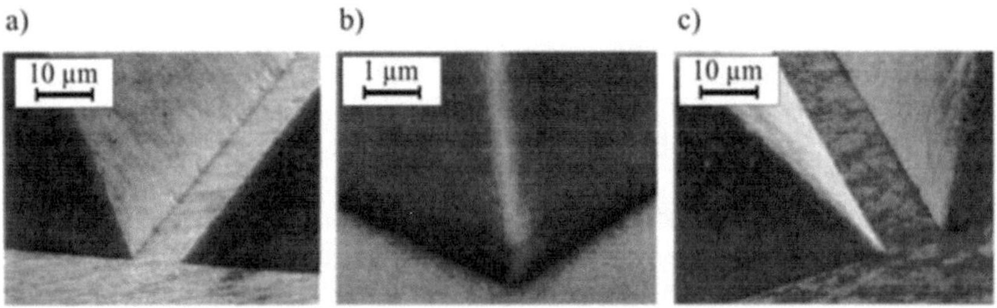


**Fig. 13.** Microgeometry after cold embossing a) aluminium, b) brass, c) stainless steel (IWU)

The aim of investigating coining processes at IWU Chemnitz was to establish a technology for the production of geometrically defined micro structures for applications in micro-fluidics, micro-optics and information technology [58,59]. In the first set of experiments, tools were made from a single crystal silicon with an almost smooth surface (Fig. 13). Thus, it was possible to create structures with lateral dimensions of 100 microns and radii of some 100 nm in aluminium, copper, brass and steel.

The analysis of the coined material shows that the shape of the edges strongly depends on the material. While aluminium shows a bulging of max. 30 % of the coining depth, steel shows only max. 5 %. Using a fine grained ZnAl alloy in a superplastic state at 250 °C (Fig. 14), high precision and high surface quality are achieved at low compressive stresses of 25 MPa.

a)                              b)                              c)

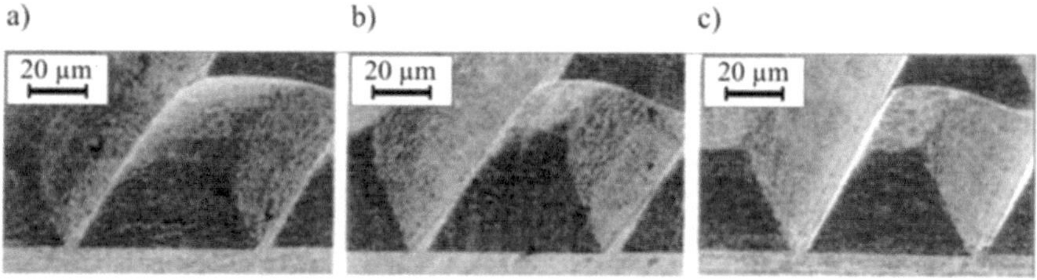


**Fig. 14.** Microgeometry in ZnAl after superplastic coining (IWU)

As surface functionalization has gained an increasing importance in the past years, coining processes have been increasingly used to manufacture smallest structures on specimen surfaces [60]. In the investigations done by Ike, aluminium sheet has been structured by using WC-dies. Additional investigations have confirmed the influence of microstructure on the shape evolution in coining [61].

In addition to bulk metal forming, micro sheet metal forming plays an increasingly important role, especially in the electronics industry. Smallest leadframes, characterized by the lead-width of about 150 microns, or the smallest connectors are made by punching, blanking and bending.

Basic investigations on the effects of miniaturization on the blanking process have been first performed by Kals []. It has been shown that, due to a lack of free surface, the normalized forming force is constant despite scaling down sheet thickness. However, when the sheet thickness is below a certain value, the forming force and the ultimate shear strength is increasing.

A particular blanking process, so-called dam-bar cutting has been investigated by [62, 63]. This is a mechanical trimming process removing the dam-bar between the leads after the IC package is encapsulated. Due to the shape of the bars, which is rectangular around the IC, investigations have to be performed considering the anisotropic behaviour of the material in the shearing line which has a direction of either 0° or 90° to the rolling direction of the strip the lead frame is made of. The punch forces are a function of the direction of the shearing line, depending on the texture of the strip material. Cuts of long and narrow slots (that means that the shearing line has a preferred direction) with a direction of 45° to the rolling direction showed a higher difference of the forces relative to cuts with 0°.

### 2.1.3 Tools

Other investigations have shown that there is a dependency between the tool and the process parameters and the accuracy in case of lead frames. The deflection of leads in the plane of the sheet increases with the decreasing width of the lead. The deflection is also influenced by different clearances between tool and die in a progressive tool. Furthermore, increasing the strip holder pressure has a positive effect on the accuracy in most cases. Also the dynamic behaviour of the tool affects the accuracy [64], e.g. increasing the blanking speed results in a decreasing accuracy. Experiments have additionally shown that increasing the strip holder force clearly improves the product quality. Other investigations on the deflection during the blanking process [65] have shown an increase of the deflection when the punch is eccentric relative to the die.

Manufacturing of tools for microforming is the key for satisfactory forming processes. An overview of the state of the art of technologies suited for the production of micro tools is given in [66, 67, 68]. If tools are to be produced with high geometrical accuracy and surface quality, two kinds of technologies should be considered: conventional technologies like machining or EDM, adapted to the needs of the micro world, or energy assisted technologies like micro laser ablation.

Production technologies, frequently used at a conventional length scale for complex 2.5 dimensional or true three-dimensional geometries are end milling processes. The geometrical accuracy of the tools produced by micro milling processes is mainly limited by the material and shape of the milling tool [69] as well as the machining strategy [70]. Using diamond cutting tools, surface roughness Ra in the range of 10 nm and smallest structures down to 100 micron are achievable. However, diamond tools are restricted to non-ferrous materials due to the high affinity of iron to carbon. To solve this problem, two solutions are applicable: ultrasonic vibrations applied to the tool [71] reduce the contact time between the diamond tool and the workpiece. The second possibility is to apply nitriding processes to the workpiece which will reduce the chemical reaction between tool and workpiece and thus reduce wear [72].

Other materials for cutting tools like cemented carbides achieve surface roughness between Rz 0.1 – 0.3 µm and are commercially available in diameters down to 50 micron. As shown in Fig. 15, structures with high geometrical accuracy are achievable.

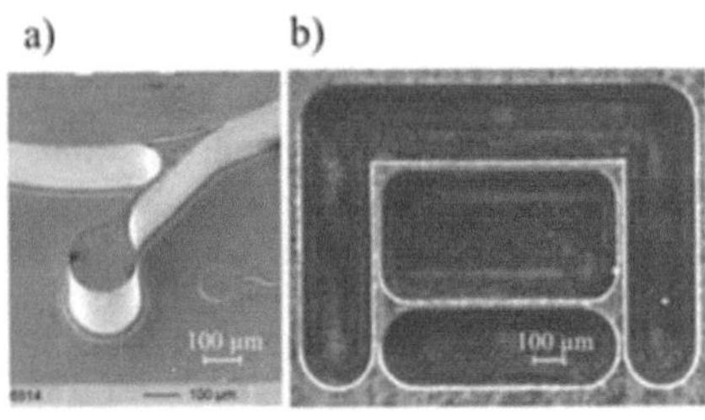


**Fig. 15.** a) Detail of a turbine inlet structure in brass (wbk)  b) small channels with feature size of 20 micron (MEC)

Wire EDM is well known for the production of tools. It is also applied in the field of micro tool manufacturing. The width of achievable contours is down to 300 µm. Investigations on this topic are carried out at the LFU Dortmund. For the production of micro tools it is necessary to use very thin wires, because of the very small contours and radii. The possibility of using wires of 10 µm in diameter has been confirmed [73], wires with diameters down to 30 µm have been applied in [74]. In order to obtain sufficient strength, tungsten with tensile strength of 2400 N/mm² is used as the wire material. It enables cutting of steel 5 mm thick and WC 3 mm thick. The smallest possible bars and slots are 20 μm to 40 μm and 50 μm to 60 μm, respectively [75]. The tolerances are typically in the range of ± 2 μm and the surface roughness reaches Ra values between 0.1 μm and 0.25 μm [,76]. Limits of μ-EDM, i.e. the non-capability of reproducing sharp edges and surface finishing as required for forming tools were recently discussed in [77].

The example of a new technology is tool manufacturing by laser ablation. It enables the production of shapes and smallest structures that can not be made using other methods e.g. by EDM [78,79]. This is because the material of the irradiated are evaporates while the material around remains cool. The very high energy density of a laser beam facilitates machining of even very brittle materials like tungsten carbide or ceramics. Micro blanking and embossing tools or extrusion dies are successfully made by laser ablation. The smallest contours are in the range of 10 μm. The surface quality of laser machined tools is another point of interest. Investigations of tools made of tungsten carbide have shown that laser machined surfaces often need some additional treatment. The best surface quality was achieved by mechanical polishing the roughness Ra could be reduced from 0.75 μm to 0.07 μm. Unfortunately, mechanical polishing is usually not possible inside small structures. Chemically polished surfaces show no improvement of the roughness, but electrolytically polished surfaces have clearly better values of Ra (0.36 μm) [, 80].

### 2.1.4 Machine and equipment

Recently, several new machines have been developed to fit the needs of micro applications. Either existing technologies have been optimized for the production of microparts or new small (micro) machines have been developed. Especially in the area of sheet and foil forming, high speed stamping or blanking machines like the BSTA series from Bruderer AG have to be mentioned. The BSTA are conventional mechanical high speed presses that are equipped with high precision guides and some more special features to increase the precision. A counter balance system acting inversely to the movement of the ram keeps the machine free of vibrations, even at a production rate of 2000 strokes per minute. The guides and levers that control the ram movement are arranged in a way that a tilting of the ram due to an eccentric load does not influence the position of the punch. This is achieved by locating the theoretical centre of rotation of the ram at the tip of the punch. For bulk metal forming Wafios AG [81] built up a multi-station former for parts down to 0.5 mm in diameter and remarkable output rate up to 400 parts per minute. Similar machine tools (multi-station former) have been developed in Japan by Daido [82].

Several machines, especially for micro forming processes, have been developed in close cooperation between academics and industry. A piezo-electric driven press has been realized by the Zentrum für Fertigungstechnik (Stuttgart) [83] which can be driven force or velocity controlled with approx. 420 strokes per minute and a maximum force of 3.5 kN. In cooperation with Schuler AG [84], a linear motor driven press has been developed for punching and blanking operations but is also applicable for micro forming processes with remarkably high number of strokes per minute (1200) and accuracy in positioning of 5.6 micron. Based on the same principle, IPU Denmark has built up a test machine for bulk microforming. Using the largest, commercially available drive providing a maximum force of 5.3 kN, the setup

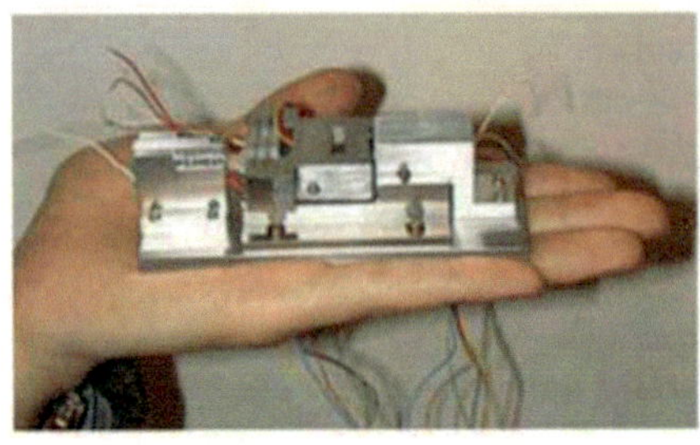

**Fig. 16.** Micromachine for superplastic extrusion (Gunma University)

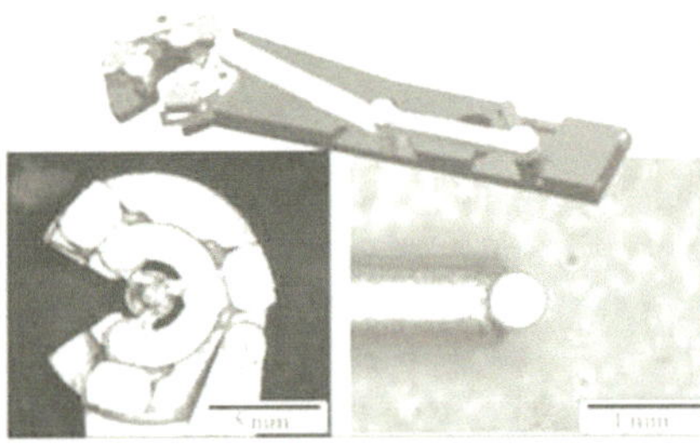

**Fig. 17.** Mechanical and vacuum microgrippers (LFT)

**Fig. 18.** Clamps and contact pins, still connected to the sheet strip (Unimet)

will facilitate two stage forming processes, punch strokes up to 25 mm with a reproducibility of 1 μm [85,86]. The latest micro forming machine has been developed by BIAS Bremen [87]. By using linear motors, air bearings and two independently controllable axes, high precision (below 3 micron position accuracy), 1250 strokes per minute and 15 kN maximum force can be achieved.

An additional goal for the future should be the reduction in the machine size itself - a micromachine for microparts. Examples are the desktop factory, one of the machines of which being a micropress [88]. A similar prototype has been presented by Gunma University. Figure 16 shows the machine tool that has been developed for superplastic forward extrusion of a microgear shaft [89].

Cold forming processes are, in general, characterized by a high production output. Using a multi-stage micro former requires the parts to be transferred and positioned in a very short time and with a few microns of positioning tolerance between the dies of the different forming stations. Another problem is caused by the small size of the parts and the high ratio of surface to volume. Due to the small weight and the adhesion forces, parts tend to stick to conventional grippers, which make correct positioning nearly impossible. The current research investigates the influence of humidity on adhesion, the use of vibrations for the purpose of decreasing adhesion and the development of grippers for micro part transfer systems. Fig. 17 shows a mechanical gripper, which makes it possible to place a part with a repeating accuracy between 5 μm and 15 μm and a vacuum microgripper, which is used in a transfer system. This prototype enables the transport of 4.3 parts per second; the transfer length is 25 mm, the part diameter 0.85 mm and the accuracy 5 μm [90].

Adhesion is not always unwanted, it can also be used to handle flat parts. The IPA Stuttgart developed an adhesive gripper for flat micro systems. The use of alcohol as adhesive results in high flexibility regarding the shape, size and material of the units handled. Furthermore, the adhesive gripping using fluids of low viscosity provides a centering effect of the unit towards the gripper as well as a compliance effect when joining parts with a multi spot contact [91].

In sheet metal forming, handling is less difficult, since the parts usually remain connected to the sheet strip (Fig. 18). Thus, only the entire strip has to be positioned and transferred. The feed rate is determined by notches that are stamped in the edge of the sheet strip during the process. The notch runs against a stopper and is enlarged with each punching stroke. A precise positioning is achieved by pins that enter into positioning holes that are punched into the strip.


## 2.2 Basic Research on Nanomaterials

On the microscopic level, plastic deformation tends to take the form of localised shear bands. Many systems of such bands subdivide the original coarse grains into so-called subgrains or dislocation cells. Initially, these subgrains have low misorientation angles, however, for larger strains, the crystallographic orientation of some subgrains changes to the extent which justifies treating them as entirely new grains, distinct from their neighbours (misorientation angle > 15°). The equivalent plastic strain at which those advanced changes occur depends on the material and the process conditions; the lowest value can be about 3 but often it is a lot more. Such a large deformation is referred to as severe plastic deformation (SPD). The process of grain subdivision and orientation evolution can be further complicated by recovery and recrystallisation taking place at elevated temperatures and phase transformation or precipitation in some metals. Nevertheless the resulting structure is extremely fine (Fig. 19), with grain sizes below 1 micrometer in the case of so-called ultrafine grained (UFG) metals or even below 0.1 micrometer for nano-crystalline (NC) metals. These structural changes result in unusual mechanical and physical properties of UFG/NC metals such as 3-4 fold increase in yield strength (compared to annealed metals), doubled ultimate tensile strength, increased hardness, toughness and high cycle (stress controlled) fatigue (HCF) life. The low cycle (strain controlled) fatigue (LCF) life is not improved because of the reduced ductility usually accompanying higher strength. Comprehensive studies on LCF-behaviour of UFG/NC metals as well as a proposal of a simple rationale explaining the difference between LCF and HCF is

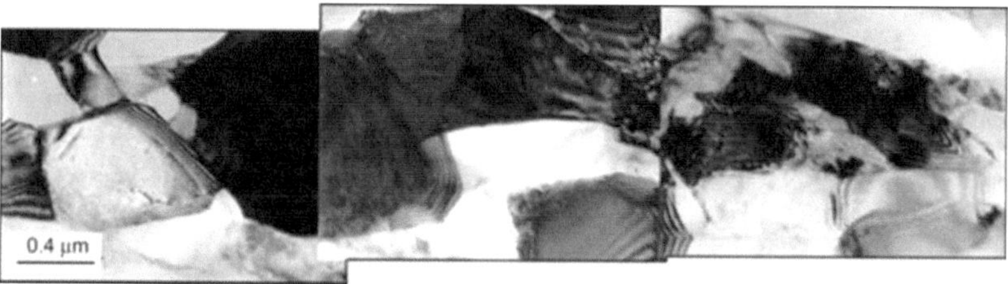


**Fig. 19.** UFG structure of Al 1070 subjected to SPD (strain 6.9)

presented in [92,93]. Despite generally lower ductility, there are examples of impressive strength gains as well as retained ductility. These were observed in pure UFG copper and titanium and UFG magnesium alloys. UFG/NC metals have a superior combination of strength and ductility at low (cryogenic) temperatures. At elevated temperatures, those UFG/NC metals which are thermally stable, exhibit enhanced superplastic properties such as increased elongation, higher strain rate (so-called high strain rate superplasticity) and/or lower temperature.

### 2.2.1 Scope for Forming Research

All these changes in structure and properties of UFG/NC metals attracted attention mainly of materials scientists and metallurgists. As a result, the majority of publications on UFG/NC metals appear at conferences and in journals serving this community. However, this situation slowly changes because SPD appears to be just another metal forming technology, with the only difference being that it usually preserves the shape and dimensions of the processed billets. Thus SPD is similar to classical metal forming with all its problems related to heating, friction, material flow, tooling, machine performance, quality, productivity and costs. These aspects of SPD have become more important recently as the increasing interest in UFG/NC metals makes them good candidates for industrial applications. Such applications require that UFG metals are produced in a wide range of shapes and dimensions and in high quantity. There is also a question of post-SPD forming of UFG/NC metals into required components by, for example, machining, spark erosion or classical metal forming. All these technologies may have to be redefined or at least tested with reference to new UFG/NC materials available. Thus, there is a lot to do for metal forming researchers and engineers starting, perhaps, from devising industrially viable methods of SPD.

### 2.2.2 SPD Processes

There are an increasing number of experimental SPD processes available (Fig. 20), however, none of them has reached the level of maturity required for mass production of UFG/NC materials. We can mention high pressure torsion (HPT) [94] and cyclic

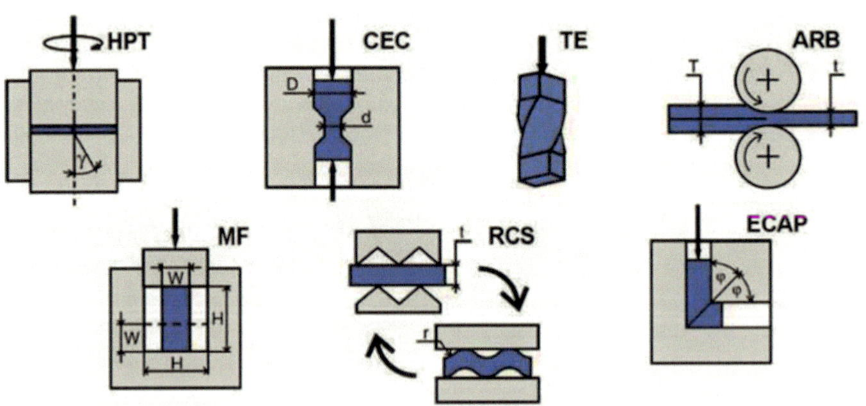


Fig. 20. SPD processes

extrusion compression (CEC) [95], twist extrusion (TE) [96] and accumulative roll bonding (ARB) [97,98], multi-axial forging (MF) [99] and repetitive corrugation and straightening (RCS) [100]. However, the most popular process is equal channel angular pressing (ECAP) [101] whose both the batch and continuous versions are close to commercialization. In this process, the billet material is forced through a sharply bent channel of a constant cross section (square/rectangular or round). When the material passes through the bend diagonal, it undergoes simple shear. For the bend angle of 90°, the amount of equivalent plastic strain generated in one pass of the billet is 1.15. Since this is less than the strain required for advanced structural changes to occur, the process is repeated several times (usually from 4 to 8 times). This gives the opportunity to change the material shear plane in each pass by rotating the billet about its axis between consecutive passes by 90° or 180°.

### 2.2.3 Modifications of ECAP

Classical ECAP, as described above, is a lengthy and cumbersome process. Therefore, there have been attempts to increase ECAP's productivity by increasing the number of channel bends in the die (Fig. 21). This has led to the concept of an S-shape channel (as opposed to L-shape channel in classical ECAP), which has two bends. In this way, it is possible to double the plastic strain generated and realize a 180° billet rotation "in die" [102]. To achieve the same for a 90° rotation, a three-dimensional, two-bend channel is required [103]. The number of channel bends can be increased to three and more, however, the force required to push the billet through such a channel becomes very high [104]. Also, the complexity of a split die that is used to form a multi-bend channel may cause problems.

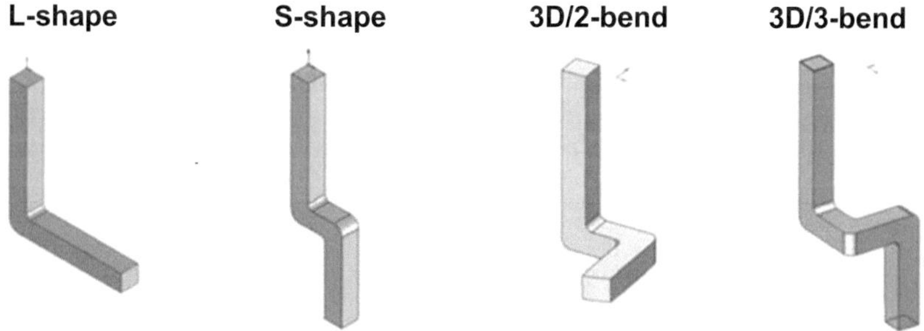


**Fig. 21.** Channel configurations in ECAP

Other issues with classical ECAP are the length of billets and material utilization. Like in forward extrusion, the length/diameter ratio of the billet is limited to about 6 because of the process force depending on friction, which increases with the length of the input channel. The inherent feature of ECAP is an undeformed material at the billet's ends. This, together with other flow related problems, can limit material utilization to 50% [105]. A partial solution to these problems can be a die with movable walls, which reduce friction [106]. However, such a die is difficult to design and manufacture and may not be suitable for all types of billets. The ideal solution would be continuous ECAP (Fig. 22). However, as the continuous SPD (CSPD) process has shown, it is not easy to feed the billet into an ECAP die using only a friction force [107]. Another process based on friction-assisted feeding uses a modified "Conform" wheel (ECAP-Conform) [108]. Yet another recent development is incremental ECAP (I-ECAP), based on an original tool kinematics [109]. In this solution, one part of the ECAP die deforms the billet by moving against it in a reciprocating manner, while the other part provides a channel for the incrementally fed billet. Synchronization of deformation and the feeding movement enables incremental shearing of the material and makes the feeding force largely reduced.

**Fig. 22.** Versions of continuous ECAP

The above examples of the SPD process development are indicative of typical research problems, which are solved using theory of plasticity, finite element simulation, tool technology, experimental techniques, etc. There is still plenty of room for an original research in this new exciting discipline of metal forming.

### 2.2.4 Post-SPD Forming

Traditional metal forming, or more generally shaping, will also require attention in order to make the best use of UFG/NC metals. The known and envisaged applications of UFG/NC metals include aluminium and copper sputtering targets (for efficiency and quality of PVD), pure titanium medical implants (for biocompatibility and strength), light armour (for ballistic performance), armour penetrators (for self sharpening effect), aerospace parts (for strength to weight ratio) and sports equipment (for performance). There are also a number of technological processes which can benefit from using UFG/NC metals such as warm forging (for better material flow and energy savings) [107], superplastic forming (for increased elongation, increased strain rate and reduced temperature) [110] and micro-manufacturing of MEMS (for avoiding scaling down problems). All these applications will require different shaping processes and, consequently, billets of different forms (bars, rods, plates, sheets, wire, foil). The preliminary results obtained for post-SPD extrusion [111] and post-SPD rolling [112] show evidence of further changes in materials properties. Very often these properties will depend on an appropriate heat treatment [113].

### 2.2.5 UFG/NC Metals for Micro-Manufacturing

In the first part of this paper, the scaling down problems were attributed to the coarse grains of metals used in microforming operations. Reducing the grain size to the UFG/NC level seems to be a very good way of alleviating these problems. This has become clear for micro backward extrusion of pure aluminium [114] and forward extrusion of a magnesium alloy [115]. Spectacular effects can be achieved on free surfaces of microformed components where, by using UFG/NC metals, a much smoother surface finish, unachievable for course grained materials, can be obtained (Fig. 23) [116]. Other positive surface effects, this time resulting from the material/tool interaction, have been reported for micromilling [117] and diamond tool turning of UFG aluminium [118]. There is also evidence of roughness reduction in micro-EDMed UFG aluminiumn [119]. Micro-manufacturing usually requires small size billets. In this case, bulk UFG/NC billets have to be converted to miniature billets. Alternatively, small size billets can be produced directly during SPD, as demonstrated by the combination of forward extrusion and ECAP [120].

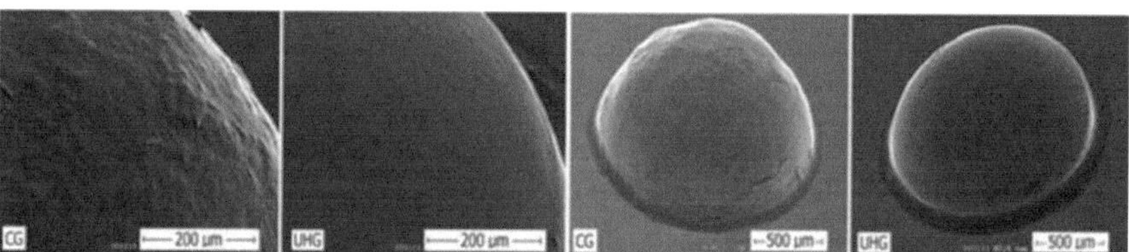


**Fig. 23.** SEM pictures of the dome part of the component made of CG and UFG Al 1070 [116]

# 3. The future

Some of the paragraphs of the last chapter recall very latest or even unpublished results. This is typical for fast progress in the area of microforming and nanostructuring of metals by SPD. We can expect a lot more research on both technologies, in particular, on the metallurgical aspects of SPD such as the mechanisms involved, thermal effects, interactions with other structural changes and the relations between structure and properties; here, achieving a good balance of strength and ductility is the main goal. There is also a research on linking the evolution of microstructure with a multiscale constitutive description of UFG/NC metals; such a model would help in process simulation and prediction of properties. The finite element technique and its modifications addressing issues specific to both technologies, will remain the main simulation tool helping to plan adequately microforming processes and to develop new, more efficient SPD techniques. There is still some way to go before we will see an industrial breakthrough in microforming and the first industrial, fully automated SPD process. A lot of design effort and new technological know-how is required to achieve this state. In parallel, processes accompanying SPD such as heat treatment and post-SPD forming will also be developed. Eventually, when UFG/NC bars/plates/sheets become readily available, more applications, including microforming, are likely to use UFG/NC materials, which will create even more research work to achieve the best match between materials and applications.

Additionally to the materials related tasks, research on process chains, process interaction and tolerance chains has to be considered. This is necessary because of the increasing complexity of products to be manufactured at microscale and thus increasing demands for accurate, reliable and reproducible manufacturing processes. With so many current interdisciplinary activities, running projects, and international networks, most of them cited in this review, as well as ideas and visions to be realized, e.g. in FP7, the research on both technologies seems to be well set for all the future tasks with their exiting challenges.

# Links

- 4M                 (www.4m.net.org)
- Masmicro            (www.masmicro.net)
- Manufuture          (www.manufuture.org)
- MINAM               (www.micronanomanufacturing.eu)
- NanoSPD             (www.nanospd.org)

**Acknowledgement:** Part of this work was carried out within the European network 4M "Multi-Material Micro Manufacturing".

# References

[1]     R. Wechsun et al., Nexus Market Analysis for microsystems, 2000-2005. Nexus Task Force Market Analysis. Wicht Technologie Consulting. München, 2002

[2]     Maeda, T.: A proposal of developing of super micro-precision press-machine. In: The Japan Society for Technology of Plasticity (publ.): 1989 Annual Report 11(1990) 12-13.

[3]     Geiger, M., Kleiner, M., Eckstein, R., Tiesler, N., Engel, U.: Microforming. CIRP Annals 50(2001)2, 445-462

[4]     Geiger, M.; Engel, U.: Microforming - a challenge to the plasticity research community. Addressed to the 40th anniversary of the JSTP. Journal of the JSTP 43(2002)494, 5-6

[5]     Lowe, T.C.; Outlook for manufacturing materials  by severe plastic deformation. In Proc. of the 3rd International Conference on Nanomaterials by Severe Plastic Deformation, SPD3, Fukuoka, Japan, September 22-26, 2005; Materials Science Forum 503-504(2006) 355-361

[6]     Engel, U.; Eckstein, R.: Microforming - from basic research to its realization. J. Mater. Process. Technol. 125-126(2002) 35-44

[7]     Vollertsen, F.; Hu, Z., Schulze Niehoff, H., Theiler, C.: State of the art in micro forming. J. Mater. Process. Technol. 151(2004) 70-79

[8]     Zhu, Y.T.; Langdon, T.G.: The fundamentals of nanostructured materials processed by severe plastic deformation.  JOM 56(2004)10, 58-63

[9]     Lowe, T.C.; Valiev, R.Z.: The use of severe plastic deformation techniques in grain refinement. JOM 56(2004) 10, 64–68

[10]    www.4m-net.org

[11]    www.masmicro.net

[12]  www.wtec.org

[13]  Pawelski, O.: Beitrag zur Ähnlichkeitstheorie der Umformtechnik. Arch. Eisenhüttenwes. 35(1964)1, 27-37

[14]  Kals, R.T.A.: Fundamentals on the Miniaturization of Sheet Metal Working Processes. Meisenbach, 1999

[15]  Kocanda A.; Prejs, T.: The effect of miniaturization on the final geometry of the bent products. In: Pietrzyk et al. (Eds.): Metall Forming 2000, Proceedings of th 8th International Conference on Metal Forming, Rotterdam, A.A.Balkema, 2000, 375-378

[16]  Raulea, L.V.; Govaert, L.E.; Baaijens, F.P.T.: Grain and Specimen Size Effects in Processing Metal Sheets. In: Geiger, M. (Ed.): Advanced Technology of Plasticity 1999. Proceedings of the 6th International Conference on Technology of Plasticity Nuremberg, September 19-24, 1999, Berlin, Springer, Vol. II, 939-944

[17]  Geiger, M.; Engel, U.; Vollertsen, F.; Kals, R.; Meßner, A.: Metal Forming of Microparts for Electronics. Production Engineering 2(1994)1, 15-18

[18]  Geiger, M.; Meßner, A.; Engel, U.: Production of Microparts - Size Effects in Bulk Metal Forming, Similarity Theory. Production Engineering 4(1997)1, 55-58

[19]  Picard, P.; Michel, J.F.: Characterization of the Constitutive Behaviour for Very Small Components in Sheet Metal Forming. Proceedings of the 2nd ESAFORM Conference on Material Forming, Guimarães, Portugal, 1999, 169-170

[20]  Geiger, M.; Meßner, A.; Engel, U.; Kals, R.; Vollertsen, F.: Design of Micro-Forming Processes - Fundamentals, Material Data and Friction Behaviour. In: Standring, P. (edtr.): Chipless 2000 - Advancing Chipless Component Manufacture. Proc. of the 9th International Cold Forging Congress, Solihull, England, May 1995. Redhill UK: FMJ Int. Publ. Ltd. 1995. pp 155-164

[21]  Engel, U.; Meßner, A.; Geiger, M.: Advanced Concept for the FE-Simulation of Metal Forming Processes for the Production of Microparts. In: Altan T. (Edtr.): Advanced Technology of Plasticity 1996, Proc. of the 5th ICTP, Oct.7-10, 1996, Columbus, Ohio USA, Vol. II, p. 903-907

[22]  Geißdörfer, S.; Engel, U.; Geiger, M.: FE-Simulation of Microforming Processes Applying a Mesoscopic Model. Int.J.Mach. Tools and Manuf. 46(2006)11, 1222-26

[23]  Geißdörfer, S.; Engel, U.: Mesoscopic Model - Simulation of Size Effects in Microforming. In: Jan Kusiak et al (Edtrs): Proceedings of the 10th International Conference on Metal Forming 2004, Krakow, 19.-23.09.2004, GRIPS' Sparkling World of Steel Vol. 1: Proceedings, No. 3, GRIPS media GmbH, 2004 pp 699-703

[24]  Geißdörfer, S.; Diehl, A.; Eichenhüller, B.; Engel, U.: Microforming - Application and 3D Simulation. In: Teti, R. (Edtr.): CIRP ICME'06 - 5th CIRP Int. Sem. on Intelligent Computation in Manufacturing Engineering, 25.-28.07.2006, Ischia, Italy, pp 521-526

[25]  Engel, U.; Geißdörfer, S.; Geiger, M.:Simulation of Microforming Processes - An Advanced Approach Applying a Mesoscopic Model (invited paper). In: Bariani, P. F. (Ed.): Proceedings of the 8th International Conference on Technology of Plasticity, 9. 13.10.2005, Verona, Italy. Padova, Italy: Edizioni Progretto 2005, pp. 447-48

[26]  Mehnert, K.; Klimanek, P.: Grain growth in metals with strong textures: Three-dimensional Monte-Carlo simulations, Computational Materials Science 9(1997) 261-266

[27]  Ashby M.F.: The deformation of Plastically Non-homogeneous Materials. Philosophical Magazine 21(1970) 399-424.

[28]  M.A. Meyers, E. Ashworth, Phil. Mag. A, 1982, 46, 737

[29]  Kals, R.T.A.: Fundamentals on the Miniaturization of Sheet Metal Working Processes. Meisenbach, 1999.

[30]   Raulea, L.V.; Govaert, L.E.; Baaijens, F.P.T.: Grain and Specimen Size Effects in Processing Metal Sheets. In: Geiger, M. (Ed.): Advanced Technology of Plasticity 1999. Proceedings of the 6th International Conference on Technology of Plasticity Nuremberg, September 19-24, 1999, Berlin, Springer, Vol. II, 939-944

[31]   Cavaliere, P.; Gabrielli, F.; Fratini, L.: Bending of very thin sheets: The influence of thickness on material characterisation. In: Pietrzyk et al. (Eds.): Metal Forming 2000. Rotterdam, A.A. Balkema, 2000, 405-410

[32]   Kals, T.A.; Eckstein, R.: Miniaturization in sheet metal working. J. Mater. Process. Technol. 103(2000) 95-101

[33]   Petch, N.J.: The Cleavage Strength of Polycrystals. J. Iron and Steel Inst. (1953) 25-28

[34]   Diehl, A.; Engel, U.; Geiger, M.: Mechanical Properties and bending behaviour of metal foils. In: Dimov, S.; Menz, W.; Fillon, B. (Eds.): Proc. of the 2nd International Conference on Multi-Material Micro Manufacture, 20.09.-22.09.2006, Grenoble, 297-300

[35]   Fleck N.A. and Hutchinson J.W.: Strain gradient plasticity. Adv. Appl. Mech. 33 (1997) 295-361.

[36]   Fleck N.A. and Hutchinson J.W.: A reformulation of strain gradient plasticity. J. Mech. Phys. Sol. 49 (2001) 2245-2271

[37]   Michel J.-F. and Picart P.: Modelling the constitutive behaviour of thin metal sheet using strain gradient theory. J. Mater. Process. Technol. 125-126(2002) 164-169

[38]   Leopold, J.: Foundations of Micro-Forming. In: Geiger, M. (Ed.): Advanced Technology of Plasticity 1999. Proceedings of the 6th International Conference on Technology of Plasticity Nuremberg, September 19-24, 1999, Berlin, Springer, Vol. II, 883-888

[39]   Raulea, L.V.; Goijaerts, A.M.; Govaert, L.E.; Baaijens, F.T.P.: Size Effects in the Processing of Thin Metal Sheets. In: Geiger, M. et al. (Eds.): Sheet Metal 1999, Proceedings of the International Conference, Bamberg, Meisenbach, 1999, 521-528

[40]   Eckstein, R.; Engel, U.: Behavior of the Grain Structure in Sheet Metal Working. In: Pitrzyk, M. et al. (Eds.) Metal Forming 2000, Proceedings of the 8th Internat. Conf. on Metal Forming, Rotterdam, A.A. Balkema, 2000, 453-459

[41]   Geiger, M.; Eckstein, R.; Engel, U.: Study of micro sheet metal working processes. Production Engng. 8(2001)1, 55-58

[42]   Taylor, G.I.: Plastic strain in metals. J. Inst. Met. 21(1938) 307-324

[43]   Chen, F.-K-, Tsai, J.-W.: A study of the size effect in micro-forming with micro-hardness tests. J. Mater. Process. Technol. 177(2006) 146-149

[44]   Engel, U.; Messner, A.; Tiesler, N.: Cold Forging of Microparts – Effect of Miniturization on Friction. In: Chenot, J.L. et al (Eds.): Proceedings of the 1st ESAFORM Conf. on Materials Forming, Sophia Antipolis, France, 77-80

[45]   Tiesler, N.; Engel, U.; Geiger, M.: Forming of Microparts – Effects of Miniaturization on Friction. In: Geiger, M. (Ed.): Advanced Technol. of Plasticity, Proc. Of the 6th Int. Conf. on Technology of Plasticity ICTP 1999, 19-24 Sep 1999, Nuremberg, Germany, Berlin, Springer, Vol. II, 889- 894

[46]   Geiger, R.: Der Stofffluß beim kombinierten Napffließpressen. Berichte aus dem Institut für Umformtechnik der Universität Stuttgart Nr. 36, Verlag Girardet Essen, 1976

[47]   Buschhausen, A.; Weinmann, K.; Lee, J.Y.; Altan, T.: Evaluation of lubrication and friction in cold forging using a double backward-extrusion process. J. Mater. Process. Technol. 33 (1992) 95-108

[48]   Engel, U.: Tribology in Microforming. Wear 260(2006) 265-273

[49]    Bay, N.: Friction stress and normal stress in bulk metal forming processes. J. Mech. Work. Technol. 14(1987)2, 203

[50]    Presz, W.: The influence of punch vibration on surface phenomena in micro-extrusion. Proceedings of ESAFORM 2006 Conference, Glasgow, 583-586

[51]    Presz, W.: Flexible Manufacturing System For Vibration Assisted Microforming. In: Proc. of the 10th Int. Esaform Conference on Material Forming, Zaragoza, Spain, April 18-20, 2007, accepted

[52]    Krishnan, N.; Cao, J.; Dohda, K.: Microforming: Study of friction conditions and the impact of low friction / high-strength die coatings on the extrusion of micropins. Proc. IMECE 2005, ASME Int. Mech. Engng. Conf., Nov 5-11, 2005, Orlando, FL, MED 16(2005)1, 331-40

[53]    Engel, U.; Tiesler, N.; Eckstein, R.: Microparts - A challenge for forming technology. In: Kuzman, K. (edtr): ICIT 2001 3rd Int. Conf. on Industrial Tools, Celje Slovenia: Tecos 2001, 31-39

[54]    Engel, U.; Egerer, E.: Basic research on cold and warm forging of microparts. Key Engng. Mater. 233-236(2002)449-56

[55]    Egerer, E.; Engel, U.: Process Characterization and Material Flow in Microforming at Elevated Temperatures. J. Manufacturing Processes 6(2004)1, 11-16

[56]    Ike, H.; Plancak, M.: Controlling metal flow of surface microgeometry in coining process. In: Geiger, M. (edtr): Advanced Technology of Plasticity, Proc. Of the 6th Int. Conf. on Technology of Plasticity ICTP 1999, 19-24 Sep 1999, Nuremberg, Germany. Vol. II, Berlin: Springer 1999, pp 907- 912

[57]    Thome, M.; Hirt, G.; Ellert, J.: Metal Flow and Die Filling in Coining of Microstructures with and without Flash, In: Proc. of the International Conference on Sheet Metal 05, 2005, pp. 631 - 639

[58]    Schubert, A.; Burkhardt, T.; Kadner, J.; Neugebauer, R.: High Precision Embossing of Metallic Parts with Microstructure. In; McKeown, P. et al (Eds.): Precision engineering – nanotechnology. Proceedings of the 1st International Euspen Conference, Aachen, Shaker, 1999, Vol. I, 530-533

[59]    Schubert, A.; Böhm, J.; Burkhardt, T.: Mikroprägen - Herstellen metallischer Mikrostukturbauteile durch Kaltprägen. wt 90(2000)11/12, 479-483

[60]    Ike, H.; Plancak, M.: Coining process as a means of controlling surface microgeometry. J. Mater. Process. Technol. 80-81(1998) 101-107

[61]    Wang, C.J.; Shan, D.B.; Zhou, J.; Guo, B.; Sun, L.N.: Size effects of the cavity dimension on the microforming ability during coining process. J. Mater. Process. Technol. (2006), doi: 10.1016/j.jmatprotec.2006.11.055 in press

[62]    Lee, W.B.; Cheung, C.F.; Chan, L.K.; Chiu, W.M.: An investigation of process parameters in the dambar cutting of integrated circuit packages. J. Mater. Process. Technol. 66 (1997) 63-72

[63]    Cheung, C.F.; Lee, W.B.; Chiu, W.M.: An investigation of tool wear in the dam-bar cutting of integrated circuit packages. Wear 237(2000) 274- 282

[64]    Jimma, T.; Sekine, F.; Sato, A.: Effects of Dynamic Behavior of Tools on Blanking of Electronic Machine Parts. In: Advanced Techn. of Plast., Vol. 3, 1990, 1459-1464

[65]    Doege, E.: Optimierung von Mikroschneidverfahren zur Herstellung von Kontakten für die Halbleitertechnik. Report, DO 190 / 96-1, DFG, Hannover, 1996

[66]    Brinksmeier, E.; Riemer, O.: Metal cutting of microstructures. In: Menz, W.; Dimov, S.: 4M2005, Proc. 1st Int. Conf. on Multi-Material Micro Manufacture, 29.06.- 01.07.2005, Karlsruhe. London: Elsevier 2005, pp. 19-25

[67]    Fleischer, J.; Kotschenreuther, J.: Manufacturing of Micro Molds by Conventional and Energy Assisted Processes. In: Menz, W.; Dimov, S.: 4M2005, Proc. 1st Int. Conf. on

Multi-Material Micro Manufacture, 29.06.-01.07.2005, Karlsruhe. London: Elsevier 2005, pp. 9-18

[68]   Uriarte, L.; Ivanov, A.; Oosterling, H.; Staemmler, L.; Tang, P.T.; Allen, D.: A Comparison between Microfabrication Technologies for Metal Tooling. In: Menz, W.; Dimov, S.: 4M2005, Proc. 1st Int. Conf. on Multi-Material Micro Manufacture, 29.06.-01.07.2005, Karlsruhe. London: Elsevier 2005, pp 351-354

[69]   Hesselbach, J. et al.: International State of the Art of Micro Production Technology. Annals of the WGP, Production Engineering 10(2004)1, 29-36

[70]   Dimov, S.; Pham, D.T.; Ivanov, A.; Popov, K.: Micro milling of thin features. In: Menz, W.; Dimov, S.: 4M2005, Proc. 1st Int. Conf. on Multi-Material Micro Manufacture, 29.06.-01.07.2005, Karlsruhe. London: Elsevier 2005, pp 363-366

[71]   Hesselbach, J. et al.: mikroPRO – Untersuchung zum internationalen Stand der Mikroproduktionstechnik, Schriftenreihe des Instituts für Werkzeugmaschinen und Fertigungstechnik der TU Braunschweig, 2002

[72]   Brinksmeier, E.; Dong, J.; Gläbe, R.: Diamond Turning of Steel Molds for Optical Applications. In: Proceedings of the 4th International Conference of the euspen, Glasgow, 2004, pp. 155ff

[73]   Masuzawa, T.: State of the Art of Micromachining. CIRP Annals 49(2000)2, 473-488

[74]   Gillner, A.; Hellrung, D. (Eds.): Produktionstechnik zum Mikrostanzen und Mikroprägen metallischer Bauteile. Abschlußbericht des Verbundprojektes PROMPT. Forschungszentrum Karlsruhe, 1999

[75]   Wolf, A.; Ehrfeld, W.; Gruber, H.P.: Mikrofunkenerosion für den Präzisionsformenbau. wt Werkstattstechnik 89 (1999) 11/12, 499-502

[76]   http://www.agiesales.com

[77]   Eriksen, R. S.; M. Arentoft and N.A. Paldan: Tool Design and Manufacturing for Bulk Forming of Micro Components. In: Proc. of the 10th Int. Esaform Conference on Material Forming, Zaragoza, Spain, April 18-20, 2007, accepted

[78]   Erhardt, R.; Schepp, F.; Schmoeckel, D.: Micro Forming with Local Part Heating by Laser Irradiation in Transparent Tools. In: Geiger, M. et al. (Eds.): Sheet Metal 1999, Proc. of the 7th Int. Conf. on Sheet Metal, SheMet'99. Bamberg, Meisenbach 1999, 497-504

[79]   Hellrung, D.; Scherr, S.: Zuverlässige Herstellung miniaturisierter Prägewerkzeuge. wt Werkstattstechnik 90(2000)11/12, 520-521

[80]   Saotome, Y.; Inoue, A.: New amorphous alloys as micromaterials and the processing technology. Proc. IEEE The 13th Annual Int. Conf. on Micro Electro Mechanical Systems 2000, 288-292

[81]   www.wafios.de

[82]   www.daido-kikai.co.jp/pdf/mini_casetto.pdf

[83]   Hess, A.: Piezo-hydraulischer Aktor für die Mikrobearbeitung von Metallen. In: Berichte aus dem Institut für Konstruktion und Fertigung in der Feinwerktechnik, Dissertation, Stuttgart, 2001

[84]   Schepp, F.: Linearmotorgetriebene Pressen für die Stanztechnik. In: Berichte aus Produktion und Umformtechnik, Schmoeckel, D.(Edtr.), Dissertation, Darmstadt, Shaker Verlag, 2001

[85]   M. Arentoft, N.A. Paldan: "Production equipment for manufacturing of micro metal components", 9th ESAFORM 2006, Glasgow, UK, pp 579-582

[86]   Paldan, N.A.; M. Arentoft, R.S.Eriksen: Production Equipment and Processes for Bulk Formed Micro Components. In: Proc. of the 10th Int. Esaform Conference on Material Forming, Zaragoza, Spain, April 18-20, 2007, accepted

[87]    Schulze Niehoff, H.; Vollertsen, F.: Entwicklung einer hochdynamischen, zweifachwirkenden Mikroumformpresse. In: Tagungsband des 2. Kolloquium Prozessskalierung, Bremen, 9.-10.11.2005, pp. 105-118

[88]    Ashida, K., Mishima, N., Maekawa, H. Tanikawa, T., Kaneko, K. & Tanaka, M.: Development of desktop machining microfactory. Proc. 2000 Japan-USA Flexible Automation Conference, July 23-26, 2000, Ann Arbor, Michigan, USA, pp 175-78

[89]    Saotome, Y.; Iwazaki, H.: Superplastic extrusion of microgear shaft of 10 μm in module. J. Microsystem Technol. 4(2000)6, 126-129

[90]    Geiger, M.; Egerer, E.; Engel, U.: Cross Transport in a multi-station former for microparts. Production Engng. 9(2002)1, 101-04

[91]    C. Bark, G. Vögele, Th. Weisener, "Bitte nicht berühren - Greifen mit Flüssigkeiten in der Mikrotechnik.", F&M 104(1996)5, 372-374

[92]    Mughrabi, H.; Höppel, H.W.; Kautz, M.: Fatigue and Microstructurre of ultrafine-grained metals produced by severe plastic deformation. Viewpoint set 35. Metals and alloys with a structural scale from the micrometer to the atomic dimensions. Scripta Mat. 51 (2004) 807-812

[93]    Höppel, H.W.; Kautz, M.; Xu, C.; Murashkin, M.; Langdon, T.G.; Valiev, R.Z.; Mughrabi, H.: An Overview: Fatigue lives and cyclic deformation behaviour of ultrafine grained metals and alloys. Int. J. of Fatigue 28(2006)9, 1001-10

[94]    Valiev, R.Z.: Mat. Sci. Eng. A234-236(1997) 59-66

[95]    Korbel, A.; Richert, M.; Richert, J.: The effects of very high cumulative deformation on structure and mechanical properties of aluminium. In: Proc. 2nd Risø Int. Symp. on Metallurgy and Material Science, Roskilde, Denmark, September 14-18, 1981, pp. 445-450

[96]    Varyukhin, V.; Beygelzimer, Y.; Synkov, S.; Orlov, D.: Application of twist extrusionin. In: Proc. of the 3rd International Conference on Nanomaterials by Severe Plastic Deformation, SPD3, Fukuoka, Japan, September 22-26, 2005, Materials Science Forum 503-504(2006) 335-340

[97]    Saito, Y.; Tsuji, N.; Utsunomiya, H.; Sakai, T.; Hong, R.G.: Scripta Mater. 39(1998) 1221-1227

[98]    Höppel, H.W.; May, J.; Goeken, M.: Enhanced strength and ductility in ultrafine-Grained Aluminium Produced by Accumulative Roll Bonding. Adv. Eng. Mat. 6(2004)9, 781-784

[99]    Ghosh, A.K.; Huang, W.: Severe deformation based process for grain subdivision and resulting microstructures. In: Proc. of the NATO Advanced Research Workshop on Investigations and Applications of Severe Plastic Deformation, Moscow, Russia, August 2-7, 1999, pp. 29-36, Kluwer Academic Publishers, 2000

[100]   Huang, J.Y.; Zhu, Y.T; Jiang, H.; Lowe, T.C.: Acta Mater. 49(2001) 1497-1505

[101]   Segal, V.M.; Reznikov, V.I.; Drobyshevskiy, A.E.; Kopylov, V.I.: Russ. Metall. (Engl. Transl.) 1(1981) 99-105

[102]   Rosochowski, A.; Olejnik, L.: J. Mater. Process. Technol. 125-126(2002) 309-316

[103]   Rosochowski A.; Olejnik, L; Richert, M.: 3D-ECAP of square aluminium billets. In: Proc. of the 8th Int. Esaform Conference on Material Forming, Cluj-Napoca, Romania, April 27-29, 2005, pp. 637-640

[104]   Rosochowski, A.; Olejnik, L.; Balendra, R.: FEM analysis of two-turn equal channel angular extrusion of cylindrical billets In: Proc. of the 7th Int. Esaform Conference on Material Forming, Trondheim, Norway, April 28-30, 2004, pp. 207-210

[105]   Barber, R.E.; Dudo, T.; Yasskin, P.B.; Hartwig, K.T.: Product yield of ECAE processed material. In: Ultrafine Grained Materials III, 2004 TMS Annual Meeting, Charlotte, North Carolina, U.S.A., March 14-18, 2004, pp. 667-672

[106]  Segal, V.M.:  Mat. Sci. Eng. A386(2004) 269-276

[107]  Srinivasan, R.; Chaudhury, P.K.; Cherukuri, B.; Han, Q.; Swenson, D.; Gros, P.: Continuous Plastic Deformation Processing of Aluminum Alloys. Final Technical Report (2006)

[108]  Raab, G.I.; Botkin, A.V.; Raab, A.G.; Valiev, R.Z.: New schemes of ECAP processes for producing nanostructured bulk metallic materials. In: Proc. of the 10th Int. Esaform Conference on Material Forming, Zaragoza, Spain, April 18-20, 2007

[109]  Rosochowski, A.; Olejnik, L.: FEM simulation of incremental shear. In: Proc. of the 10th Int. Esaform Conference on Material Forming, Zaragoza, Spain, April 18-20, 2007

[110]  Xu, C.; Furukawa, M.; Horita, Z.; Langdon, T.G.: Achieving a superplastic forming capability through severe plastic deformation. In: Proc. of the 2nd Conference "Nanomaterials by Severe Plastic Deformation - NANOSPD2", Vienna, Austria, December 9-13, 2002, pp. 701-710, WILEY-VCH, 2004

[111]  Salimgareeva, G.H.; Semenova, I.P.; Latysh, V.V.; Valiev R.Z.: Nanostructuring of Ti in long-sized Ti rods by severe plastic deformation. In: Proc. of the 8th Int. Esaform Conference on Material Forming, Cluj-Napoca, Romania, April 27-29, 2005

[112]  Islamgaliev, R.K.; Yunusova, N.F.; Bardinova, M.A.; Soschnikova, E.P.; Krasilnikov, N.A.; Valiev, R.Z.; Schoberth, A.: Microstructure and mechanical properties of UFG Al alloy sheets produced by equal channel angular pressing and warm rolling. In: Proc. of the 9th Int. Esaform Conference on Material Forming, Glasgow, United Kingdom, April 26-28, 2006, pp. 547-550

[113]  Mughrabi, H.; Höppel, H.W.; Kautz, M.; Valiev, R.Z.: Annealing treatments to enhance thermal and mechanical stability of ultrafine-grained metals produced by severe plastic deformation. Int. J. Mater. Res. 94(2003) 1079-1083

[114]  Rosochowski, A.; Presz, W.; Olejnik, L.; Richert, M.: Micro-extrusion of UFG aluminium. In: Menz, W.; Dimov, S. (eds): 4M2005, Proc. 1st Int. Conf. on Mult-Material Micro Manufacture, 29.06.-01.07.2005, Karlsruhe. London: Elsevier 2005, pp. 161-164

[115]  Kim, W. J.; Sa, Y.K.: Scripta Mater., 54(2006) 1391-1395

[116]  Presz, W.; Rosochowski, A.: The influence of grain size on surface quality of microformed components. In: Proc. of the 9th Int. Esaform Conference on Material Forming, Glasgow, United Kingdom, April 26-28, 2006, pp. 587-590

[117]  Popov, K.; Dimov, S.; Pham, D.T.; Minev, R.; Rosochowski, A.; Olejnik, L.; Richert, M.: Micro-extrusion of UFG aluminium. In: Menz, W.; Fillon, B.; Dimov, S. (eds): 4M2006, Proc. 2nd Int. Conf. on Multi-Material Micro Manufacture, Sep 20-22, 2006, Grenoble, France. London: Elsevier 2006, pp. 127-130

[118]  Brinksmeier, E.; Osmer, J.; Rosochowski, A.; Olejnik, L.; Richert, M.: Diamond turning of ultrafine grained aluminium alloys. In: Proc. 7th Euespen International Conference, Bremen, May 20-24, 2007

[119]  Rosochowski, A.; Olejnik, L.; Roginski, S.; Richert, M.: Micro-EDM of UFG aluminium. In: 4M2007, Proc. 3rd Int. Conf. on Multi-Material Micro Manufacture, Borovets, Bulgaria, Oct 3-5, 2007

[120]  Valiev, R.Z.; Estrin, Y.; Raab, G.I.; Janecek, M.; Zi, A.: Processing ultra fine grained net-shaped MEMS parts using severe plastic deformation. In: Proc. of the 10th Int. Esaform Conference on Material Forming, Zaragoza, Spain, April 18-20, 2007

# Multiscale Approaches

Laurent Duchêne[1] and Anne Marie Habraken[1]

[1] Department ArGEnCo, Division MS$^2$F, University of Liège,
Chemin des Chevreuils 1, 4000 Liège, Belgium
{l.duchene, anne.habraken}@ulg.ac.be

**Abstract.** This paper presents a review of the main families of multiscale models. A first group of models is interested in an accurate modelling of the texture induced anisotropy of the material during numerical simulations. The differences between the proposed models are mainly due to different choices concerning the necessary compromise between the importance of the microscopic roots of the model and the maximum admissible computation time. The length scale of the investigated process is also an important parameter. The second group of micro-macro models is based on an analysis of the dislocation densities linked to the plastic deformations. A discussion concerning the past evolution, the recent achievements and the future trends concerning multiscale models is also provided.

**Keywords:** Micro-macro, texture, homogenization techniques, dislocation densities, finite element modelling.

## 1 Introduction

The microscopic mechanisms involved during plastic deformation of metals are various and very complex, depending on the material, the forming process and the experimental conditions investigated. Nowadays, numerous complex constitutive laws are developed in order to improve the accuracy of the finite element (FE) technique.

Sophisticated macroscopicphenomenological models (dedicated to yield locus anisotropy or/and hardening behaviour) with an increasing number of parameters are proposed. This chapter mainly focuses on another approach: the multiscale material models which are based on the physics and include more and more refined microscopic mechanisms.

Macroscopic phenomenological models are efficient for the numerical simulations of industrial forming processes (e.g. automotive industry, can forming) due to the low computation time required. However, multiscale models are very helpful for the identification of the material properties. For instance, while experimental tests provide some points on the yield locus, micro-macro models are able to compute the shape of the yield locus useful for the setting of the phenomenological models.

Accurate models based on the physics are also required for a deep understanding of the material behaviour, which is a crucial point for the development of new materials with optimum texture and microstructure and their associated forming processes. For instance, the springback modelling after any metal sheet forming process and the earing prediction during deep drawing are still under progress.

Even if the multiscale models generally require large computation time, the continuous improvement of the computer performances permits to account for more and more refined microscopic plastic deformation mechanisms.

Section 2 presents an overview of the multiscale models where the anisotropy of the material is deduced from the crystallographic texture. The efficiency of these models generally results from a compromise between computation time considerations and the refinement of the microscopic models.

Section 3 is devoted to models taking into consideration the dislocation densities inside the crystals constituting the material. Microscopic models dedicated to small length scale modelling as well as macroscopic models with microscopic roots are presented.

Discussions, future trends and conclusions end this current review.

# 2  Multiscale models based on texture

## 2.1  General features

An important point when dealing with multiscale models is the micro-macro transition (or homogenization technique), which is necessary to deduce the macroscopic material behaviour from microscopic considerations. This step must be achieved with care because it determines the accuracy of the multiscale model.

At this stage, different scales for the analysis of the problem are defined:

• The scale of the sample: only the macroscopic stress field and strain rate field are important at that scale. They are computed by the finite element code on the basis of the lower scale analyses.

• The scale where the macroscopic fields are assumed to be constant in order to be able to achieve the micro-macro transition which is the topic of section 2.2. From a finite element point of view, this scale is the scale of one integration point of one finite element. From a physical point of view, this scale should be the smallest representative volume of the polycrystal behaviour. A sufficient number of crystals must be included in this volume to correctly represent the material texture.

• The scale of the crystal or the microscopic scale: a microscopic material model (based e.g. on dislocation sliding) must be developed at this scale. It is the starting point for the micro-macro transition.

• The dislocation pattern scale: this scale is larger than a single dislocation but smaller than the crystal scale. The interaction between dislocations, the presence of obstacles or substructures inside one grain are analysed at this scale. The microscopic events linked to this length scale are more deeply analyzed in section 3.

• The atomistic scale: this scale is not considered here even if dislocations dynamic simulations have given lots of new insights into the elementary processes of the dislocation motions and their interactions that define the dislocation patterning [1-3].

The smallest scale considered in this section is the scale of the crystal called the microscopic scale. The micro-macro transition results from the averaging of the microscopic values over the representative volume element (RVE):

$$\underline{\underline{\sigma}}^{macro} = \frac{1}{V} \int_{RVE} \underline{\underline{\sigma}}^{micro} dV \; , \tag{1}$$

$$\underline{\underline{\dot{\varepsilon}}}^{p,macro} = \frac{1}{V} \int_{RVE} \underline{\underline{\dot{\varepsilon}}}^{p,micro} dV \; . \tag{2}$$

With the use of $f(g)$, the orientation distribution function (ODF), one has:

$$\underline{\underline{\sigma}}^{macro} = \int \underline{\underline{\sigma}}^{micro}(g)\, f(g) dg \; , \tag{3}$$

$$\underline{\underline{\dot{\varepsilon}}}^{p,macro} = \int \underline{\underline{\dot{\varepsilon}}}^{p,micro}(g)\, f(g) dg \; , \tag{4}$$

the integration being done over all the possible orientations of the crystals.

Even if the relation between microscopic stress and plastic strain rate is known, the corresponding relation at the macroscopic level is not straightforward. The averaging over all the crystallographic orientations seems a very simple concept but it must be done simultaneously on the stress and the plastic strain rate because of their interaction in the single crystal. A solution to this problem is then very hard to find and is discussed in section 2.2.

## 2.2 Homogenization techniques

Several methods have successively been proposed in order to solve the micro-macro transition. The main ones are:

<u>Sachs' model</u>. Based on [4], one has derived the assumption of a homogeneous stress distribution throughout the whole polycrystal. The stress in each crystal is then chosen equal to the macroscopic stress. The averaging must then only be achieved on the plastic strain rate. This model is not very satisfactory and gives rise, in the general case, to a contradiction. Indeed, each crystal having its own orientation, imposing a common stress expressed in the sample coordinate system to all the crystals consists in imposing a different stress to each crystal in its reference system. The yield locus of each crystal being quite anisotropic, the imposed stress cannot fall on the yield locus for each crystal. One can understand that no compatibility of the strain rate will be fulfilled between neighbouring grains with such an approach. This model is generally not implemented in FE method.

<u>The full constraints (FC) Taylor's model</u> [5]. According to literature, Taylor type models are the most widely used for the computation of the constitutive response of

polycrystal aggregates. Taylor's model assumes a homogeneous distribution of the velocity gradient trough the polycrystal. This assumption is expressed for each crystal by:

$$\underline{\underline{L}}^{micro} = \underline{\underline{L}}^{macro} \ . \tag{5}$$

Within a finite element code when elasticity is neglected, the plastic strain rate is the symmetric part of the velocity gradient $\underline{\underline{L}}$ :

$$\underline{\underline{\dot{\varepsilon}}}^{p,macro} = \frac{1}{2}\left( \underline{\underline{L}} + \underline{\underline{L}}^{T} \right) \text{, with } \underline{\underline{L}} = \frac{\partial \underline{v}}{\partial \underline{x}} \ , \tag{6}$$

where $\underline{v}$ is the velocity field and $\underline{x}$ is the spatial coordinates.

Thanks to Taylor's assumption, the micro-macro transition can easily be implemented into a finite element code. The velocity gradient is given by the finite element code at each integration point and the corresponding stress is sought for. The plastic strain rate is deduced from equation (6) in a macroscopic point of view; the microscopic one is identical for all the crystals of the RVE associated with one integration point (see equation (5)), however it must be rotated to be expressed in each crystal lattice reference system. Each crystal in its own reference system sustains a different plastic strain rate. The microscopic stress is computed for each crystal from its yield locus. The macroscopic stress of the polycrystal is obtained by a weighted averaging of the microscopic ones with the use of the ODF as shown by equation (3).

This procedure consumes computation time if a large number of crystals is considered. Anyway, it offers the advantage that each crystal is treated successively and independently of the other crystals. So, multiprocessor computation can easily be applied. As the Voigt's model is an upper bound for the elastic stiffness matrix, the Taylor's model is an upper bound for the yield stress. The stress equilibrium between individual grains is generally violated with the Taylor's model.

According to literature (see e.g. [6]), in spite of its limitations, this model appears to be quite successful in the prediction of the stress-strain response of the polycrystal and of the texture during plastic deformations. It is however widely recognized that the textures predicted by Taylor type models are much stronger than the actual measurements.

<u>The relaxed constraints (RC) models [7-9]</u>. Taylor's assumption of a homogeneous plastic strain rate in the RVE is, in some applications, too restrictive. From a physical point of view, nothing ensures that the plastic strain rate is constant in a crystal and identical in all the neighbouring crystals; it is not even sure that the plastic strain rate is close from one crystal to another. According to the considered forming process, it is interesting to modify the assumption of a homogeneous plastic strain rate. In the case of rolling, the X-axis being the rolling direction, the Y-axis being the transverse direction and the Z-axis being the normal direction, the lath model relaxes the XZ component of the plastic strain rate. The relaxed component is no more identical for all the crystals but is free. The pancake model relaxes in addition the YZ component. The lath and pancake methods seem to be more satisfactory for the modelling of the roll-

ing process than the FC Taylor's model (particularly for the prediction of the deformation textures), they take into account the elongated shape of the crystals.

Note that the compatibility conditions disappearing on the relaxed plastic strain rate components must be replaced by equilibrium conditions on the corresponding stress components.

The main disadvantage of these relaxed models is that they are dedicated to a particular forming process; they cannot be used for an arbitrary strain history. The generalised relaxed constraints method (see [10]) overcomes this drawback by choosing automatically the plastic strains that should be relaxed.

The multiple points models: as it has been explained, due to their assumptions, the previous models treat each crystal in turn. Interesting from a computational point of view, this choice increases the difficulty to take into account the effect of the interaction between adjacent crystals. That is the reason why multiple points models have been investigated. For instance, the Lamel model [11-13] examines the interaction between 2 grains still assuming rolling simulation. The flattening and the elongation of the rolled crystals led to the idea of considering 2 grains having the same size and shape and which lie exactly on top of each other. The boundary between these 2 grains being parallel to the plane of the steel sheet. The FC Taylor's condition of uniform plastic strain rate is here applied to the set of the 2 grains and not to each grain. Under such conditions, the crystals have 2 relaxations: the XZ and the YZ components of the velocity gradient are relaxed (using the same coordinates as above). Now, the relaxation of the top grain must be the opposite of the relaxation of the bottom grain, which is different from the RC models. The stress equilibrium between the 2 grains must be verified. Table 1 of [13] shows that the Lamel model predicts more accurately the deformation texture than the RC or the FC Taylor's models. A stack of 3 grains instead of 2 is also proposed by [13]. A more recent Lamel version with, in addition, the relaxation of the XY component has also been investigated [12].

Figure 1 schematically compares the FC Taylor's, the Lamel and the pancake models. Opposite shear appears with the Lamel model (computed by minimization of the total plastic work rate in the two grains). No shear appears with the FC Taylor's model because it is prescribed in a macroscopic point of view while the pancake model violates the compatibility between both grains.

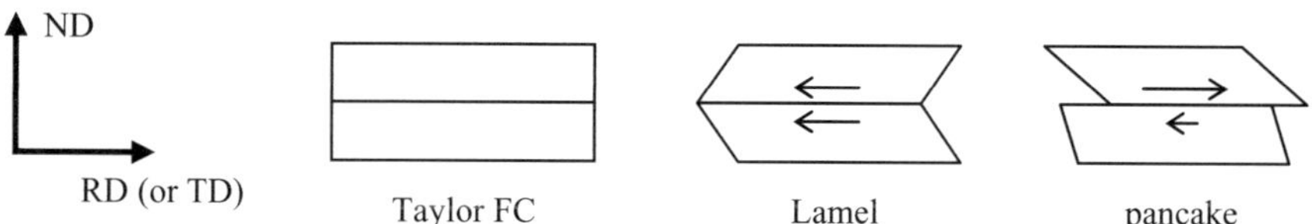


**Figure 1:** Comparison of the FC Taylor's, the Lamel and the Pancake models (adapted from Fig. 1 of [13])

The GIA (Grain Interaction model) [14] is similar but more general than Lamel model. It considers the interaction between 8 grains. The relaxed shear components are as in the Lamel model. This model is not limited to the rolling deformation mode; but it requires larger computation time.

In the Lamel model, the interfaces between two adjacent grains must always be parallel to the RD – TD plane. This constraint limits this model to rolling processes. In the recent Advanced Lamel (ALAMEL) model [15], the orientations of the interfaces are defined either by the user as a function of the material and the process investigated or they can be randomly chosen with a rule taking into account the grain shape. Therefore, the ALAMEL model is suitable for any deformation mode.

<u>The self-consistent model</u> (see [16-17]). It is a generalisation of the multiple points models in the sense that all the crystals are treated as if they would act simultaneously. The self-consistent model considers each grain in turn as an inclusion embedded in an ideal homogeneous plastic matrix. As this plastic matrix is expected to represent the polycrystal behaviour, it is obtained by an averaging of the single crystal behaviour. An iterative procedure must be used. From a computational point of view, Taylor's model, for instance, is used to obtain a first approximation for the matrix behaviour. Each grain is then computed as the inclusion into that matrix. The new matrix behaviour is obtained by averaging on all the grains. The computation is repeated until convergence is obtained on the matrix behaviour. With this model, the stress and the plastic strain rate are allowed to be different from one grain to another (the assumption of uniform stress and plastic strain rate inside each crystal is however kept). This model should conceptually be more accurate than the previous ones but requires larger computation time. However, on a local point of view, neither equilibrium nor displacement compatibility are fulfilled.

Note that a first trial of a self-consistent approach was Kröner's model [18]. On the basis of Eshelby's work, Kröner treats the problem of a spherical plastic inclusion representing one crystal embedded in an elastic matrix. The polycrystal behaviour is the average on all the crystals. The plastic incompatibility between the crystals is accommodated elastically. As a consequence, this model overestimates the yield stress just like Taylor's model. Moreover, as the matrix is assumed to be purely elastic, it cannot correctly represent the polycrystal behaviour. In that sense, Kröner's model is not literally a self-consistent model.

A visco-plastic self-consistent polycrystalline model [19] implemented in a 3D FE code extends previous versions limited to elasto-plasticity towards actual elasto-viscoplastic behaviour (see also [20]).

An original micro-macro model based on a solid volume fraction internal variable approach and a self-consistent approximation is presented in [21] to describe the isothermal steady state flow behaviour of semi-solid material (thixoforming) in a large range of strain rates. A specific self-consistent model based on the integral equation for the translated visco-plastic strain rate field is proposed in [22].

<u>Pilvin's model</u> [23]: in order to be able to simulate complex path loading, starting from Kröner's model, Pilvin added accommodation variables (β-law). A 2 stage complex mathematical formulation is used for the micro-macro transition. The physical meaning of this model is not clear; nevertheless, it compares favourably to other models.

## 2.3  Micro-macro models without macroscopic yield locus

A large class of models assumes that a set of representative crystals is associated to each integration point of a macroscopic FE mesh. For each crystal, an elasto-plastic or elasto-visco-plastic crystal plasticity model is chosen and the homogenization technique takes care of both the average process to provide the macroscopic answer and the identification of the microscopic quantities [24-25].

The crystal plasticity finite element method (CPFEM) [6;26-27]: In this method, each finite element represents one grain of the polycrystal. The constitutive law for one particular element is then the microscopic law governing the crystal behaviour. In order to model the polycrystal behaviour, each finite element corresponds to a particular crystal orientation so that the ODF of the material is correctly represented by the whole finite element mesh. A particular procedure must be used to assign one specific orientation to each finite element of the mesh.

Note that some variants of this model are proposed: one finite element can represent more than one grain (for instance, each integration point is assigned a different lattice orientation; if each element contains 8 integration points, it has 8 different lattice orientations, i.e. each element is assumed to be made of 8 grains). On the other hand, each element of the finite element mesh can represent a region smaller than a grain; one grain is then modelled by several finite elements. Other variants have also been investigated (by e.g. [6]).

The main advantage of this model is that it simultaneously ensures stress equilibrium and deformation compatibility between grains. It is basically the goal of the finite element technique. This point has been proved to be a significant improvement compared to the Taylor's model. Large deformation heterogeneities between grains have indeed been observed with the finite element technique (as shown by Figure 2).

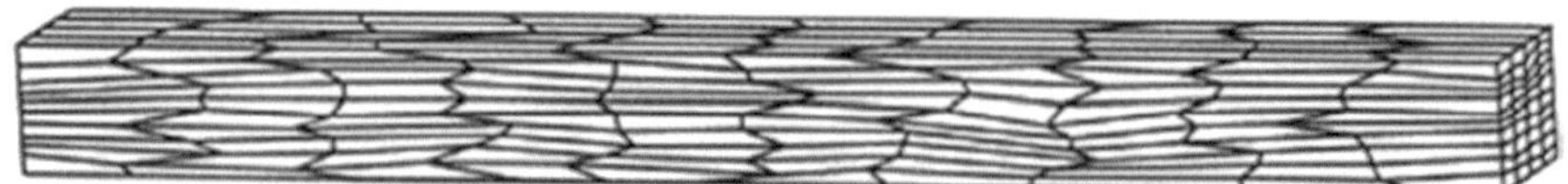

**Figure 2:** Deformed finite element mesh from the simulation of plane strain compression on model 3D-400E-400g2 to a 70% reduction level ($\varepsilon$=-1.2). From [6].

Unfortunately, this method requires larger computation time than the Taylor's model. For the finite element simulation of a complex forming process, this technique can hardly be used. Indeed, for this method, one finite element has a length scale of the order of the size of one crystal. While the global mesh has the size of the sample. A very large number of elements should be used in most cases of actual forming processes.

Nevertheless, a multi-level FEM approach or FEM$^2$ method [28-31] would allow to avoid (or reduce) the problem of the previous approach. Two distinct finite element meshes are used. The first one is a macroscopic mesh (at the scale of the sample) representing the forming process. The second finite element mesh is a microscopic one and is used to achieve the micro-macro transition. The size of that mesh is of the order of the representative volume element, i.e. the number of elements of this second mesh must be such that the ODF is correctly represented. Such a microscopic finite element mesh is then supposed to be placed at each integration point of the macroscopic mesh. One complete computation at the micro level must then be achieved each time an integration point of the macro level is treated. Anyway, due to the large length scale difference between the two meshes, the overall number of elements and nodes in the FEM$^2$ method (on both scales) is expected to be lower than for the CPFEM.

The multisite model [32-34] of Delannay adopts a different strategy. It assumes (i) that the deformation of each grain is predominantly influenced by short-range interaction with adjacent grains, (ii) that local strains deviate from their macroscopic average according to specific relaxation modes, and (iii) that the macroscopic strain is achieved on average by every pair of adjacent grains.

The model bears some resemblance with the Lamel model. However, as it is a severe limitation of the Lamel model, the relaxation modes in the multisite model do not assume that grain boundaries are aligned with the rolling plane. Instead, as in the Alamel procedure, the grains interact across a planar interface that is not necessarily parallel to the rolling plane. The relaxation modes in the multisite model are defined by the user, so that the model can either reduce to the full constraints Taylor's model, the pancake, Lamel or Alamel models. According to the finite element mesh and the way to represent the texture of the material, multisite model is able to achieve a CPFEM analysis.

The large flexibility of the multisite model allows generating different micro-macro models; on the other hand, large computation time is generally required during FE simulations.

The method proposed by Dawson and Kumar [35-38] is based on two coupled FEM analyses. A classical mechanical analysis is applied at the scale of the forming process. The texture computation relies on a second FEM analysis in the crystal orientation space. The ODF is described by a FE mesh in order to compute the texture evolution by solving the ODF conservation equation [39].

## 2.4  Micro-macro models with macroscopic yield locus

The computation time for the models presented in previous section is generally very large when industrial forming processes are investigated. Therefore, micro-macro models with a macroscopic yield locus were developed. The macroscopic yield locus is employed during the FE computation in order to reduce computation time. This yield locus is determined from the micro-macro model either in a pre-processor or during the FE computation.

To define the macroscopic yield locus, Van Houtte and Van Bael [40-41] proposed to use the rate of plastic work per unit volume as plastic potential in the strain rate space. The yield stress (expressed as a vector in 5D space) is the first derivative of the potential:

$$\underline{S} = \frac{\partial \Psi}{\partial \underline{D}} \ . \tag{7}$$

A $\underline{6^{th}\ order\ series}$ expansion is used to approximate the plastic potential:

$$\Psi\left(\underline{D}\right) = \tau D\sqrt{\frac{2}{3}}\overline{M} \approx \tau D F_{pqrstu}\frac{D_p}{D}\frac{D_q}{D}\frac{D_r}{D}\frac{D_s}{D}\frac{D_t}{D}\frac{D_u}{D} \quad , \tag{8}$$

$$with\ 1 \le p \le q \le r \le s \le t \le u \le 5$$

where $D$ is the norm of the strain rate $\underline{D}$ defined in 5D space; $\tau$ characterizes the hardening and $\overline{M}$ is the average Taylor factor. The computation of the 210 series expansion coefficients $F_{pqrstu}$ is achieved by a least square fitting based on average Taylor factor or, more efficiently, using a linear method based on texture C-coefficients with pre-calculated libraries.

As the convexity of the $6^{th}$ order formulation of (8) is not guaranteed, according to the modelled material, so called fish-tails, particularly inconvenient during FE simulations, are sometimes encountered. Therefore, a new version of the 6th order potential was proposed [42]:

$$\Psi\left(\underline{D}\right) = \tau D\sqrt{\frac{2}{3}}\overline{M} \approx \tau \sqrt[6]{\alpha'_{pqrstu}D_p D_q D_r D_s D_t D_u} \quad . \tag{9}$$

$$with\ 1 \le p \le q \le r \le s \le t \le u \le 5$$

The calculation of the 210 series expansion coefficients $\alpha'_{pqrstu}$ is achieved by a three step iterative procedure that ensures strict convexity.

Similarly, Arminjon and co-workers [43] used a $4^{th}$ order series expansion to define a macroscopic yield locus. This formulation was implemented in a finite element code [44-45] to simulate deep drawing of steel and aluminium.

Darrieulat and Montheillet [46] proposed a methodology to derive a quadratic yield function for orthotropic aggregates of cubic crystal from the associated texture components. The parameters of Hill's quadratic yield locus are determined by averaging the functions corresponding to the individual texture components of the investigated material's texture.

Maudlin and co-workers [47] have investigated such an approach with a yield locus approximated by a set of hyperplanes (plane in 5 dimensional deviatoric stress space). They called the fitting procedure "tessellation", highlighting the fact that the continuity between the hyperplanes must be fulfilled, i.e. the yield locus must be a continuous surface. The tessellation is achieved on 647 stress points that are calculated on the basis of the texture of the material using a visco-plastic self-consistent model [19] assuming a combination of prismatic and pyramidal slip systems [48]. 1226 linear functions defining the hyperplanes are then obtained. The yield locus is described with a continuous mathematical formulation with sufficient detail to be integrated in a finite element code.

The $\underline{Minty}$ model developed by Duchêne and Habraken [49] is a crystal plasticity law adapted to macroscopic forming processes simulations. This law is based on a local yield locus approach able to predict texture evolution during FE modelling of industrial forming processes. With this model, only a small zone of the yield locus is computed. This zone is updated when its position is no longer located in the part of interest in the yield locus or when the yield locus changes due to texture evolution.

This model is specific in the sense that it does not use a yield locus formulation either for plastic criterion or in the stress integration scheme. A linear stress-strain interpolation in the 5-dimensional (5D) stress space is used at the macroscopic scale:

$$\underline{\sigma} = \tau \underline{\underline{C}} . \underline{u} \ . \tag{10}$$

In this equation, $\underline{\sigma}$ is a 5D vector containing the deviatoric part of the stress; the hydrostatic part being computed according to an isotropic elasticity law. The 5D vector $\underline{u}$ is the deviatoric plastic strain rate direction (it is a unit vector). $\tau$ is a scalar describing an isotropic work hardening.

The macroscopic anisotropic interpolation is included in matrix $\underline{C}$. Its identification relies on 5 directions: $\underline{u}_i$ ($i$=1...5) advisedly chosen in the deviatoric strain rate space and their associated deviatoric stresses: $\underline{\sigma}_i$ ($i$=1...5) computed by the polycrystal plasticity model. This micro-macro model uses Taylor's assumption of equal macroscopic strain and microscopic crystal strain. It computes the average of the response of a set of representative crystals evaluated with a microscopic model taking into account the plasticity at the level of the slip systems. Two versions of this Full Constraints (FC) Taylor's model are investigated: one coupled with a rate insensitive crystal plastic model and one coupled with a visco-plastic crystal model in [50].

Texture evolution is computed using Taylor's model on the basis of the strain history for each integration point every 10 FE time steps. This law is very modular and the principle of this approach can be coupled with any microscopic model and homogenization technique. Further details and properties of Minty law can be found in [49].

Minty and the 6th order yield locus (fitted on Taylor and Lamel models) are compared during deep drawing simulations in [51].


## 3 Multiscale models based on dislocation densities


### 3.1 Description and main features

Multiscale models based on dislocation densities can have a strong macroscopic character: the FE mesh has macroscopic scale and the constitutive law relies on balance equation of different mechanisms of generation and annihilation of dislocations, see for instance the models of Pietrzyk [52-53] or Kopp [54]. Microstructure evolution can also be followed by such an approach, when the link between dislocation densities and phase transformation [55-56] or recrystallization [57] is clearly identified.

Another type of dislocation models looks at a smaller scale. The assumption of uniform stress and strain distribution inside each crystal can yield to inaccurate results depending on the material and the process investigated. In particular, when the dimensions of the sample are of the same order than the length scale of the microstructure (the grains size), the new generation of strain-gradient crystal plasticity models are of great interest [58].

Hereafter only a few examples are provided.

## 3.2  Microscopic models

The internal variable model of Pietrzyk [52-53] is based on the computation of the stress during plastic deformation as a function of the evolution of dislocation populations controlled by competition between storage and annihilation of dislocations. Various models for the description of the dislocation density evolution were proposed. In this respect, a simplified analytical solution of differential equation is proposed in [59]. Models with only one internal variable: the average dislocation density [60] or with two internal variables: the densities of mobile and trapped dislocations [61] were investigated. Finally, a complex model based on the distribution of dislocation density function was developed [53]. This full model accounting for the distribution of dislocation density provides accurate predictions during hot forming of metals [57] but requires larger computation time during FE simulations.

Geers and co-workers proposed a crystal plasticity model [62] on the polycrystal scale which considers each grain as a single crystal core surrounded by flat bi-crystals representing the grain boundaries. When the polycrystal plastically deforms, due to the bi-crystal interface conditions (stress equilibrium and deformation compatibility), the bi-crystals and the crystal core will behave differently. This heterogeneous behaviour gives rise to the generation of geometrically necessary dislocations (GND) in order to fulfill crystallographic lattice compatibility. These GND's act as obstacles to the motion of the statistically stored dislocations (SSD), which carry the plastic deformation.

In order to improve the model, the heterogeneities within each grain were afterwards considered at the single crystal level [63-64]. Therefore, each grain had to be modelled with a sufficient number of finite elements (around 20 in [64]). GND's and SSD's are still accounted for, while grain boundary dislocations (GBD) densities are added in order to consider the initial lattice mismatch between adjacent grains. Application of this model to constrained simple shear [63] and plane stress tension [64] are presented.

## 3.3  Macroscopic models with microscopic physical roots

The Teodosiu and Hu's hardening model [65-66] is a physically-based microstructural model. Basically, it is able to describe both kinematic and isotropic hardening taking into account the influence of the dislocation structures and their evolutions, at a macroscopic scale. It allows to describe complex hardening behaviours induced by strain-path changes.

The model is described by 13 material parameters and depends on four state variables: $\underline{\underline{P}}, \underline{\underline{S}}, \underline{\underline{X}}, R$. The variable $\underline{\underline{P}}$ is a second order-tensor that depicts the polarity of the persistent dislocation structures (PDS) and $\underline{\underline{S}}$ is a fourth-order tensor that describes the directional strength of the PDS's. The scalar $R$ represents the isotropic hardening due to the randomly distributed dislocations and the second-order tensor $\underline{\underline{X}}$ is the back stress. These state variables evolve with respect to the equivalent plastic strain rate $\dot{p}$ with the form $\dot{Y} = f(Y)\dot{p}$. A precise description of these evolution equations can be found in [67]. The yield condition is given by

$$\bar{\sigma} = \sigma_y = Y_0 + R + f\,|\underline{\underline{S}}|\,, \qquad (11)$$

where $\bar{\sigma}$ is the equivalent stress, function of $\underline{\underline{\sigma}} - \underline{\underline{X}}$, $\sigma_y$ is the current elastic limit, $Y_0$ is the initial size of the yield locus and $R + f\,|\underline{\underline{S}}|$ represents the isotropic hardening. The expression of $\bar{\sigma}$ depends on the definition of the associated yield locus.

For instance, a validation of the Teodosiu and Hu's hardening model coupled with the Minty constitutive law during deep drawing simulations is presented in [68].

Levkovitch and co-workers [69] proposed a similar phenomenological model based on microstructural material behaviour. The evolution of the polarized dislocation structure on the grain level, representing the main cause of the induced flow anisotropy at the macroscopic level is taken into account. Besides the isotropic and kinematic hardening, the model also accounts for the change of the yield locus shape (distortional hardening). The model is validated thanks to metal forming simulations inducing complex strain path changes.

The 3IVM model [54;70] considers three internal variables for the description of the microstructure. These variables are three dislocation type densities: the mobile dislocations, the immobile dislocations in cell walls and the immobile dislocations in cell interiors. This flow stress model is linked to a Taylor-type model in a finite element code to simulate forming processes: stretch forging of an austenitic steel and hot rolling of aluminium alloys are reported in [54].

## 4   Discussion and future trends

Thanks to the evolution of the computer capabilities, strong progresses have been noticed concerning the multiscale models. Concerning the modelling of the materials anisotropy, three periods can be distinguished.

Before 1900, only a few research teams are involved in this field but clear tendencies are already initiated: the use of Taylor's model with (parallel) computation of the average behaviour of a set of representative crystals or the use of a fitted yield locus formulation based on the initial texture of the material.

Due to a rapid increase of the number of research teams in the polycrystal domain, the period 1990-2000 is characterized by a strong development of sophisticated homogenization techniques in order to supersede the Taylor's model. The first FEM$^2$

models are proposed so as to take into account texture evolution and adjacent grains interaction. Crystal plasticity finite element method is still rarely developed.

Since 2000, FEM$^2$ and CPFEM techniques are more and more investigated. Concurrently, the robustness of the texture based yield loci was improved.

Concerning the microstructure modelling, the progress origins are different:
- A better knowledge of the microscopic mechanisms linked to the phase transformations and the recrystallizations yielded to more and more realistic physical models.
- The increasing computation power allows better analyzing dislocation densities associated to different mechanisms, it allows for instance to study the induced microstructure during not only monotonic strain path but also two-stage strain paths[71].

The current evolution of multiscale models concerns the modelling of multiple phase materials (while single phase models were initially developed). Body centred cubic (bcc) and face centred cubic (fcc) material were first investigated because of their rather continuous behaviour. Several models are now devoted to hexagonal closed packed (hcp) lattice material [72,73]. The main complications come from the lower number and the particular orientations of the slip systems (compared to fcc and bcc materials).

The number and the variety of the polycrystalline models rapidly increase. These new models are very helpful for the identification of more simple models. For instance CPFEM models can be used for the validation of micro-macro models.

## 5 Conclusion

Even without analyzing the atomistic scale, a lot of research work is still required to improve the accuracy of the numerical results compared to the experiment. Several studies concerning forming processes have already been achieved. A lower effort was devoted to the structure toughness and consequently to the fatigue analysis.

The rapidly increasing number of micro-macro research teams prefigures future improvements of the numerical models concerning both their accuracy and their rapidity.

Anyway, the complexity of the numerical model must always be adapted to the material and the process investigated. For instance, if a simple constitutive law (e.g. Von Mises yield locus with isotropic hardening) provides accurate results for the studied process, multiscale models should be avoided. It is important to correctly analyze the complexity of the involved deformation mechanisms in order to choose the most adequate model (multiscale or phenomenological model). Macroscopic phenomenological models based on and identified from micro-macro models should still be taken into consideration as an interesting intermediate solution.

## References

1. Devincre, B., Kubin, L.P. Philos. Trans. R. Soc. Lond. Ser. A 355 (1997) 2003

2. Schiotz, J., Di Tolla, F.D., Jacobsen, K.W. Nature 391 (1998) 561
3. Van Swygenhoven, H., Derlet, P.M., Froseth, A.G.: Nucleation and propagation of dislocations in nanocrystalline fcc metals. Acta Materialia 54 (2006) 1975-1983
4. Sachs, G.: Zur ableitung einer fliessbedingung. Z. Verein Deutscher Ing, 72 (1928) 734-736
5. Taylor, G.I.: Plastic strain in metals. J. Inst. Met 62 (1938) 307-324
6. Bachu, V., Kalidindi, S.R.: On the accuracy of the predictions of texture evolution by the finite element technique for fcc polycrystals. Materials Science and Engineering A257 (1998) 108-117
7. Honeff, H., Mecking, H.: Analysis of the deformation texture at different rolling conditions. In: Nagashima, S. (Ed.), Proceedings of ICOTOM 6, vol. 1. The Iron and Steel Institute of Japan, Tokyo (1981) 347–355
8. Kocks, U.F., Chandra, H.: Slip geometry in partially constrained deformation. Acta Metall. 30 (1982) 695
9. Van Houtte, P.: On the equivalence of the relaxed Taylor theory and the Bishop–Hill theory for partially constrained plastic deformation of crystals. Mater. Sci. Eng. 55 (1982) 69–77
10. Van Houtte, P., Rabet, L.: Generalisation of the relaxed constraints models for the prediction of deformation textures. Revue de Métallurgie-CIT/Science et Génie des matériaux (1997) 1483-1494
11. Van Houtte, P., Delannay, L., Samajdar, I.: Quantitative prediction of cold rolling textures in low-carbon steel by means of the LAMEL model. Textures and Microstructures 31 (1999) 109-147
12. Van Houtte, P., Delannay, L., Kalidindi, S.R.: Comparison of two grain interaction models for polycrystal plasticity and deformation texture prediction. Int. J. Plasticity 18 (2002) 359-377
13. Liu, Y.S., Delannay, L., Van Houtte, P.: Application of the Lamel model for simulating cold rolling texture in molybdenum sheet. Acta Materialia 50 (2002) 1849-1856
14. Crumbach, M., Pomana, G., Wagner, P., Gottstein, G.: A Taylor type deformation texture model considering grain interaction and material properties. Part I – Fundamentals. In: Gottstein, G., Molodov, D.A. (Eds.), Recrystallisation and Grain Growth, Proceedings of the First Joint Conference, Springer, Berlin (2001) 1053–1060
15. Van Houtte, P., Li, S., Seefeldt, M., Delannay, L.: Deformation texture prediction: from the Taylor model to the advanced Lamel model. Int J Plasticity 21 (2005) 589-624
16. Berveiller, M., Zaoui, A.: Modeling of the plastic behaviour of inhomogeneous media. J. Engng. Mater. Technol., 106 (1984) 295-298
17. Berbenni, S., Favier, V., Berveiller, M.: Impact of the grain size distribution on the yield stress of heterogeneous materials. Int J Plasticity 23 (2007) 114-142
18. Kröner, E.: Berechnung der elastischen konstanten des vielkristalls aus den konstanten des einskristalls. Z. Phys. (1958) 151
19. Lebensohn, R.A., Tome, C.N.: A self-consistent anisotropic approach for the simulation of plastic deformation and texture development of polycrystals: application to zirconium alloys. Acta Metall. Mater. 41 (1993) 2611-2624
20. Molinari, A., Toth, L.: Tuning a self-consistent viscoplastic model by finite element results I: Modelling. Acta Metall. Mater. 42 (1994) 2453-2458
21. Rouff, C., Favier, V., Bigot, R., Berveiller, M., Robelet, M.: Micro-macro modeling of the steady-state semi-solid behavior. The 5th ESAFORM Conference on Material Forming (2002) Krakow, Poland
22. Sabar, H., Berveiller, M., Favier, V., Berbenni, S.: A new class of micro-macro models for elastic-viscoplastic heterogeneous materials. Int. J. of Solids and Structures 39 (2002) 3257-3276
23. Frénois, S.: Modélisation polycristalline du comportement mécanique du tantale. Application à la mise en forme par hydroformage. Ph. D. Thesis (2001) Ecole Centrale des Arts et Manufactures, Ecole Centrale Paris

24. Anand, L., Balasubramanian, S., Kothari, M.: Constitutive modelling of polycrystalline metals at large strains: Application to deformation processing, large plastic deformation of crystalline aggregates. International Centre for Mechanical Sciences, Courses and Lectures n°376, Springer Verlag, 109-172

25. Beaudoin, A.J., Dawson, P.R., Mathur, K.K., Kocks, U.F., Korzekwa, D.A.: Application of polycrystal plasticity to sheet forming. Comp. Methods Appl. Mech. Eng. 117 (1994) 49-70

26. Bate, P.: Modelling deformation microstructure with the crystal plasticity finite-element method. Philos. T. Roy. Soc. A 357 (1999) 1589

27. Kalidindi, S.R., Bronkhorst, C.A., Anand, L.: Crystallographic texture evolution in bulk deformation processing of FCC metals. J. Mech. Phys. Solids 40 (1992) 537–569

28. Smit, R.J.M., Brekelmans, W.A.M., Meijer, H.E.H.: Prediction of the mechanical behavior of nonlinear heterogeneous systems by multi-level finite element modeling. Comp. Meth. Appl. Mech. Eng. 155 (1998) 181-192

29. Miehe, C., Schröder, J., Schotte, J.: Computational homogenization analysis in finite plasticity, simulation of texture development in polycrystalline materials. Comp. Meth. Appl. Mech. Eng. 171 (1999) 387-418

30. Geers, M.G.D., Kouznetsova, V., Brekelmans, W.A.M.: Constitutive approaches for the multi-level analysis of the mechanics of microstructures. 5th National Congress on Theoretical and Applied Mechanics, Louvain-la-Neuve, May 23-24 (2000)

31. Feyel, F., Chaboche, J.L.: Multiscale non linear FE analysis of composite structures: damage and fiber size effects. Euromech 417, October 2-4 (2000), University of Technology of Troyes, France

32. Delannay, L., Logé, R.E., Signorelli, J.W., Chastel, Y.: Evaluation of a multisite model for prediction of rolling textures in hcp metals. Int. J. of Form. Proc. 8 (2005) 131

33. Delannay, L., Loge, R.E., Chastel, Y., Van Houtte, P.: Prediction of intergranular strains in cubic metals using a multisite elastic–plastic model. Acta Mater. 50 (2002) 5127–5138

34. Delannay, L., Kalidindi, S.R., Van Houtte, P.: Quantitative prediction of texture in aluminium cold rolled to moderate strains. Mater. Sci. Eng. A 336 (2002) 233–244

35. Kumar, A., Dawson, P.R.: The simulation of texture evolution during bulk deformation processes using finite elements over orientation space. Simulation of Materials Processing: Theory, Methods and Applications, Shen & Dawson, Balkema (1995)

36. Kumar, A., Dawson, P.R.: Polycrystal plasticity modeling of bulk forming with finite elements over orientation space. Comp. Mech. 17 (1995) 10-25

37. Kumar, A., Dawson, P.R.: The simulation of texture evolution with finite elements over orientation space. I. Development, II. Application to planar crystals. Comp. Methods Appl. Mech. Eng. 130 (1996) 227-261

38. Dawson, P.R., Kumar, A.: Deformation process simulations using polycrystal plasticity. Large plastic deformation of crystalline aggregates, International Centre for Mechanical Sciences, Courses and Lectures n°376, Springer Verlag (1997) 247

39. Clement, A.: Prediction of deformation texture using a physical principle of conservation. Mater. Sci. Eng. 55 (1982) 203-210

40. Van Houtte, P.: Application of Plastic Potentials to Strain Rate Sensitive and Insensitive Anisotropic Materials. Int. J. Plasticity 10 (1994) 719-748

41. Van Bael, A., Van Houtte, P.: Convex fourth and sixth-order plastic potentials derived from crystallographic texture. J.Phys. IV France 105 (2003) 39-46

42. Van Houtte, P., Van Bael, A.: Convex plastic potentials of fourth and sixth rank for anisotropic materials. Int. J. Plasticity 20 (2004) 1505-1524

43. Arminjon, M., Bacroix, B., Imbault, D., Raphanel, J.L.: A fourth-order plastic potential for anisotropic metals and its analytical calculation from the texture function. Acta Mech. 107 (1994) 33–51

44. Bacroix, B., Gilormini, P.: Finite element simulations of earing in polycrystalline materials using a texture-adjusted strain-rate potential. Model. Simulation Mater. Sci. Eng. 3 (1995) 1-21

45. Zhou, Y., Jonas, J.J., Szabo, L., Makinde, A., Jain, M., MacEwen, S.R.: Incorporation of an anisotropic (texture-based) strain-rate potential into three-dimensional finite element simulations. Int. J. Plasticity 13 (1997) 165-181

46. Darrieulat, M., Montheillet, F.: A texture based continuum approach for predicting the plastic behaviour of rolled sheet. Int. J. Plasticity 19 (2003) 517-546

47. Maudlin, P.J., Wright, S.I., Kocks, U.F., Sahota, M.S.: An application of multisurface plasticity theory : yield surfaces of textured materials. Acta Mater. 44 (1996) 4027-4032

48. Tomé, C.N., Maudlin, P.J., Lebensohn, R.A., Kaschner, G.C.: Mechanical response of zirconium—I. Derivation of a polycrystal constitutive law and finite element analysis. Acta Mater. 49 (2001) 3085-3096

49. Habraken, A.M., Duchêne, L.: Anisotropic elasto-plastic finite element analysis using a stress-strain interpolation method based on a polycrystalline model. Int J Plasticity 20 (2004) 1525-1560

50. Duchêne, L., El Houdaigui, F., Habraken, A.M.: Length changes and texture prediction during free end torsion test of copper bars with fem and remeshing techniques. International Journal of Plasticity (2007) doi: 10.1016/j.ijplas.2007.01.008

51. Van Bael, A., He, S., Van Houtte, P., Delannay, L., Duchêne, L., Habrakne, A.M.: Finite element simulations of cup drawing using the Taylor and the Lamel model. The 7th ESAFORM Conference on Material Forming (2004) Trondheim, Norway

52. Madej, L., Kuziak, R., Pietrzyk, M.: Validation of the history dependant constitutive law under varying conditions of hot deformation. The 6th ESAFORM Conference on Material Forming (2003) Salerno, Italy

53. Pietrzyk, M.: Numerical aspects of the simulation of hot metal forming using internal variable method. Metall. Foundry Eng. 20 (1994) 429-439

54. Luce, R., Aretz, H., Kopp, R., Goerdeler, M., Gottstein, G.: Microstructure simulation of multistep hot forming processes. The 4th ESAFORM Conference on Material Forming (2001) Liège, Belgium

55. Van Rompaey, T., Lani, F., Pardoen, T., Blanpain, B., Wollants, P.: Micromechanical study of the martensitic transformation in TRIP-assisted multi-phase steels. Solid State Transformation and Heat Treatment, ed. A. Hazotte, Wiley-VCH (2005) 87-94

56. Van Rompaey, T., Furnemont, Q., Jacques, P.J., Pardoen, T., Blanpain, B., Wollants, P.: Micromechanical modelling of TRIP steels. Steel Research International 74-6 (2003) 365-369

57. Pietrzyk, M., Roucoules, C., Hodgson, P.D.: Dislocation model for work hardening and recrystallization applied to the finite element simulation of hot forming. Proc. Conf. NUMIFORM (1995) 315-320

58. Cheong, K.S., Busso, E.P., Arsenlis, A.: A study of microstructural length scale effects on the behaviour of FCC polycrystals using strain gradient concepts. Int. J. Plast. 21 (2005) 1797-1814

59. Madej, L., Pietrzyk, M.: Analysis of possibilities of material behaviour modelling during hot plastic deformation. Metall. Foundry Eng. Special issue (2001) 143-149

60. Ordon, J., Kuziak, R., Pietrzyk, M.: History dependant constitutive law for austenitic steels. Proc. Metal Forming Conference (2000) Krakow, Poland, 747-753

61. Ordon, J., Pietrzyk, M., Kedzierski, Z., Kuziak, R.: Constitutive model based on two internal variables for constant and changing deformation conditions. Thermomechanical Processing Conference (2002) Sheffield, UK

62. Evers, L.P., Parks, D.M., Brekelmans, W.A.M., Geers, M.G.D.: Crystal plasticity model with enhanced hardening by geometrically necessary dislocation accumulation. Journal of the Mechanics and Physics of Solids 50 (2002) 2403-2424

63. Evers, L.P., Brekelmans, W.A.M., Geers, M.G.D.: Non-local crystal plasticity model with intrinsic SSD and GND effects. Journal of the Mechanics and Physics of Solids 52 (2004) 2379-2401

64. Evers, L.P., Brekelmans, W.A.M., Geers, M.G.D.: scale dependent crystal plasticity framework with dislocation density and grain boundary effects. Int. J. of Solids and Structures 41 (2004) 5209-5230

65. Teodosiu, C., Hu, Z.: Evolution of the intergranular microstructure at moderate and large strains: modelling and computational significance. In Simulation of Materials Processing: Theory, Methods and Applications, NUMIFORM, Shen SF, Dawson PR (Eds.), Balkema, Rotterdam (1995) 173-182

66. Haddadi, H., Bouvier, S., Banu, M., Maier, C., Teodosiu, C.: Towards an accurate description of the anisotropic behaviour of sheet metals under large plastic deformations: Modelling, numerical analysis and identification. Int. J. Plasticity 22 (2006) 2226-2271

67. Bouvier, S., Haddadi, H.: Modelling the behaviour of a bake-hardening steel using a dislocation structure based model. The 4th ESAFORM Conference on Material Forming (2001) Liège, Belgium

68. Duchêne, L., de Montleau, P., El Houdaigui, F., Bouvier, S., Habraken, A.M.: Analysis of texture evolution and hardening behaviour during deep drawing with an improved mixed type FEM element. Proc. Conf. NUMISHEET (2005)

69. Levkovitch, V., Svendsen, B., Wang, J.: Micromechanically motivated phenomenological modelling of induced flow anisotropy and its application to metal forming processes with complex strain path changes. The 9th ESAFORM Conference on Material Forming (2006) Glasgow, UK

70. Roters, F., Raabe, D., Gottstein, G.: Work hardening in heterogeneous alloys – a microstructural approach based on three internal state variables. Acta Mater. 48 (2000) 4181-4189

71. Franz G.,Abed-Meraim, Ben Zineb T. Lemoine X., Berveiller M.: A multiscale model based on intragranular microstructure – Prediction of dislocation patterns at microscopic scale, The 10th ESAFORM Conference on Material Forming (2007) Zaragoza, Spain.

72. Plunkett B., Lebensohn R. A., Cazacu O., Barlat F.: Anisotropic yield function of hexagonal materials taking into account texture development and anisotropic hardening , Acta Mater., 54 (2006) 4159-4169

73. Dunlop J.W., Bréchet Y.J.M., Legras L.,Estrin Y.: Dislocation density-based modelling of plastic deformation of Zircaloy-4, Materials Science and Engineering A, 443 (2007) 77-86

# Anisotropy and Formability

Dorel Banabic[1], Frédéric Barlat[2], Oana Cazacu[3] and Toshihiko Kuwabara[4],

[1] CERTETA - Research Centre on Sheet Metal Forming Technology, Technical University of Cluj Napoca, 400020 Cluj Napoca, Romania,

[2] Alloy Technology and Materials Research Division, Alcoa Technical Center, 100 Technical Drive, Alcoa Center, PA 15069-0001,USA,

[3] Department of Mechanical and Aerospace Engineering, University of Florida/REEF, Shalimar, FL 32579, USA

[4] Institute of Symbiotic Science and Technology, Tokyo University of Agriculture and Technology, 2-24-16, Nakacho, Koganei-shi, Tokyo 184-8588, Japan

**Abstract.** The chapter presents synthetically the most recent models of the anisotropic plastic behavior. The first section gives an overview of the classical models. In the next step, the discussion is focused on the anisotropic formulations developed on the basis of the theories of linear transformations and tensor representations, respectively. Those models are applied to different types of materials: body centered, faced centered and hexagonal-close packed metals. A brief review of the experimental methods used for observing and modeling the anisotropic plastic behavior of metallic sheets and tubes under biaxial loading is presented together with the models and methods developed for predicting and establishing the limit strains. The capabilities of some commercial programs specially designed for the computation of forming limit curves (FLC) are also analyzed.

**Keywords:** Anisotropy, Yield criteria, Biaxial tensile tests, Forming Limit Diagrams.

# 1    Introduction

Given the current trend of globalization and active competition on the world market, especially on the automotive one, the reduction of the lead time can be decisive. Virtual manufacturing using finite element analyses may contribute to the reduction of the lead time. Finite element analysis has been applied extensively to compare design options, understand the influence of process conditions on both formability and structural performance and to reduce the trial and error in the development of tools for optimum performance. Key in improving the accuracy of these analyses is the use of appropriate constitutive and formability models.

## 2  Anisotropic plastic potentials

Aspects of the constitutive models for metal forming applications were discussed by Barlat [34], [37]. A number of reviews concerning plastic yielding in metals can be found in the literature [10], [33], [218], [221], [222] and this section is only a brief summary from the author's perspective.

### 2.1  Classical anisotropic yield functions

The most popular isotropic yield conditions, verified for many metals, were proposed by Tresca and von Mises and may be expressed in terms of the principal values of the stress ( $\sigma_i$ ) or the deviatoric stress ( $S_j$ ) tensors as [90]

$$\phi = \left|\sigma_1 - \sigma_2\right|^a + \left|\sigma_2 - \sigma_3\right|^a + \left|\sigma_3 - \sigma_1\right|^a = \left|S_1 - S_2\right|^a + \left|S_2 - S_3\right|^a + \left|S_3 - S_1\right|^a = 2\bar{\sigma}^a \qquad (1)$$

where $\bar{\sigma}$ defines the effective stress. For an exponent $a = 2$ or $a = 4$, Eq. (1) reduces to von Mises, whereas for $a = 1$ and in the limiting case $a \to \infty$, it leads to Tresca yield condition. Its main advantage is that it leads to a good approximation of yield loci computed using the Bishop-Hill crystal plasticity model by setting $a = 6$ for BCC and $a = 8$ for FCC materials, respectively [90], [147].

For isotropic materials, yield criteria have the same form in any reference frame. For anisotropic materials, yielding properties are directional, thus the expression of the yield criterion depends on the reference frame. The simplest form of the yield criterion is with respect to a coordinate system associated with the axes of symmetry of the material. For instance, due to the symmetry of their thermo-mechanical processing history, sheet metals exhibit orthotropic symmetry characterized by three mutually orthogonal planes of symmetry. These are denoted **x**, **y** and **z** and correspond to the rolling, transverse and normal directions of the sheet, respectively. Hill [91] proposed an extension of the isotropic Mises criterion to orthotropic materials

$$\phi = F\left(\sigma_{yy} - \sigma_{zz}\right)^2 + G\left(\sigma_{zz} - \sigma_{xx}\right)^2 + H\left(\sigma_{xx} - \sigma_{yy}\right)^2 + 2\left(L\sigma_{yz}^2 + M\sigma_{zx}^2 + N\sigma_{xy}^2\right) = \bar{\sigma}^2 \qquad (2)$$

In this equation, $F, G, H, L, M$ and $N$ are material constants. The validity of this yield function has been explored in numerous experiments, the consensus being that it is well suited to specific metals and textures, especially steel [95], [156]. Hill [93] proposed a non-quadratic yield criterion to describe materials other than steel and derived four special cases from the general form. The general expression of Hill's (1979) yield criterion accounts for planar anisotropy, provided that the directions of the principal stresses are superimposed with the anisotropy axes. The most widely used expression of this yield criterion is the so-called "Special Case IV," which applies to materials exhibiting planar isotropy (with an average Lankford coefficient $\bar{r}$) for plane stress states

$$\phi = \left|\sigma_1 + \sigma_2\right|^a + \left(1 + 2\bar{r}\right)\left|\sigma_1 - \sigma_2\right|^a = 2\left(1 + \bar{r}\right)\bar{\sigma}^a \tag{3}$$

Hill proposed other non-quadratic plane stress yield criteria [95], [96]. Independently from Hill, Hosford [107] used Hershey's isotropic criterion [90], Eq. (1), to describe crystal plasticity results and proposed the following generalization [108] for materials exhibiting orthotropic symmetry

$$\phi = F\left|\sigma_{yy} - \sigma_{zz}\right|^a + G\left|\sigma_{zz} - \sigma_{xx}\right|^a + H\left|\sigma_{xx} - \sigma_{yy}\right|^a = \bar{\sigma}^a \tag{4}$$

An important drawback of this as well as of Hill's (1979) criteria is that they do not involve shear stresses. Thus, these criteria cannot account for the continuous variation of plastic properties between the sheet's **x** and **y** axes. Barlat and Lian [27] successfully extended Hosford's (1979) [108] criterion to capture the influence of the shear stress

$$\phi = \bar{c}\left|k_1 + k_2\right|^a + \bar{c}\left|k_1 - k_2\right|^a + \left(2 - \bar{c}\right)\left|2k_2\right|^a = 2\bar{\sigma}^a \tag{5}$$

where $\bar{a}, \bar{c}, \bar{h}$ and $\bar{p}$ are material coefficients and

$$k_1 = \frac{\sigma_{xx} + \bar{h}\sigma_{yy}}{2}, \quad k_2 = \sqrt{\left(\frac{\sigma_{xx} - \bar{h}\sigma_{yy}}{2}\right)^2 + \bar{p}^2\sigma_{xy}^2} \tag{6}$$

Barlat et al. [28] proposed a yield criterion for a general stress state, denoted Yld91, which extends the isotropic Hershey's criterion, Eq. (1), to orthotropic symmetry as well. Anisotropy is introduced by replacing the principal values of the stress tensor by those of a stress tensor modified with weighting coefficients. Karafillis and Boyce [120] proposed a generalization Hershey's criterion

$$\phi = \left(1 - c\right)\left(\left|S_1 - S_2\right|^a + \left|S_2 - S_3\right|^a + \left|S_3 - S_1\right|^a\right) + \frac{3^a c}{2^{a-1} + 1}\left(\left|S_1\right|^a + \left|S_2\right|^a + \left|S_3\right|^a\right) = 2\bar{\sigma}^a \tag{7}$$

where $c$ is a constant, and extended it to orthotropic materials, thus generalizing Yld91.

## 2.2  Theories of linear transformations and tensor representations

The procedure used to introduce anisotropy by Barlat et al. [28] and Karafillis and Boyce [120] is equivalent with the application of a fourth order linear transformation operator on the stress tensor $\tilde{\sigma} = \mathbf{L}\sigma$ or on the stress deviator $\tilde{s} = \mathbf{L}s$, for pressure independent plasticity. Thus, an anisotropic yield function is obtained from an isotropic function by substituting the principal value of the stress tensor (or deviator) by the principal value of $\tilde{\sigma}$ or $\tilde{s}$. The coefficients characterizing anisotropy are the components of the fourth order tensor $\mathbf{L}$. Similar to the stiffness or compliance tensor in elastic anisotropy, the symmetry of a material is reflected by the symmetry of the tensor $\mathbf{L}$. The advantage of this theory is that if the associated isotropic yield function is convex in principal stress space, a property that is relatively easy to check in this space, then the anisotropic version is automatically convex. This property ensures stability in numerical simulations where the potential is used.

Some authors realized that the approaches that describe anisotropic behavior proposed thus far did not contain enough coefficients to provide a good description of plastic anisotropy, even in the case of uniaxial tension (e.g., Barlat et al. [30], [31]). Therefore, two or multiple ($n$) linear transformations were proposed in order to increase the number of anisotropy coefficients, e.g., for $n = 2$ and pressure independent plasticity

$$\tilde{s}' = \mathbf{C}'s\,,\ \tilde{s}'' = \mathbf{C}''s \tag{8}$$

where $\mathbf{C}'$ and $\mathbf{C}''$ are the two tensors containing the anisotropy coefficients. Examples of such yield functions are given in the next section.

An alternative approach to extend any isotropic yield criterion such as to describe any type of material symmetry was proposed by Cazacu and Barlat [62], [63]. Within the framework of the theory of representation of tensor functions, they developed generalizations of the stress deviator invariants $J_2$ and $J_3$. These generalized invariants were required to be homogeneous functions of degree two and three in stresses, respectively, that reduce to $J_2$ and $J_3$ for isotropic conditions, are insensitive to pressure, and are invariant to any transformation belonging to the symmetry group of the material. The anisotropic yield criterion is obtained by substituting in the expression of the isotropic criterion the stress deviator invariants by their respective anisotropic generalization. Example of such yield functions are given in the next section.

## 2.3  Advanced anisotropic yield criteria for body and faced centered metals

These advanced yield criteria includes more coefficients and are able to described anisotropy of the tensile properties (yield stress and Lankford coefficient). BCC and FCC materials deform by dislocation glide on certain slip planes and directions. Because glide can occur indifferently in a direction or the reverse, tension and compression flow stresses are identical at the same amount of deformation. Therefore, the yield function must exhibit the inversion symmetry ($\phi(\sigma_{ij}) = \phi(-\sigma_{ij})$). Moreover, the associated isotropic function must be a good approximation of the crystal plasticity response for an aggregate containing a random distribution of grain orientations.

For plane stress, the anisotropic Yld2000-2d yield function $\phi$ was defined as [32]

$$\phi = \left|\tilde{S}_1' - \tilde{S}_2'\right|^a + \left|2\tilde{S}_2'' + \tilde{S}_1''\right|^a + \left|2\tilde{S}_1'' + \tilde{S}_2''\right|^a = 2\bar{\sigma}^a \tag{9}$$

$\tilde{S}_1', \tilde{S}_2'$ and $\tilde{S}_1'', \tilde{S}_2''$ are the principal values of $\tilde{\mathbf{s}}'$ and $\tilde{\mathbf{s}}''$, respectively. It can be shown that eight independent coefficients are available in Yld2000-2d, i.e., three $C_{ij}'$ and five $C_{kl}''$.

Independently, Banabic et al. based on the Barlat'89 criterion [27] have developed since 2000 [9] the so-called BBC family criteria [14], [18]. The last version, BBC2003 criterion, was defined as

$$\alpha|\Gamma + \Psi|^a + \alpha|\Gamma - \Psi|^a + (1-\alpha)|2\Lambda|^a = \bar{\sigma}^a \tag{10}$$

where $\Gamma, \Psi$ and $\Lambda$ are expressions of the three plane stress components and anisotropy coefficients. It was shown that this yield function contains eight independent coefficients and that, in fact, it was the same as Yld2000-2d written is a different form [38].

For a pressure-independent material under a general stress state, Bron and Besson [49] extended Karafillis and Boyce yield function with two linear transformations to a formulation containing 12 coefficients. Under the same conditions, Barlat et al., [36] proposed the yield function Yld2004-18p, which extends Eq. (1)

$$\phi(s_{\alpha\beta}) = \Phi\left(\tilde{S}_i', \tilde{S}_j''\right) = \sum_{i,j}^{1,3}\left|\tilde{S}_i' - \tilde{S}_j''\right|^a = 4\bar{\sigma}^a \tag{11}$$

The two transformed stress deviators $\tilde{\mathbf{s}}'$ and $\tilde{\mathbf{s}}''$ can be written in a matrix form as

$$\tilde{\mathbf{s}} \equiv \begin{bmatrix} \tilde{s}_{xx} \\ \tilde{s}_{yy} \\ \tilde{s}_{zz} \\ \tilde{s}_{yz} \\ \tilde{s}_{zx} \\ \tilde{s}_{xy} \end{bmatrix} = \begin{bmatrix} 0 & -c_{12} & -c_{13} & 0 & 0 & 0 \\ -c_{21} & 0 & -c_{23} & 0 & 0 & 0 \\ -c_{31} & -c_{32} & 0 & 0 & 0 & 0 \\ 0 & 0 & 0 & c_{44} & 0 & 0 \\ 0 & 0 & 0 & 0 & c_{55} & 0 \\ 0 & 0 & 0 & 0 & 0 & c_{66} \end{bmatrix} \begin{bmatrix} s_{xx} \\ s_{yy} \\ s_{zz} \\ s_{yz} \\ s_{zx} \\ s_{xy} \end{bmatrix} \tag{12}$$

with the appropriate symbols (prime and double prime) for each transformation, i.e., $C'_{ij}$ for $\tilde{\mathbf{s}}'$ and $C''_{ij}$ for $\tilde{\mathbf{s}}''$. In this formulation, 18 parameters (nine per linear transformation) are available to characterize anisotropy.

As explained above, using the generalized invariants, any isotropic yield criterion can be extended to describe orthotropic materials. In [62], this approach was used to extend Drucker's 1949 isotropic yield criterion [72], which provides a yield surface shape intermediate between that on von Mises and Tresca. The proposed orthotropic criterion is

$$\phi = \left(J_2^o\right)^3 - c\left(J_3^o\right)^2 = k^2 \tag{13}$$

where $J_2^o$ and $J_3^o$ are the orthotropic generalizations of the principal stress tensor invariants. For full 3-D stress conditions, the criterion involves 18 anisotropy coefficients.

### 2.4  Anisotropic yield criteria for hexagonal-close packed metals

Metals with hexagonal close packed (HCP) crystal structure deform plastically by slip and twinning. As opposed to slip, twinning is a directional shear mechanism: shear in one direction can produce twinning while shear in the opposite direction cannot. For example, in magnesium alloys sheets twinning is not active in tension along any direction in the plane of the sheet, but is easily activated in compression. As a result the average in-plane compressive yield stress is about half the average in-plane tensile yield stress (e.g. see Lou et al. [148]). Thus, the yield surfaces are not symmetric with respect to the stress free condition. Since hcp metals sheets exhibit strong basal textures ($c$-axis oriented predominantly perpendicular to the thickness direction), a pronounced anisotropy in yielding is observed. To account for both strength differential (SD) effects and the anisotropy displayed by HCP metals, Hosford [108] proposed the following modification of Hill's (1948) orthotropic yield criterion [91]:

$$A\sigma_{xx} + B\sigma_{yy} - \left(A+B\right)\sigma_{zz} + F\left(\sigma_{yy} - \sigma_{zz}\right)^2 + G\left(\sigma_{zz} - \sigma_{xx}\right)^2 + H\left(\sigma_{xx} - \sigma_{yy}\right)^2 = 1. \tag{14}$$

where *A, B, F, G, H* are material coefficients and **x, y, z** are normal to the mutually orthogonal planes of symmetry of the material . Since the criterion does not involve shear stresses, it cannot account for the continuous variation of the plastic properties between the material axes of symmetry. Liu et al. [146] have proposed an extension of Hill (1948) yield criterion in the form which involves shear terms, but SD effects are due to pressure (first invariant of stress). To develop macroscopic formulation that account for *both the asymmetry and anisotropy in yielding exhibited by hcp materials,* the approach adopted by Cazacu and Barlat [64] was to develop an isotropic yield function that could describe SD effects and then extend the isotropic formulations such as to account for orthotropy. Thus, based on the polycrystalline simulations of randomly oriented polycrystals deforming solely by twinning, the following yield function was proposed:

$$f \equiv \left(J_2\right)^{\frac{3}{2}} - cJ_3 = \tau_Y^3 \, ,$$

(15)

where $\tau_Y$ is the yield stress in pure shear and $c$ is a material constant expressible solely in terms of the uniaxial yield stresses in tension and compression, respectively. For equal yield stresses in tension and compression $c = 0$, hence the proposed criterion reduces to the von Mises yield criterion. Anisotropy was introduced in the formulation using the generalized invariants approach, i.e. in the expression of the isotropic yield criterion (3) the invariants of the stress deviator were replaced with the generalized invariants $J_2^o$ and $J_3^o$, respectively. For full stress state (3D) conditions, anisotropy is described by 18 coefficients. This orthotropic yield function is homogeneous of degree three in stresses, yet for certain HCP materials such as titanium the yield surfaces are quadratic (see Liu et al. [146] , Cazacu et al. [65] etc). To overcome this limitation, Cazacu et al. [65] proposed the following isotropic yield criterion for which the degree of homogeneity $a$ is not fixed

$$\phi = \left\|S_1\right| - kS_1\right|^a + \left\|S_2\right| - kS_2\right|^a + \left\|S_3\right| - kS_3\right|^a$$

(16)

and $k$ is a strength differential parameter. To capture simultaneously anisotropy and tension/compression asymmetry, this isotropic yield criterion was extended to orthotropy by applying a fourth-order linear operator on the stress deviator. The anisotropic yield criterion (denoted CPB05) is of the form:

$$F = \left(\left|\Sigma_1\right| - k\Sigma_1\right)^a + \left(\left|\Sigma_2\right| - k\Sigma_2\right)^a + \left(\left|\Sigma_3\right| - k\Sigma_3\right)^a$$

(17)

with $\Sigma_1, \Sigma_2, \Sigma_3$ are the principal values of the transformed stress tensor. Thus, for full 3-D stress states, nine anisotropy coefficients are involved in the criterion. It was shown that the criterion describes with great accuracy the yielding behavior of

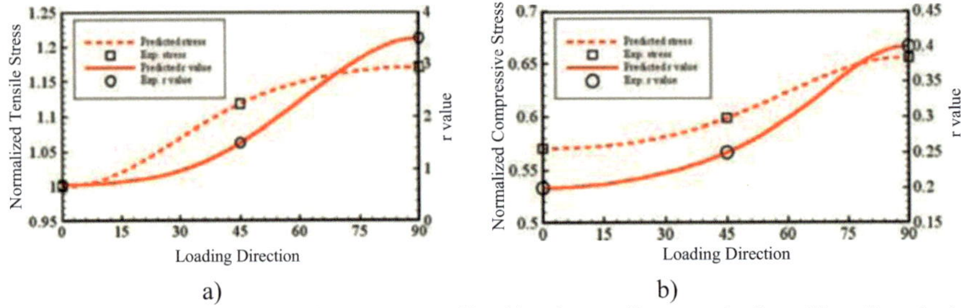

**Fig. 5.1.** Anisotropy of the yield stress (normalized by the tensile stress in the rolling direction) and the r-values for AZ31B Mg, measured and predicted with the extension of the yield criterion (16) involving 2 linear transformations for (a) tensile loading; (b) compressive loading. Experimental data after Lou et al. [148].

magnesium alloys (see Cazacu et al. [65]). Additional linear transformations can be incorporated into the CPB05 criterion for an improved representation of the anisotropy (see Fig. 5.1).

## 2.5  Strain rate potentials

In classical flow theory of plasticity, the yield function serves as a potential for the plastic strain rates (increments), i.e.,

$$\dot{\varepsilon}_{ij}^{p} = \dot{\lambda}\, \partial\phi / \partial s_{ij} \tag{18}$$

Hill [94] proved the existence of the dual conjugate $\psi$ to a stress potential $\phi$ for rate independent perfect plasticity, i.e., its inverse expressed in terms of the dual strain rate variables. In this case, the stress is given by the gradient of this potential, i.e.,

$$s_{ij} = \mu\, \partial\psi / \partial\dot{\varepsilon}_{ij} \tag{19}$$

For instance, Fortunier [78] proposed a dual formulation for the single crystal stress potential obeying the Schmid law and Van Houtte [193], [194] for a polycrystal. In general, it is difficult to find the dual conjugate of stress potentials analytically. However, it is possible to define strain rate potentials that describe the material anisotropy as an independent definition of the material behavior. This is formally identical to deriving yield functions. Because a strain rate potential and the associated stress potential are dual of each other, their mathematical properties are similar. Strain rate potentials are useful for rigid plasticity finite elements (FE) analysis and design codes. Stress and strain rate potentials can be used with equal degree of success (e.g., Li et al. [143]) but stress potentials have received more attention in the literature. The strain rate potential  proposed by Barlat et al. [29] has a structure similar to the stress potential Yld91 but in strain rate space. Recently, Barlat and Chung [35] and Kim et al. [121] developed a strain rate potential

$$\psi = \left|\tilde{E}'_1\right|^b + \left|\tilde{E}'_2\right|^b + \left|\tilde{E}'_3\right|^b + \left|\tilde{E}''_2 + \tilde{E}''_3\right|^b + \left|\tilde{E}''_3 + \tilde{E}''_1\right|^b + \left|\tilde{E}''_1 + \tilde{E}''_2\right|^b = (2^{2-b} + 2)\dot{\bar{\varepsilon}}^b \qquad (20)$$

where $\dot{\bar{\varepsilon}}$ is the corresponding effective strain rate and $b$ is an exponent recommended to be 3/2 for BCC and 4/3 for FCC materials, respectively. $\tilde{E}'_i$ and $\tilde{E}''_i$ are the principal values of two tensors $\tilde{\varepsilon}'$ and $\tilde{\varepsilon}''$ resulting from two linear transformations of the strain rate tensor $\dot{\varepsilon}$

$$\tilde{\varepsilon}' = \mathbf{B}'\mathbf{T}\dot{\varepsilon}, \quad \tilde{\varepsilon}'' = \mathbf{B}''\mathbf{T}\dot{\varepsilon} \qquad (21)$$

The two $4^{\text{th}}$ order tensors $\mathbf{B}'$ and $\mathbf{B}''$ contain the anisotropy coefficients and $\mathbf{T}$ is necessary to ensure that the gradient of the strain rate potential is deviatoric (see Kim et al. [121]).

## 2.6 Discussion

In the past, until the late 1980's yield functions were generic for anisotropic materials. Therefore, they were seldom accurate enough. They did not contain enough anisotropy coefficients and could not describe the anisotropies of the Lankford coefficient and yield stress in uniaxial tension simultaneously. The inversion symmetry was always assumed. To the exception of Hill's 1948, the yield functions were applicable to restricted stress cases, for instance formulated with normal stress components only, in order to avoid a number of formulation issues [109]. Therefore, they were not suitable for implementation in finite element codes.

At present, yield functions (or strain rate potentials) are developed for specific materials, FCC, BCC or HCP, and take into account the distinctive features of the response of these materials. For instance, for HCP materials, the inversion symmetry is not imposed anymore. Because of the higher number of anisotropy coefficients, they are more accurate than the yield functions used in the past. In particular, they are able to capture the anisotropy of uniaxial properties (Lankford coefficients and yield stress) simultaneously. The formulations are developed for either plane stress or general stress states without restrictions and can be implemented in finite element codes with relative ease (e.g., Yoon et al. [210], [211], [212]). Some efforts are now initiated to formulate macroscopic level models that account for the evolution of anisotropy due to evolving texture. The simulation results (Plunkett et al. [167]) suggest that a computationally efficient macroscale model that incorporates relevant information about the behavior of HCP metals at different length-scales describes with high fidelity the quasi-static deformation of initially textured HCP materials for single path loadings.

In the future, the evolution of yield function coefficients will be self-contained in the formulation for any linear and non-linear loading. Therefore, as in the case of crystal plasticity, it will be possible to describe the evolution of anisotropy as deformation proceeds. However, the computation time will be much lower than what it is required for calculations in which crystal plasticity models are directly linked to finite element (FE) codes.

## 3    Experimental validation of the anisotropic models

In sheet or tube forming processes, materials are generally subjected to multiaxial loads. Therefore, multiaxial loading experiments are infinitely preferable for checking the plasticity models to be used in simulations. Servo-controlled testing machines are necessary for such experiments. This section is a brief review of experimental methods for observing and modeling the anisotropic plastic behavior of metal sheets and tubes under biaxial loading. The reader is also referred to the excellent reviews of earlier work on experimental plasticity by Michno and Findley [158], Hecker [89], Ikegami [114], Bell [40], Phillips [166], Szczepiński [185], Stout and Kocks [177], McDowell [154] and Kuwabara [136].

### 3.1    Biaxial compression

Biaxial compression tests are effective in observing yielding behavior in the $\pi$-plane [139], [151], [190]. One of the disadvantages of the biaxial compression test is the difficulty in obtaining accurate stress-strain relations because of friction between the specimen and tool. Moreover, when the plastic deformation mechanism of the material is influenced by the hydrostatic component of stress [149], [176], the yield locus shapes obtained from the biaxial compression experiment may differ from those obtained from the biaxial tension test [149].

### 3.2    Biaxial tension using a cruciform specimen

A variety of cruciform specimens have been proposed for the biaxial tension experiments on sheet metals (Shiratori and Ikegami [175]; Kreißig and Schindler [125]; Ferron and Makinde [76]; Makinde et al. [150]; Demmerle and Boehler [70]; Boehler et al. [42]; Lin and Ding [145]; Müller and Pöhlandt [160]; Kuwabara et al. [130]; Hoferlin et al. [100]; Borsutzki et al. [44]). Cruciform specimens are suitable for observing the behavior of sheet metals in small plastic strain ranges of less than several percent.

Fig. 5.2 shows examples of experimental work contours for steel alloys with different r-values and the calculated yield loci based on the Taylor-Bishop-Hill (TBH) model and yield functions. The level of plastic work, $W$, is represented by the corresponding uniaxial plastic strain, $\varepsilon_0^p$, observed for the uniaxial tension test in the rolling direction of the material. A group of stress points, $(\sigma_x, \sigma_y)$, consisting of a work contour for a specific value of $\varepsilon_0^p$ are normalized by the uniaxial tensile flow

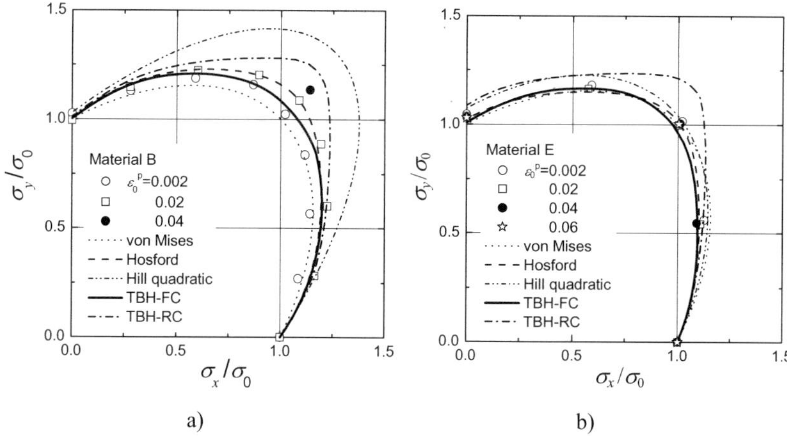

a)                                           b)

**Fig. 5.2.** Contours of plastic work and the yield loci calculated on the basis of the full constraints (FC) and relaxed constraints (RC) Taylor-Bishop-Hill (TBH) model, and yield functions [133]. (a) Ultralow-carbon IF steel ($r_0 = 2.01$, $r_{45} = 1.52$, $r_{90} = 2.42$) and (b) cold-rolled dual-phase steel ($r_0 = 0.82$, $r_{45} = 1.08$, $r_{90} = 1.08$).

stress, $\sigma_0$, corresponding to the $\varepsilon_0^p$. Experimental work contours corresponding to $\varepsilon_0^p = 0.002$, effectively viewed as a yield locus of the material, are closer to the theoretical yield loci based on the TBH-FC model. Von Mises and Hosford's yield criteria agree more closely with the experimental work contours than Hill's quadratic yield criterion for both materials.

## 3.3  Biaxial tension test on metal tubes

Multiaxial testing of thin-walled tubular specimens loaded in combined tension-torsion or tension-internal pressure modes is suitable for measuring the behavior of metals subjected to large plastic strain. Anisotropic plastic deformation behavior of brass tube [97], aluminum tube [135] and steel tube [217] was observed up to fracture and formulated on the basis of work contours. A yield vertex was successfully observed at the point of loading [134], [131], using the abrupt strain path change method proposed by Kuroda and Tvergaard [126].

## 3.4  Stress reversal test on sheet metals

Stress reversal tests are effective in observing the Baushinger effect of sheet metals: cyclic bending-unbending (Weinmann et al. [203]; Yoshida et al. [213]); cyclic simple shearing (G'sell et al. [85]; Miyauchi [159]; Hu et al. [111]); and in-plane tension/compression stress reversals tests (Iwata et al. [117]; Kuwabara et al. [132]; Yoshida et al. [214], Boger et al. [43]).

## 3.5  Discussion

In the past, hydraulic bulging tests or biaxial compression tests were widely used. The major limitations are the difficulties in observing accurate stress strain curves in small plastic strain ranges less than several percent and changing stress/strain path during loading.

At present, biaxial tension tests using cruciform specimens are becoming popular for observing the yield locus in the first quadrant of stress space. If one wishes to observe the anisotropic plastic deformation behavior of sheet metals for large plastic strain ranges up to fracture, one can make thin-walled tubular specimens from a flat sheet and test them using combined tension-torsion or tension-internal pressure testing machines [217]. According to biaxial tension experiments under linear and bilinear stress paths, it has been found that phenomenological plasticity models based on the isotropic hardening assumption, in conjunction with an appropriate anisotropic yield function, are still useful for describing anisotropic plastic deformation behaviors of metal sheets and tubes [136]. If stress reversal plays a dominant roll in a forming operation, such as bending-unbending, the Bauschinger effect must be measured accurately using a stress reversal test (see section 3.4) and formulated to make more sophisticated material models.

In the future, material testing based on non-linear loading and multistage loading will be of importance in aid of checking the validity of material models for such loading conditions as those occurring in real forming processes. Measurement and formulation of the non-linear stress-strain relations of prestrained sheet metals subjected to biaxial unloading will be necessary from the viewpoint of improving the accuracy of springback simulations of three dimensional sheet metal parts, such as automotive body panels.

# 4  Formability of metallic materials

The formability describes the capability of a sheet metal to undergo plastic deformation in order to get some shape without defects. During the last decades different assessment methods of the metals sheets formability have been developed. The most useful tool used to asses the formability is the forming limit diagram (FLD). This method meets both manufacturer and user's requirements and is widely used in factory and research laboratories. One of the major advantages of the FLD concept consists of the fact that the plastic instability can also be described by theoretical models. A detailed presentation of this method can be found in the literature [11], [104], [202], [208].

## 4.1  Prediction of the FLC

Various theoretical models have been developed for the calculation of forming limit curves (FLC).

The first ones were proposed by Swift [184] and Hill [92] assuming homogeneous sheet metals (the so-called models of diffuse necking and localized necking, respectively). The Swift model has been developed later by Hora (so-called Modified Maximum Force Criterion - MMFC) [102], [103]. Marciniak and Kuczynski (MK model) [152] proposed a model taking into account that sheet metals are non-homogeneous from both the geometrical and the structural point of view. Stören and Rice [181] have been developed a model based on the bifurcation theory. Dudzinski and Molinari [73] used the method of linear perturbations for analyzing the strain localization and computing the limit strains. Bressan and Williams [48] have introduced so-called "Through Thickness Shear Instability Criterion" in order to take into account the shear fracture mode. Since the theoretical models are rather complex and need a profound knowledge of continuum mechanics and mathematics while their results are not always in agreement with experiments, some semi-empirical models have been developed in recent years.

At present, the most widely used models for the computation of the limit strains are those proposed by Marciniak and Kuczynski [152] and Hora [102], [103], respectively. As a consequence, the models previously mentioned will be briefly discussed in the following (see also [11]). The Forming Limit Stress Diagram (FLSD) proposed by Arrieux et al. [4] has been also intensively studied during the last decade.

On the basis of the experimental investigations concerning the strain localization of some specimens subjected to hydraulic bulging or punch stretching [152] has concluded that the necking is usually initiated by a geometrical or structural non-homogeneity of the material. This non-homogeneity may be associated to a variation of the sheet thickness (geometrical non-homogeneity) or some defects of the lattice (structural non-homogeneity).The analysis of the necking process have been performed assuming a geometrical non-homogeneity in the form of a thickness variation. This variation is usually due to some defects in the technological procedure used to obtain the sheet metal. The thickness variation is generally gentle. However, the theoretical model assumes a sudden variation in order to simplify the calculations (Fig. 5.3.a). The theoretical model proposed by Marciniak assumes that the specimen has two regions: region "$a$" having a uniform thickness $t_0^a$, and region "$b$" having the thickness $t_0^b$. The initial geometrical non-homogeneity of the specimen is described by the so-called "coefficient of geometrical non-homogeneity", $f$, expressed as the ratio of the thickness in the two regions: $f=t_o^b/t_o^a$. In the MK model, the strain and stress states in the two regions are analyzed watching the principal strain $\varepsilon_1^b$ in region "$b$" in relation with the principal strain $\varepsilon_1^a$ in region "$a$". When the ratio of these strains $\varepsilon_1^b/\varepsilon_1^a$ becomes too large (infinitely large in theory, but greater than 10 in practice), one may consider that the entire straining of the specimen is localized in region "$b$". The shape and position of the curve $\varepsilon_1^a$-$\varepsilon_1^b$ depend on the value of the $f$-coefficient. If $f=1$ (geometrically homogeneous sheet), the curve becomes coincident with the first bisectrix. Thus this theory cannot model the strain localization for geometrically homogeneous sheets. The value of the principal strain $\varepsilon_1^a$ in region "$a$" corresponding to non-significant straining of this region as compared to region "$b$" (the straining being localized in region "$b$") represents the limit strain $\varepsilon_1^{a*}$. This strain together with the second principal strain $\varepsilon_2^{a*}$ in region "$a$" define a point belonging to the FLC. Assuming different strain ratios $\rho = d\varepsilon_2/d\varepsilon_1$, one obtains different points on the FLC. By scrolling the range $0 < \rho < 1$, one gets the FLC for biaxial tension ($\varepsilon_1 > 0, \varepsilon_2 > 0$). In this domain, the orientation of the geometrical non-homogeneity with respect to the principal directions is assumed to be the same during the entire forming

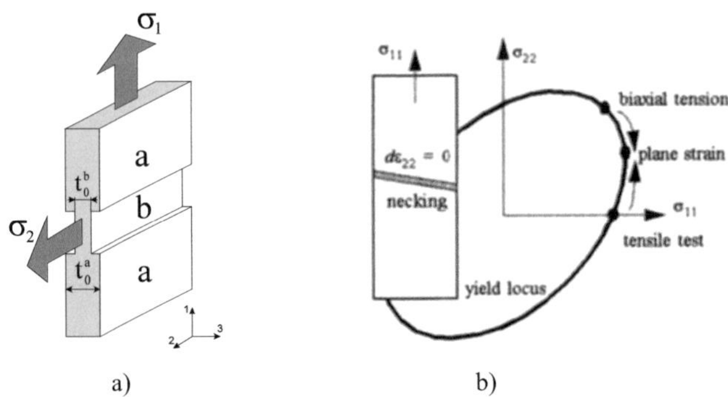



**Fig. 5.3.** (a) Geometrical model of the MK theory; (b) basic principle of the MMFC criterion.

process.

The 'Modified Maximum Force Criterion' (MMFC) for diffuse necking proposed by Hora et al. [102] is based on Considère's maximum force criterion. The idea behind the MMFC-model is to factor in an additional increase in hardening, which is triggered by the deviation from the initial, homogeneous stress condition—e.g. uniaxial tension—to the stress condition of local necking and with this to the point of plane strain (Fig. 5.3.b) [103]. In order to take into account the influence of the thickness and the strain rate sensitivity index on the limit strains, an enhanced MMFC (eMMFC) has been proposed by Hora et al. [103] and by Brunet [50], [54], respectively. The advantage of the MMFC criteria can be found in their independence of the inhomogeneity assumption. This criterion could be used to calculate FLC for non-linear strain path. A drawback of the MMFC models is the fact that it contains a singularity that emerges if the yield locus contains straight line segments [2].

During the last decade the research in the field of the forming limits prediction have been focused mainly on the following aspects:

**Implementation of new constitutive equations in the models used for the computation of the limit strains.** The effect of the shape of the yield locus on the limit strains has been analyzed in detail by Barlat et al. [27]. As we have emphasized in section 2, a lot of new yield criteria have been developed during the last decade. Many of those criteria have been already implemented in the computational models of the limit strains, in order to improve the predictive capabilities. Banabic et al. have implemented various yield criteria in the MK model (Hill'93 [8], [13], BBC yield criteria [15], [16], [165] and Cazacu-Barlat [17], [164], [165]. In the Figure 5.4.a is presented the theoretical FLC predicted using BBC2003 criterion versus experimental data for AA5182-0 aluminium alloy [16]. Butuc et al. have used the Barlat '97 [56], [58], [59] and BBC2000 [56] yield criteria. Cao [60], [209] used the Karafillis and Boyce [120] yield criterion in the MK model to analyze the effect of changing strain-paths on the FLC. Kuroda and Tvergaard [127] used four different yield criteria to fit a set of experimental data. The Yld 2000 [32] formulation has been included by Aretz [2] in the MK model for studying the influence of the biaxial coefficient of plastic anisotropy on the FLCs. Kim et al. [122] used the Yld2000 [32] criterion to analyze the formability of a sandwich sheets.

Vegter et al. have implemented their own yield criterion [195], [196] in the MK model. Ganjiani and Assempour [80] have improved the analytical approach for determination of FLC considering the effects of yield functions (Hosford [108] and BBC2000 [14]) The Teodosiu hardening model [189] associated with different yield criteria has been implemented by Butuc et al. [57] in a MK theory for studying the influence of the loading path change on the limit strains.

**Implementation of the polycrystalline models.** The adaptability of the texture based models to the MK theory of the strain localisation has been proved in the 1980's by Bate [39], Assaro [7], Barlat [25], [26] and later by Van Houtte [192] and Neale [115], [172], [204], [205],[206] teams. Later on, Viatkina et al. [197] have used such models for the computation of FLCs. The texture-based yield criterion developed by Van Houtte et al. [194] has been implemented in FLC models, the results being compared both with those provided by phenomenological models and with experimental data [17]. Van Houtte model [193] coupled with a dislocation based hardening model [189] have been implemented by Hiwatashi et al. [99], [195] in order to predict the forming limits corresponding to change strain paths. A microstructural model developed for the description of the aluminium alloy hardening (ALFLOW) has been used by Berstad et al. [41] to predict the forming limits of the AA3103-0 alloy. Boudeau [45], [46] used the linear stability analysis combined with a polycrystalline model to predict and to analyze the influences on the FLC. A polycristal plasticity model has been used by McGinty [155] to conduct parametric studies of FLC. Knockaert et al. [123] have used a rate-independent polycrystalline plasticity to predict the limit strains. The influence of the texture on the FLCs has been studied by Kuroda [129] and Fjedbo et al. [77].

**Implementation of the ductile damage models.** Several types of ductile damage models have been developed during the time, e.g. Gurson, Kachanov, Chaboche,

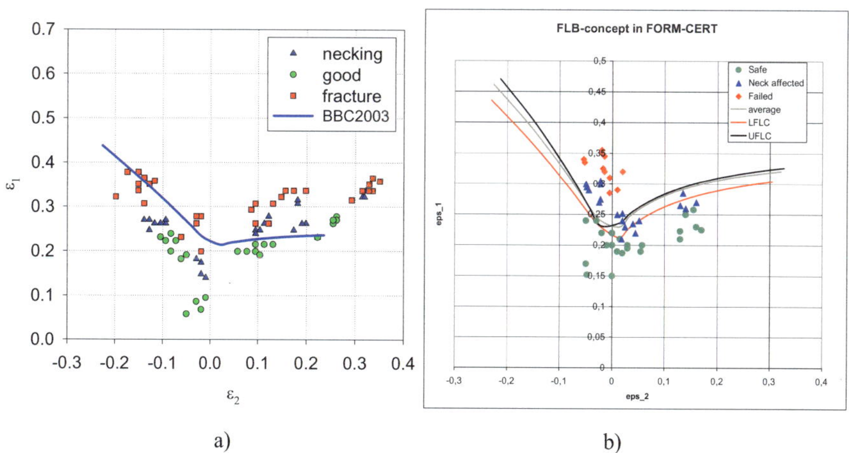



a)                                        b)

**Fig. 5.4.** (a) Theoretical FLC versus experimental data for AA5182-0 aluminium alloy; (b) Predicted Forming Limit Band versus experimental data for AA6111-T43 aluminium alloy (LFLC-lower FLC, UFLC-upper FLC).

Gologanu (see details in [141], Chapter 6). Those models have been frequently used during the last decade for the computation of the limit strains. Brunet et al. have used the Gologanu model [83] for calculating such limit strains [51], [52], [53]. The effects of texture and damage evolution on the limit strains have been studied by Hu et al. [112]. Chow et al. have developed a ductile damage model and implemented it into the MK theory both for linear [67] and complex load paths [68], [69]. An anisotropic model of Gurson type have been used by Huang et al. [113] for the computation of the FLCs. Ragab et al., [170] use a new model to predict the FLC for kinematically hardened voided sheet metals. Han and Kim [86] used an original ductile fracture criterion to calculate the FLC. Lemaitre's ductile damage model has been also implemented by Teixeira [186].

**Enhancing the existing models to take into account new material or process parameters.** The influence of different parameters on the limit strains has been analyzed since the end of the 1960's. More recently, several new introduced parameters have been included in the MK model: the shape of the yield locus [14], the forming temperature [1] and the coefficient of biaxial anisotropy [2]. The influence of the different effects on the limit strains have been studied: the effect of the surface defects [98], the effect of the void growth [169], the effect of grain size [173], the effect of the tangential-strain rate. Brunet and Clerc [54] have extended Hora's model by including the strain rate sensitivity exponent and used it for studying the influence of that parameter on the limit strains. Chan [66] has developed a model of forming limits prediction for the superplastic forming.

**Extending the FLC models for non-linear strain-paths.** During the sheet metal forming processes, the material is usually subjected to complex strain patterns. Nakazima [161] has proved that complex loads modify the shape and position of the FLC's. This fact imposes the determination of the limit strains for complex strain-paths. The development of the computational models for complex strai-paths in the frame of the MK theory has become an active research field in the early 1980's (see Barata et al. [23], [24] and also [202]). The refinement of those models has been intensively approached only during the last period. Butuc [55], [57], [58], [59] has developed a general computer code for the FLC computation in the case of complex load paths using various hardening models (both phenomenological – Swift, Voce, and miocrostructural ones – Teodosiu-Hu). Rajarajan et al. [168] have validated the CRACH model for the case of complex strain-paths. Cao [60], [209] analyzed the influence of the changing strain paths on the limit strains. Hiwatashi et al. [99] have used Teodosiu's model for studying the influence on the strain-path change on FLCs. Kuroda and Tvergaard [128] have studied the effect of the strain-path change on the limit strains using four anisotropic models.

**Using advanced numerical methods for the solution of the limit strain models.** Wagoner and his co-workers have used the finite element method for the numerical determination of the limit strains in the frame of the MK theory [163], [219]. Later on, FEM has been also used by Horstemayer [105], Tai and Lee [186], Nandedkar [162], Gänser et al. [81], Evangelista [74], Van der Boogarard [191], Lademo [137], [138], Berstad [41], Brunet [53], Paraianu [164], Teixeira [186], Hopperstad [101]. The results reported by the researchers previously mentioned are promising.

**Modeling the Forming Limit Band concept.** The analysis of the variability of the sheet metal parameters and their effects on the formability has been recently performed by [61], [171]. On the basis of the variability of the limit strains established by experiments, Janssens et al. [118] introduced the Forming Limit Band concept. This is a strip containing almost all of the limit strain states. The concept has been extended by Strano and Colosimo[182], [183]. Asuming the variability of the mechanical parameters of the sheet metal, Banabic and Vos [22], [201] have developed a computational method of the Forming Limit Band. In the Figure 5.4.b is presented the predicted Forming Limit Band versus experimental data for AA6111-T43 aluminium alloy.

**Forming Limit Stress Diagram (FLSD).** The concept of Forming Limit Stress Diagram (FLSD) proposed by Arrieux et al. [4] has been intensively studied recently. The FLSD is reportedly independent of strain paths [4], [5], [6], [84], [178], [179], [180], [207], [216]. Biaxial stress experiments on aluminum alloy tube [215] and steel tube [217] for many linear and bilinear stress paths revealed that forming limit stresses are effectively path independent, provided that unloading is included between the loading paths and that the material work hardens isotropically. The FLSD concept therefore appears to be useful, particularly in multistage forming, for predicting the failure of metal tubes and sheets.

**Developing commercial codes for FLC computation (see section 4.3).**


## 4.2  Advanced methods to determine the FLC

Since the proposal of the FLC concept, many researchers have been actively involved in the development of experimental methods for the accurate and objective determination of the limit strains. These experimental aspects have been the most important obstacles limiting the practical use of the FLC's. During the last years, the digital processing of the images has allowed the development of refined methods for the experimental measurement of the limit strains. These methods aim to remove the subjective perturbations induced by the human operator in the process of image analysis. More precisely, new algorithms for the detection of the defect occurrence on the formed part have been developed. They have contributed to the increased accuracy of the limit strain measurement and to the reduction of the discrepancy between the experimental data obtained in different laboratories. In the following, we shall present some of these methods. Further details related to this problem can be found in [11], [12], [104].

Takashina and his co-workers [187] first proposed a simply method to determine the limit strains (so-called "three circle method"). The method has been improved by Veerman [196]. Bragard [47] developed in 1972 a more precise method of determining the limit strains based on interpolation. This method is later improved by D'Haeyer and Bragard [71] using the name of "the double profile method". In 1972 Hecker [88] proposed a method based on the determination of three types of ellipses around the fracture: fractured, necked and acceptable.  The method consists in determining the major and minor strains of the different types of ellipses in the neighborhood of the fracture on the deformed piece and transposing them on FLD. The limit curve is traced between the point corresponding to the ellipses affected by necking and the acceptable ones. The method has been used on a large scale because of simplicity. Kobayashi defines the limit strain based on the accelerated increase of

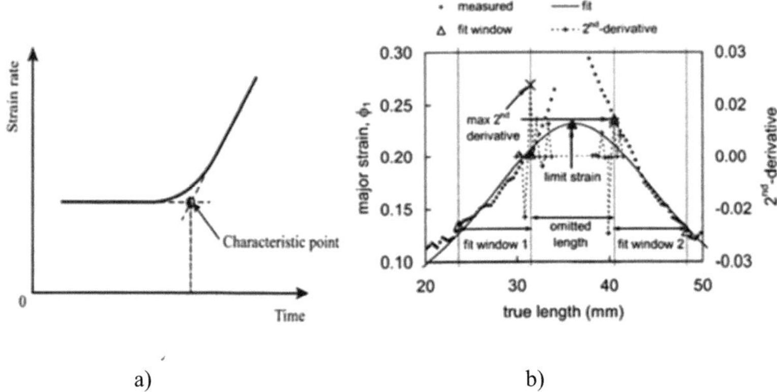


**Fig. 5.5.** Methods to determine the limit strains: a) Strain-rate versus time; b) IDDRG method

the roughness in the necking area. The Zürich meeting in 1973 of the IDDRG workgroup, following an analysis of several versions of limit strain determination, recommends using an improved version of the Bragard method. This is known as the "Zürich Nr.5 method" [157]. You find a review of these methods in [11].

Together with the development "of online" video strain measuring methods, new methods of determining the limit strains have been proposed in the last years. A new criterion based on the evolution of the strain rate as a function of time during the forming process has been proposed by the SOLLAC team [153]. The method is based on the observation than the beginning of the necking is accompanied by a considerable increase of the strain rate (see Fig.5.5.a).

According with this method the start necking point corresponds to the dramatically changing in the strain-rate versus time variation (characteristic point). This point could be determined by the intersection of the two straight lines corresponding to the first and the last sector of the curve. The strain – rate evolutions are automatically determined by images analysis. The strain-rate method has been used recently by Volk [200]. He used the idea to identify a regular grid for the optical measurement as a typical mesh of a finite element method.

The Nakajima workgroup of the IDDRG has developed a new method [110], the so-called "in-process measurement" method (see Fig.5.5.b). A guideline for the determination of FLC based on this method is presented in the paper [144]. The method is similar with the Bragard one. Using a video camera system, a film of the forming process is made. Based on the film of the forming process, the development of the strain distribution starting from the onset of necking and finally up to the fracture is analyzed. The method is a very robust one and gives a very good repeatability of the results. Base of this achievements, the expert group of Nakajima workgroup propose a revision of the ISO 12004 standard "Metallic materials-sheet and strip-Determination of the forming limit curves" [116].

Based on the video camera measurement some systems have been developed by the commercial company to determine automatically the FLC. CAMSYS company has developed the first automatically system (ASAME - Automated Strain Analysis and Measurement Environment) used on the large scale, both in research laboratory and industry [142]. The INSA Lyon developed a FLC determination system (IcaForm) based on the spray of a random pattern of paint at the surface of the sample to determine strain distribution [50]. An opto-mechanical device adaptable allows determining easily the FLC. An objective criterion to identify the start of local necking automatically has been proposed recently [174]. The "Autogrid" system developed by Vialux company offer the possibility to determine the limit strains automatically and independent of any operator. The methodology used to define the limit strain is presented in details in the paper [75]. GOM Company has developed for the FLC determination so-called ARAMIS system [79]. The methodology used is according with the Nakajima workgroup recommendation. For an FLC are used five different geometries, for each geometry three specimens and for each specimen three to five parallel sections. The FLC determination procedure can be done automatically.


## 4.3  Commercial programs

In the last decade have been developed more commercial programs for the limit strains prediction. In this section will be presented more significant ones.

Based on a Marciniak-Kuczynski model [152], Jurco and Banabic [31], [119] have developed so-called FORM-CERT commercial code. The BBC 2003 yield criterion [18] is implemented in this model. This yield criterion can be reduced to simpler formulations (Hill'48, Hill'79, Barlat'89, etc). In this way, the yield criterion can be also used in the situations when only 2, 4, 5, 6, or 7 mechanical constants are available. The program consists in four modules: a graphical interface for input, a module for the identification and visualization of the yield surfaces, of the strain hardening laws and a module for calculating and visualizing the forming limit curves. The numerical results can be compared with experimental data, using the import/export facilities included in the program. The FORM-CERT code can be directly coupled with the finite element codes.

Hora and his co-workers [103] have developed MATFORM code based on the MMFC model. This code is able to calculate and plot the limit strains and also the visualization of the strain hardening curve and yield loci using Hill'48, Hill'79, Hill'90 and Barlat'89 criteria. The program is useful for evaluation of most common experiments like tensile, bulge, Miyauchi, torsion dilatometer and tube hydroforming tests. The program is very well documented and is able to export the constitutive models in FEM specific form for the application in the mostly spread FEM-codes like Autoform or PamStamp.

Using the CRACH algorithm (based on the Marciniak-Kuczynski model), Gese and Dell [82] have developed two software: CrachLAB, a product for prediction of the initial FLC and CrachFEM a product for coupling with the FEM codes. Criteria for ductile and shear fracture have been included in CrachFEM to cover the whole variety of fracture modes for sheet materials. The material model used to calculate instability describes: the initial anisotropy (using Hill'48 criterion), the combined isotropic-kinematic hardening and the strain rate sensitivity. CrachFEM is now included in the FEM codes PamStamp and PamCrash of ESI Group.

## 4.4  Discussion

In the past, the FLC models provided an approximate description of the experimental results. Such models were used especially for obtaining qualitative information concerning the necking/tearing phenomena.

At present, the FLC models allow a sufficiently accurate prediction of the limit strains, but each model suffers from its own limitations (see section 4.1). There is no model that can be applied to any sort of sheet metal, any type of crystallographic structure, any strain- path or any variation range of the process parameters (strain rate, temperature, pressure, etc.).

The future research will be focused on a more profound analysis of the phenomena accompanying the necking and fracture of the sheet metals. On the basis of the analysis, more realistic models will be developed in order to obtain better predictions of the limit strains. New models will be developed for prediction of the limit strains for special sheet metal forming processes: superplastic forming, forming at very high pressure, incremental forming etc. Commercial codes allowing the quick and accurate calculation of the FLC's both for linear and complex strain-paths will be developed. The texture models will be also implemented in such commercial programs. The FLC computation will be included in the finite element codes used for the simulation of the sheet metal forming processes. The aim is to develop automatic decision tools (based on artificial intelligence methods) useful in the technological design departments. The stochastic modeling of the FLC's will be developed in order to increase the robustness of the sheet metal forming simulation programs. More refined, accurate and objective experimental methods for the experimental determination of the limit strains (e.g. methods based on thermal or acoustic effects) will be also developed.

# References


1.  Abedrabbo, N., Pourboghrat, F., Carsley, J.: Forming of aluminum alloys at elevated temperatures. Int. J. Plasticity 22 (2006) 314-373
2.  Aretz, H.: Numerical restrictions of the modified maximum force criterion for prediction of forming limits in sheet metal forming. Modelling Simul. Mater. Sci. Eng. 12 (2004) 677-692
3.  Aretz, H.: Impact of the equibiaxial plastic strain ratio on the FLD prediction. In: Juster, N., Rosochowski, A. (eds.): Proc. 9th ESAFORM Conference on Material Forming. Glasgow, April 2006. AKAPIT, Krakow (2006) 311-314
4.  Arrieux, R., Bedrin, C., Boivin, M.: Determination of an intrinsec Forming Limit Stress Diagram for isotropic sheets. In: Proc. of the 12th IDDRG Congress. S-ta Margerita Ligure (1982) 61-71
5.  Arrieux, R., Bedrin, C., Boivin, M.: Determination of the Strain Path Influence of the Forming Limit Diagrams, from the Limit Stress Curve. Annals of the CIRP. 34 (1985) 205-208
6.  Arrieux, R.: Determination of the Forming Limit Stress Curve for Anisotropic Sheets. Annals of the CIRP. 36 (1987) 195-198
7.  Asaro, R.J., Needleman, A.: Texture development and strain hardening in rate-dependent polycrystals. Acta Metall. 33 (1985) 923-953

8.  Banabic, D.: Limit strains in the sheet metals by using the 1993 Hill's yield criterion. J. Materials Process. Techn. 92-93 (1999) 429-432

9.  Banabic, D., Comsa, D.S., Balan, T., A new yield criterion for orthotropic sheet metals under plane –stress conditions. In: Banabic, D. (ed.): Proc. of the 7th Conf. „TPR2000". Cluj Napoca, (2000) 217-224

10.  Banabic, D.: Anisotropy of Sheet Metals. In: Banabic, D. (ed.): Formability of Metallic Materials, Springer-Verlag, Berlin Heidelberg New York (2000) 119-172

11.  Banabic, D.: Forming Limits of Sheet Metals. In: Banabic, D. (ed.): Formability of Metallic Materials, Springer-Verlag, Berlin Heidelberg New York (2000) 173-215

12.  Banabic, D.: Theoretical Models of the FLD's. In: Banabic, D. (ed.): Formability of Metallic Materials, Springer-Verlag, Berlin Heidelberg New York (2000) 317-327

13.  Banabic, D., Dannenmann, E.: The influence of the yield locus shape on the limits strains, J. Materials Process. Techn. 109 (2001) 9-12

14.  Banabic, D., Kuwabara, T., Balan, T., Comsa, D.S., Julean, D.: Non-quadratic yield criterion for orthotropic sheet metals under plane-stress conditions. Int. J. Mech. Sci. 45 (2003) 797-811

15.  Banabic, D. et al.: FLD theoretical model using a new anisotropic yield criterion, J. Materials Process. Techn. 157-158 (2004) 23-27

16.  Banabic, D.: Anisotropy  and formability of AA5182-0 aluminium alloy sheets. Annales of CIRP. 53 (2004) 219-222

17.  Banabic, D. et al.: Prediction of FLC from two anisotropic constitutive models. In: Stören, S. (ed.): Proc. 7th ESAFORM Conference on Material Forming. Trondheim, (2004) 455-459

18.  Banabic, D., Aretz, H., Comsa, D.S., Paraianu, L. : An improved analytical description of orthotropy in metallic sheets. Int. J. Plasticity 21 (2005) 493–512

19.  Banabic, D., Aretz, H., Paraianu, L., Jurco, P.: Application of various FLD modelling approaches. J. Modelling Simul. Materials Science Eng. 13 (2005) 759-769

20.  Banabic, D., Cazacu, O., Paraianu, L., Jurco, P.: Recent Developments in the Formability of Aluminum Alloys. In: Smith, L.M., Pourboghrat, F., Yoon, J.-W., Stoughton,T.B. (eds): Proc. of the NUMISHEET 2005 Conference. AIP (2005) 466-472

21.  Banabic, D.: Numerical prediction of FLC using the M-K-Model combined with advanced material models. In: Hora, P. (ed): Numerical and experimental methods in prediction of forming limits in sheet forming and tube hydroforming processes. ETH Zürich, Zürich (2006) 37-42

22.  Banabic, D., Vos, M.: Increasing the Robustness in the Simulation of Sheet Metal Forming Processes using a new Concept - Forming Limit Band. Annales of CIRP. 56 (2007) (in press)

23.  Barata da Rocha, A., Jalinier, J.M.: Plastic instability of sheet metals under simple and complex strain path. Trans. Iron Steel Inst. Japan 24 (1984) 133-140

24.  Barata da Rocha, A., Barlat, F., Jalinier, J.M.: Prediction of the forming limit diagrams of anisotropic sheets in linear and non-linear loading. Mat. Sci. Eng. 68 (1985) 151-164

25.  Barlat, F., Richmond, O.: Prediction of tricomponent plane stress yield surfaces and associated flow and failure behavior of strongly textured FCC sheets. Mat. Sci. Eng. 95 (1985) 15-29

26.  Barlat, F.: Crystallographic texture, anisotropic yield surfaces and forming limits of sheet metals. Mat. Sci. Eng. 91 (1987) 55-72

27.  Barlat, F., Lian, J.: Plastic Behavior and Stretchability of Sheet Metals. Part I: Yield Function for Orthotropic Sheets under Plane Stress Conditions. Int. J. Plasticity 5 (1989) 51–66

28. Barlat, F., Lege, D.J., Brem, J.C.: A six-component yield function for anisotropic materials. Int. J. Plasticity 7 (1991) 693-712
29. Barlat, F., Chung, K.: Anisotropic Potentials for Plastically Deforming. Metals, Model. Simul. Mater. Sci. Eng. 1 (1993) 403–416
30. Barlat, F., Becker, R.C., Hayashida, Y., Maeda, Y., Yanagawa, M., Chung, K., Brem, J.C., Lege, D.J., Matsui, K., Murtha, S.J., Hattori, S.: Yielding description of solution strengthened aluminum alloys. Int. J. Plasticity 13 (1997)185-401
31. Barlat, F., Maeda, Y., Chung, K., Yanagawa, M., Brem, J.C., Hayashida, Y., Lege, D.J., Matsui, K., Murtha, S.J., Hattori, S., Becker, R.C., Makosey, S.: Yield function development for aluminum alloy sheets. J. Mech. Phys. Solids 45 (1997) 1727-1763
32. Barlat, F., Brem, J.C., Yoon, J.W., Chung, K., Dick, R.E., Lege, D.J., Pourboghrat, F., Choi, S.-H., Chu, E.: Plane stress yield function for aluminum alloy sheet-Part I: Theory. Int. J. Plasticity 19 (2003) 1297-1319
33. Barlat, F., Cazacu, O. Życzowski, M., Banabic, D., Yoon, J.W.: Yield surface plasticity and anisotropy. In: Raabe, D., Roters, F., Barlat, F., Chen, L.-Q. (eds): Continuum Scale Simulation of Engineering Materials–Fundamentals–Microstructures – Process Applications. WileyVCH Verlag Manheim (2004) 145-177
34. Barlat, F.: Constitutive modeling for metals. In: Banabic, D., (ed.): Proc. 8th ESAFORM Conference on Material Forming. Cluj-Napoca April 2005. The Publishing House of the Romanian Academy, Bucharest (2005) 3-10
35. Barlat, F., Chung, K.: Anisotropic strain rate potential for aluminum alloy plasticity. In: Banabic, D. (ed.): Proc. 8th ESAFORM Conference on Material Forming. Cluj-Napoca April 2005. The Publishing House of the Romanian Academy, Bucharest (2005) 415-418
36. Barlat, F., Aretz, H., Yoon, J.W., Karabin, M.E., Brem, J.C., Dick, R.E.: Linear transformation-based anisotropic yield functions. Int. J. Plasticity 21 (2005) 1009–1039
37. Barlat, F.: Constitutive modeling for metals. In: Banabic D. (ed.): Advanced Methods in Material Forming. Springer-Verlag, Berlin Heidelberg New York (2007) 5-25
38. Barlat, F., Yoon, J.W., Cazacu, O.: On linear transformations of stress tensors for the description of plastic anisotropy. Int J. Plasticity 23 (2007) 876-896.
39. Bate, P.: The prediction of limit strains in steel sheet using a discrete slip plasticity model. Int. J. Mech. Sci. 26 (1984) 373-384
40. Bell, J.F.: The Experimental Foundations of Solid Mechanics. In: Truesdell, C. (ed.): Mechanics of Solids. Vol. I Springer-Verlag, Berlin Heidelberg New York (1984)
41. Berstad, T. et al.: FEM and a microstructure based work-hardening model used to calculate FLCs. In: Stören, S. (ed.): Proc. 7th ESAFORM Conference on Material Forming. Trondheim, (2004) 131-134
42. Boehler, J.P., Demmerle, S., Koss, S.: A new direct biaxial testing machine for anisotropic materials. Exp. Mech. 34 (1984) 1-9
43. Boger, R.K., Wagoner, R.H., Barlat, F., Lee M.G., Chung, K.: Continuous, large strain, tension/compression testing of sheet material. Int. J. Plasticity 21 (2005) 2319-2343
44. Borsutzki, M., Keßler, L., Sonne, H-M.: Kennzeichnung des Verfestigungsverhaltens von Werkstoffen mit der Biaxialprüfung. In: Werkstoffprüfung 2002, Proc. DVM-Conference, Bad Nauheim (2002) 186 (in German)
45. Boudeau, N., Gelin J.C., Salhi S.: Computational prediction of the localized necking in sheet forming based on microstructural material aspects. Computational Materials Science 11 (1998) 45-64
46. Boudeau, N., Gelin, J.C.: Necking in sheet metal forming. Influence of macroscopic and microscopic properties of materials. Int. J. Mech. Sciences 42 (2000) 2209-2232

47. Bragard, A., Baret, J.C., Bonnarens, H.: A simplified technique to determine the FLD at onset of necking. CRM 33 (1972) 53-63
48. Bressan, J.D., Williams, J.A.: The use of a shear instability criterion to predict local necking in sheet metal deformation. Int. J. Mech. Sciences 25 (1983) 155-168
49. Bron, F., Besson, J.: A yield function for anisotropic materials. Application to aluminum alloys. Int. J. Plasticity 20 (2003) 937-963
50. Brunet, M., Morestin, F.: Experimental and analytical necking studies of anisotropic sheet metals. J. Materials Process. Techn. 112 (2001) 214-226
51. Brunet, M., Morestin, F., Walter, H.: Damage modeling in sheet metals forming processes with experimental validations. In: Habraken, A.M. (ed.): Proc. 4th ESAFORM Conference on Material Forming. Liege (2001) 209-212
52. Brunet, M., Morestin, F., Walter, H.: Anisotropic ductile fracture in sheet metal forming processes using damage theory. In: Pietrzyk, M., Mitura, Z., Kaczmat, J. (eds.): Proc. 5th ESAFORM Conference on Material Forming. Krakow (2002) 135-138
53. Brunet, M., Morestin, F., Walter-Laberre, H.: Failure analysis of anisotropic sheet metals using a non-local plastic damage model. J. Materials Process. Techn. 170 (2005) 457-470
54. Brunet M., Clerc P.: Two prediction methods for ductile sheet metal failure, In: Proc. 10th ESAFORM Conference on Material Forming, Zaragoza (2007) (in press)
55. Butuc, C. et al.: A more general model for FLD prediction. J. Materials Proc. Techn. 125-126 (2002) 213-218
56. Butuc, C. et al.: Influence of constitutive equations and strain-path change on the forming limit diagram for 5182 Aluminum Alloy. In: Pietrzyk, M., Mitura, Z., Kaczmat,J. (eds.): Proc. 5th ESAFORM Conference on Material Forming, Krakow (2002) 715-719
57. Butuc, C., Gracio, J.J., Barata da Rocha, A.: A theoretical study on forming limit diagrams prediction. J. Material Proc. Techn., 142 (2003) 714–724
58. Butuc, C., Gracio, J.J., Barata da Rocha, A.: Application of the YLD 96 yield criterion on describing the anisotropy and formability of the BCC materials, In: Banabic, D. (ed.): Proc. 8th ESAFORM Conference on Material Forming. Cluj-Napoca April 2005. The Publishing House of the Romanian Academy, Bucharest (2005) 391-394
59. Butuc, C. et al.: An experimental and theoretical analysis on the application of stress-based forming limit criterion. Int. J. Mech. Sci. 48 (2006) 414–429
60. Cao, J. et al.: Prediction of localized thinning in sheet metal using a general anisotropic yield criterion. Int. J. Plasticity 16 (2000) 1105-1129
61. Carleer, B., Sigvant, M.: Process Scatter with Respect to Material Scatter. In: Liewald, M. (ed.): New Developments in Sheet Metal Forming. Institute for Metal Forming Technology, University of Stuttgart (2006) 225-239
62. Cazacu, O., Barlat, F.: Generalization of Drucker's yield criterion to orthotropy. Mathematics and Mechanics of Solids. 6 (2001) 613-630
63. Cazacu, O., Barlat, F.: Application of representation theory to describe yielding of anisotropic aluminum alloys. Int. J. of Engng. Sci. 41 (2003) 1367-1385
64. Cazacu, O., Barlat, F.: A criterion for description of anisotropy and yield differential effects in pressure-insensitive metals. Int. J. Plasticity 20 (2004) 2027-2045
65. Cazacu, O., Plunkett, B., Barlat, F.: Orthotropic yield criterion for hexagonal close packed metals. Int. J. Plasticity 22 (2006) 1171-1194
66. Chan, K.C., Tong G.Q.: Formability of high-strain-rate superplastic Al-4.4Cu-1.5Mg/21SiCW, composite under biaxial tension. Material Science Eng. A340 (2003) 49-57

67. Chow, C.L. et al.: A unified damage approach for predicting FLDs. Trans ASME, J. Eng. Materials Techn. 119 (1997) 346-353

68. Chow, C.L., Yang X.J.: Prediction of the FLD on the basis of the damage criterion under non-proportional loading. Proc. Instn. Mechn. Eng. 215C (2001) 405-414

69. Chow, C.L. et al.: Prediction of FLD for AL6111-T4 under non-proportional loading. Int. J. Mech. Sciences 43 (2001) 471-486

70. Demmerle, S., Boehler, J.P.: Optimal design of biaxial tensile cruciform specimens. J. Mech. Phys Solids 41 (1983) 143-181

71. d'Hayer, R., Bragard, A.: Determination of the limiting strains at the onset of necking. CRM 42 (1975) 33-35

72. Drucker, D.C.: Relation of experiments to mathematical theories of plasticity. J. Appl. Mech. 16 (1949) 349-357

73. Dudzinski, D., Molinari, A.: Instability of visco-plastic deformation in biaxial loading. C.R. Acad. Sci. Paris 307 (1988) 1315-1321

74. Evangelista, S.H. et al.: Implementing a modified Marciniak-Kuczynki model using the FEM for the simulation of sheet metal deep drawing. J. Materials Process. Techn. 130-131 (2002) 135-144

75. Feldmann, P., Schatz, M.: Effective evaluation of FLC-tests with the optical in-process strain analysis system AUTOGRID. In: Hora, P. (ed): Numerical and experimental methods in prediction of forming limits in sheet forming and tube hydroforming processes. ETH Zürich, Zürich (2006) 69-73

76. Ferron, G., Makinde, A. J.: Design and development of a biaxial strength testing device. J. Testing Eval. 16 (1988) 253-256

77. Fjeldbo, S.K. et al.: A numerical study on the onset of plastic instability in extruded materials with strong through-thickness texture variation. In: Banabic, D. (ed.): Proc. 8th ESAFORM Conference on Material Forming. Cluj-Napoca April 2005. The Publishing House of the Romanian Academy, Bucharest (2005) 209-213

78. Fortunier, R.: Dual potentials and extremum work principles in single crystal plasticity. J. Mech. Phys. Solids 37 (1989) 779-790

79. Friebe, H., et al.: FLC determination and forming analysis by optical measurement system. In: Hora, P. (ed.): Numerical and experimental methods in prediction of forming limits in sheet forming and tube hydroforming processes. ETH Zürich, Zürich (2006) 74-81

80. Ganjiani, M., Assempour, A.: An improved analytical approach for determination of FLD considering the effects of yield functions. J. Materials Process. Techn. 182 (2007) 598-607

81. Gänser, H.P., Werner, E.A., Fisher, F.D.: FLDs: a micromechanical approach. Int. J. Mech. Sciences. 42 (2000) 2041-2054

82. Gese, H., Dell, H.: Numerical prediction of FLC with the program CRACH. In: Hora, P. (ed): Numerical and experimental methods in prediction of forming limits in sheet forming and tube hydroforming processes. ETH Zürich, Zürich (2006) 43-49

83. Gologanu, M. et al.: Recent extension of Gurson's model for porous ductile metals. In: Suquet, P. (ed.): Continuum Micromechanics. Springer-Verlag, Berlin Heidelberg New York (1997) 61-130

84. Gronostajski, J.: Application of limit stresses to determine limit strains at complex strain paths. Archiwum Hutnictwa. 30 (1985) 41-56

85. G'sell, C., Boni, S., Shrivastava, S.: Application of the plane simple shear test for determination of the plastic behaviour of solid polymers at large strains. J. Mater. Sci. 18 (1983) 903-918

86. Han, H.N., Kim, K.H.: A ductile fracture criterion in sheet metal forming process. J. Materials Process. Techn. 142 (2003) 231-238

87. Hashiguchi, K., Protasov, A.: Localized necking analysis by the subloading surface model with tangential-strain rate and anisotropy. Int. J. Plasticity 20 (2004) 1909-1930

88. Hecker, S.S.: A simple forming limit curve technique and results on aluminum alloys. In: proc. of the IDDRG Congress, Amsterdam (1972) 5.1-5.8

89. Hecker, S.S.: Experimental studies of yield phenomena in biaxially loaded metals. In: Stricklin, J.A., Saczalski, K.H. (eds.): Constitutive Equations in Viscoplasticity: Computational and Engineering Aspects. ASME, New York (1976) 1-33

90. Hershey, A.V.: The plasticity of an isotropic aggregate of anisotropic face centred cubic crystals. J. Appl. Mech. 21 (1954) 241-249

91. Hill, R.: A theory of the yielding and plastic flow of anisotropic metals. Proc. Roy. Soc. London A193 (1948) 281–297

92. Hill, R.: On discontinous plastic states, with special reference to localized necking in thin sheets. J. Mech. Phys. Sol. 1 (1952) 19-30

93. Hill, R.: Theoretical plasticity of textured aggregates. Math. Proc. Cambridge Philos. Soc. 85 (1979) 179–191

94. Hill, R.: Constitutive dual potential in classical plasticity. J. Mech. Phys. Solids 35 (1987) 23–33

95. Hill, R.: Constitutive modelling of orthotropic plasticity in sheet metals. J. Mech. Phys. Solids 38 (1990) 405–417

96. Hill, R.: A user-friendly theory of orthotropic plasticity in sheet metals. Int. J. Mech. Sci. 35 (1993) 19–25

97. Hill, R., Hecker, S.S., Stout, M.G.: An investigation of plastic flow and differential work hardening in orthotropic brass tubes under fluid pressure and axial load. Int. J. Solids Struct. 31 (1994) 2999-3021

98. Hiroi T., Nishimura H.: The influence of surface defects on the forming-limit diagram of sheet metal. J. Materials Process. Techn. 72 (1997) 102–109

99. Hiwatashi, S., Van Bael, A., Van Houtte, P., Teodosiu, C.: Prediction of the forming limit strains under strain-path changes: application of an anisotropic model based on texture and dislocation structure. Int. J. Plasticity 14 (1998) 647-669

100. Hoferlin, E., Van Bael, A., Van Houtte, P., Steyaert, G., De Maré, C.: The design of a biaxial tensile test and its use for the validation of crystallographic yield loci. Modelling Simul. Mater. Sci. Eng. 8 (2000) 423-433

101. Hopperstad, O.S. et al.: A preliminary numerical study on the influence of PLC on the formability of aluminium alloys, In: Juster, N., Rosochowski, A. (eds.): Proc. 9$^{th}$ ESAFORM Conference on Material Forming. Glasgow, April 2006. AKAPIT, Krakow (2006) 315-318

102. Hora, P., Tong, L.: Prediction methods for ductile sheet metal failure using FE-simulation. In: Proc. of the IDDRG Congress. Lisbon (1994) 363-375

103. Hora, P., Tong, L., Reissner, J.: A prediction method for ductile sheet metal failure. In: Lee, J.K., Kinzel, G.L., Wagoner, R.H. (eds): Proc. of the NUMISHEET 1996 Conference, Dearborn (1996) 252-256

104. Hora, P., Krauer, J. (eds): Numerical and experimental methods in prediction of forming limits in sheet metal forming and tube hydroforming processes. FLC-Zürich 06 Conference, Zürich (2006)

105. Horstemeyer, M.F., Chiesa, M.L., Bamman, D.J.: Predicting FLDs with explicit and implicit FE codes. In: Proc. SAE Conference, Detroit (1994) 481-495

106. Hosford, W.F.: Texture strengthening. Metals. Eng. Quarterly 6 (1966) 13-19

107. Hosford, W.F.: A generalized isotropic yield criterion. J. Appl. Mech. Trans. ASME 39 (1972) 607–609

108. Hosford, W.F.: On yield loci of anisotropic cubic metals. In: Proceedings 7th North American Metalworking Conference., SME, Dearborn MI, (1979) 191-197

109. Hosford, W.F.: Comments on anisotropic yield criteria. Int. J. Mech. Sci. 27 (1985) 423-427

110. Hotz, W.: European efforts in standardization of FLC. In: Hora, P. (ed): Numerical and experimental methods in prediction of forming limits in sheet forming and tube hydroforming processes. ETH Zürich, Zürich (2006) 24-25

111. Hu, Z., Rauch, E.F., Teodosiu, C.: Work hardening behavior of mild steel under stress reversal at large strains. Int. J. Plasticity 8 (1992) 839-856

112. Hu, J.G. et al.: Influence of damage and texture evolution on limit strain in biaxially stretched aluminium alloy sheets. Materials Science Eng. A251 (1998) 243-250

113. Huang, H.M., Pan, J., Tang, S.C.: Failure prediction in anisotropic sheet metals under forming operations with consideration of rotating principal stretch directions. Int. J. Plasticity 16 (2000) 611-633

114. Ikegami, K.: Experimental Plasticity on the Anisotropy of Metals. In: Boehler, J.P., (ed.): Mechanical Behavior of Anisotropic Solids. Proceedings of the Euromech Colloquim 115, Colloques Inter. du CNRS, Paris (1979) 201-242

115. Inal, K., Neale, K.W., Aboutajeddine, A.: Forming limit comparisons for FCC and BCC sheets. Int. J. Plasticity 21 (2005) 1255–1266

116. ISO 12004: Metallic materials-sheet and strip-Determination of the forming limit curves. (2006)

117. Iwata, N., Matsui, M., Kato, T., Kaneko, K., Tsutamori, H., Suzuki, N., Gotoh, M.: Numerical prediction of spring-back behavior of a stamped metal sheet by considering material nonlinearity during unloading. In: Mori, K. (ed.): Proc. 7th Int. Conf. Numerical Methods in Industrial Forming Processes, Balkema (2001) 693

118. Janssens, K., Lambert, F., Vanrostenberghe, S., Vermeulen, M.: Statistical evaluation of the uncertainty of experimentally characterised forming limits of sheet steel. J. Materials Process. Techn. 112 (2001) 174-184

119. Jurco, P., Banabic, D.: A user-frienldy programme for calculating Forming Limit Diagrams. In: Banabic, D. (ed): Proc. 8th ESAFORM Conference on Material Forming. Cluj-Napoca April 2005. The Publishing House of the Romanian Academy, Bucharest (2005) 423-427

120. Karafillis, A.P., Boyce, M.C.: A general anisotropic yield criterion using bounds and a transformation weighting tensor. J. Mech. Phys. Solids 41 (1993) 1859-1886

121. Kim, D., Barlat, F., Bouvier, S., Rabahallah, Balan, T., M., Chung, K.: Non-quadratic anisotropic potential based on linear transformation of plastic strain rate. Int. J. Plasticity (2007) (in press)

122. Kim, K.J. et al.: Formability of AA5182/polypropilylene/AA5182 sandwich sheets. J. Materials Process. Technol. 139 (2003) 1-7

123. Knockaert, R. et al.: Forming limits prediction using rate-independent polycrystalline plasticity. Int. J. Plasticity 16 (2000) 179-198

124. Kobayashi ,T., Ishigaki, H., Tadayuki, A.: Effect of strain ratios on the deforming limit of steel sheet and its application to the actual press forming. In: Proc. of the IDDRG Congress, Amsterdam (1972) 8.1-8.4

125. Kreißig, R., Schindler, J.: Some experimental results on yield condition in plane stress state. Acta Mech. 65 (1986) 169-179

126. Kuroda, M., Tvergaard, V.: Use of abrupt strain path change for determining subsequent yield surface: illustrations of basic idea. Acta Mater. 47 (1999) 3879-3890

127. Kuroda M., Tvergaard V.: FLD for anisotropic metal sheets with different yield criteria. Int. J. Solids Struct. 37 (2000) 5037-5059

128. Kuroda M., Tvergaard V.: Effect of strain path change on limits to ductility pf anisotropic metal sheets. Int. J. Mech. Sciences 42 (2000) 867-887

129. Kuroda, M.: Effects of texture on mechanical properties of aluminium alloys sheets and texture optimization strategy. In: Smith, L.M., Pourboghrat, F., Yoon, J.-W., Stoughton,T.B. (eds): Proc. of the NUMISHEET 2005 Conf. AIP (2005) 445-450

130. Kuwabara, T., Ikeda, S., Kuroda, T.: Measurement and Analysis of Differential Work hardening in Cold-Rolled Steel Sheet under Biaxial Tension. J. Materials Process. Technol. 80-81 (1998) 517-523

131. Kuwabara, T., Kuroda, M., Tvergaard, V., Nomura, K.: Use of abrupt strain path change for determining subsequent yield surface: experimental study with metal sheets. Acta Mater. 48 (2000) 2071-2079.

132. Kuwabara, T., Nagata, K., Nakako, T.: Measurement and analysis of the Bauschinger effect of sheet metals subjected to in-plane stress reversals. In: Torralba, J. M. (ed.): Proc. AMPT '01, Univ. Carlos III de Madrid, Madrid (2001) 407

133. Kuwabara, T., Van Bael, A., Iizuka, E.: Measurement and Analysis of Yield Locus and Work hardening Characteristics of Steel Sheets with Different R-values. Acta Mater. 50 (2002) 3717-3729

134. Kuwabara, T., Ishiki, M., Kuroda, M., Takahashi, S.: Yield Locus and Work-Hardening Behavior of a Thin-Walled Steel Tube Subjected to Combined Tension-Internal Pressure. Journal de Physique IV 105 (2003) 347-354

135. Kuwabara, T., Yoshida, K., Narihara, K., Takahashi S.: Anisotropic plastic deformation of extruded aluminum alloy tube under axial forces and internal pressure. Int. J. Plasticity 21 (2005) 101-117

136. Kuwabara, T.: Advances in experiments on metal sheets and tubes in support of constitutive modeling and forming simulations. Int. J. Plasticity 23 (2007) 385-419

137. Lademo, O.G., Berstad, T., Hopperstad, O.S., Pedersen, K.O.: A numerical tool for formability analysis of aluminium alloys. Steel Grips 2 (2004) Suppl. Metal Forming 2004, Krakow (2004) 427-437

138. Lademo O.G. et al.: Prediction of plastic instability in extruded aluminium alloys using shell analysis and a coupled model of elasto-plasticity and damage. J. Materials Process. Techn. 166 (2005) 247–255

139. Lee, D., Backofen, W.A.: An experimental determination of the yield locus for titanium and titanium-alloy sheet. Trans. TMS-AIME 236 (1966) 1077-1084

140. Lee W.B., Tai W.H., Tang C.Y.: Damage evolution and forming limit predictions of an AA2024-T3 aluminium alloy. J. Material Process. Techn. 63 (1997) 100-104

141. Lemaitre, J. (ed.): Continuous damage. In: Handbook of Materials Behavior Models, Academic Press, San Diego, CA, (2001) 411-793

142. Lewison, D.J., Lee, D.: Determination of Forming Limits by Digital Image Processing Methods. In: Proceedings of International Body Engineering Conference and Exposition (IBEC), Detroit (MI) (1999) (Paper 01-3168)

143. Li, S., Hoferlin, E., Van Bael, A., Van Houtte, P.: Application of a texture-based plastic potential in earing prediction of an IF steel. Adv. Eng. Materials (2001) 990-994

144. Liebertz,H. et al.: Guideline for the determination of forming limit curves.In: Proc. of the IDDRG Conference, Sindelfilgen (2004) 216-224

145. Lin, S.B., Ding, J.L.: Experimental study of the plastic yielding of rolled sheet metals with the cruciform plate specimen. Int. J. Plasticity 11 (1995) 583-604

146. Liu, C., Huang,Y., Stout, M.G.: On the asymmetric yield surface of plastically orthotropic materials: A phenomenological study. Acta Mater. 45 (1997) 2397-2406

147. Logan, R.W., Hosford, W.F.: Upper-bound anisotropic yield locus calculations assuming pencil glide. Int. J. Mech. Sci. 22 (1980) 419–430

148. Lou, X. Y., Li, M., Boger, R. K., Agnew, S. R., Wagoner, R. H.: Hardening evolution of AZ31B Mg sheet. Int. J. Plasticity 23 (2007) 44-86

149. Lowden, M.A.W., Hutchinson, W.B.: Texture strengthening and strength differential in titanium-6A-4V. Metall .Trans. 6A (1975) 441-448

150. Makinde, A., Thibodeau, L., Neale, K.W.: Development of an apparatus for biaxial testing using cruciform specimens. Exp. Mech. 32 (1992) 138-144

151. Maeda, Y., Yanagawa, M., Barlat, F., Chung, K., Hayashida, Y., Hattori, S., Matsui, K., Brem, J.C., Lege, D.J., Murtha, S.J., Ishikawa, T.: Experimental analysis of aluminum yield surface for binary Al-Mg alloy sheet samples. Int. J. Plasticity 14 (1998) 301-318

152. Marciniak, Z., Kuczynski, K.: Limit strains in the processes of stretch forming sheet metal. Int. J. Mechan. Sciences  9 (1967) 609-620

153. Marron, G. et al.: A new necking criterion for the Forming Limit Diagrams, IDDRG 1997 WG Meeting, Haugesund (1997)

154. McDowell, D. L.: Modeling and experiments in plasticity. Int. J. Solids Struct. 37 (2000) 293-309

155. McGinty, R., McDowell, D.L.: Application of multiscale crystal plasticity models to FLD. Trans.ASME., J. Eng. Mater. Techn. 126 (2004) 285-291

156. Mellor, P.B.: Sheet metal forming. Int. Metals Review 1 (1981) 1–20

157. Methods of determining the forming limit curve. IDDRG Meeting, Zurich (1983)

158. Michno, M. J. Jr., Findley, W. N.: An historical perspective of yield surface investigations for metals. Int. J. Non-Linear Mech. 11 (1976) 59-82

159. Miyauchi, K.: Bauschinger effect in planar shear deformation of sheet metals. In: Advanced Technology of Plasticity, Proc. 1st Int. Conf. Technology of Plasticity, The Japan Society for Technology of Plasticity, Tokyo, (1984) 623

160. Müller, W., Pöhlandt, K., J.: New experiments for determining yield loci of sheet metal. J. Materials Process.Techn. 60 (1996) 643-648

161. Nakazima, K, Kikuma, T, Hasuka, K.: Study  on  the formability of steel sheets. Yawata Tech. Rep. No. 284 (1971) 678-680

162. Nandedkar, V.M, Narashimhan, K.: Prediction of forming limits incorporating work-hardening behavior. In: Gelin, J.C., Picart, P. (eds): Proc. of the NUMISHEET 1999 Conference, Besancon (1999) 437-442

163. Narashimhan, K., Wagoner, R.H.: Finite Element Modeling simulation of in-plane FLD of sheets containing finite defects. Metallurgical Trans. 22A (1991) 2655-2665

164. Paraianu, L., Banabic, D.: Calculation of Forming Limit Diagrams Using a Finite Element Model. In: Banabic, D. (ed.): Proc. 8th ESAFORM Conference on Material Forming. Cluj-Napoca, April 2005. The Publishing House of the Romanian Academy, Bucharest (2005) 419-423

165. Paraianu, L., Comsa, D.S., Gracio, J.J., Banabic, D.:  Influence of yield locus and strain-rate sensitivity on the Forming Limit Diagrams. In: Juster N., Rosochowski A. (eds.): Proc. 9th ESAFORM Conference on Material Forming, Glasgow, April 2006, The Publishing House AKAPIT, Krakow (2006) 343-346

166. Phillips, A.: A review of quasistatic experimental plasticity and viscoplasticity. Int. J. Plasticity 2 (1986) 315-328

167. Plunkett, B., Lebensohn, R.A., Cazacu, O., Barlat, F.: Anisotropic yield function of hexagonal materials taking into account texture development and anisotropic hardening. Acta Materialia 54 (2006) 4159-4169

168. Rajarajan G. et al.: Validation of the non-linear strain –path model CRACH to enhance the interpretation of FE simulations in multistage forming operations, In: Banabic, D. (ed.): Proc. 8th ESAFORM Conference on Material Forming. Cluj-Napoca, April 2005. The Publishing House of the Romanian Academy, Bucharest (2005) 387-390

169. Ragab A.R., Saleh, C.: Effect of void growth on the predicting forming limit strains for planar isotropic sheet metals. Mechanics of Materials 32 (2000) 71-84

170. Ragab, A.R., Saleh, C., Zaafarani, N.N.: Forming limit diagrams for kinematically hardened voided sheet metals. J. Materials Process. Techn. 128 (2002) 302–312

171. Rechberger, F., Till, E.T.: Influence of scatter of materials properties on the formability of parts. In: Kergen, R. (ed): Forming the future, Proc. IDDRG 2004 Conference, Sindelfingen (2004) 236-245

172. Savoie J. et al.: Prediction of the FLD using crystal plasticity model. Materials Science Eng. A257 (1998) 128-133

173. Shakeri, M., Sadough, A., Dariani, B.M.: Effect of pre-straining and grain size on the limit strains in sheet metal forming. Proc.Instn. Mech. Engrs. 214B (2000) 821-827

174. Schatz, M., Keller, S., Feldmann, P.: Experimental determination of the FLD for sheet thickness from 2.5 to 5.0 mm (in German). UTF Science III (2005) 1-8

175. Shiratori, E., Ikegami, K.: Experimental study of the subsequent yield surface by using cross-shaped specimens. J. Mech. Phys Solids 16 (1968) 373-394

176. Spitzig, W.A., Richmond, O.: The effect of pressure on the flow stress of metals. Acta Metall. 32 (1984) 457-463

177. Stout, M. G., Kocks, U. F.: Effects of Texture on Plasticity. In: Kocks, U.F., Tomé, C.N., Wenk, H.-R. (eds.): Texture and Anisotropy, Cambridge University Press, Cambridge, (1998) 420-465

178. Stoughton, T.B.: A general forming limit criterion for sheet metal forming. Int. J. Mech. Sci. 42 (2000) 1-27

179. Stoughton, T.B., Zhu, X.: Review of theoretical models of the strain-based FLD and their relevance to the stress-based FLD. Int. J. Plasticity 20 (2004) 1463-1486

180. Stoughton, T.B., Yoon, J.W.: Sheet metal formability analysis for anisotropic materials under non-proportional loading. Int. J. Mech. Sci. (in press)

181. Stören, S., Rice, J.R.: Localized necking in thin sheets. J. Mech. Phys. Solids 23 (1975) 421-441

182. Strano, M., Colosimo, B.M.: Logistic regression analysis for experimental determination of forming limit diagrams. Int. J. Machine Tools Manuf. 46 (2006) 673-682

183. Strano, M., Colosimo, B.M.: Ordinal logistic regression analysis for statistical determination of forming limit diagrams. In: In: Juster N., Rosochowski A. (eds.): Proc. 9th ESAFORM Conference on Material Forming, Glasgow, April 2006, The Publishing House AKAPIT, Krakow (2006) 303-306

184. Swift, H.W.: Plastic instability under plane stress. J.Mech. Phys.Sol. 1 (1952) 1-16

185. Szczepiński, W. (ed.): Experimental Methods in Mechanics of Solids, Elsevier, Amsterdam (1990)

186. Tai, W.H., Lee, W.B.: Finite element simulation of in plane forming processes of sheets containing plastic damage. In: Lee, J.K., Kinzel, G.L., Wagoner, R.H. (eds): Proc. of the NUMISHEET 1996 Conference, Dearborn (1996) 257-261

187. Takashina, K. et al.: Relation between the manufacturing conditions and the average strain according to the scribed circle tests in steel sheets. La Metallurgia Italiana 8 (1968) 757-765

188. Teixeira, P. et al.: Finite element prediction of fracture onset in sheet metal forming using a ductile damage model. In: Proc. of the IDDRG 2006 Conference, Porto (2006) 239-245

189. Teodosiu, C., Hu, H.: Microstructure in the continuum modeling of plastic anisotropy. In: Shen, S., Dawson, P.R. (eds.): Proc. of the Conference, NUMIFORM'95 on Simulation of Materials Processing, Theory, Methods and Applications, Balkema, Rotterdam (1995) 173

190. Tozawa, Y.: Plastic deformation behavior under conditions of combined stress. In: Koistinen, D.P., Wang, N-.M. (eds.): Mechanics of Sheet Metal Forming. Plenum Press, New York (1978) 81-110

191. Van der Boogaard, A.H., Huetink, J.: Prediction of sheet necking with shell finite element models. In: .Brucato, V. (ed.): Proc. 6th ESAFORM Conference on Material Forming, Salerno, April 2003, Nuova Ipsa Editore, Palermo (2003) 191-194

192. Van Houtte, P., Toth L.S.: Generalization of the Marciniak-Kuczynski defect model for predicting FLD. In: Lee, W.B. (ed.): Advances in Engineering Plasticity and its Application, Elsevier, Amsterdam (1993) 1013-1020

193. Van Houtte, P.: Application of plastic potentials to strain rate sensitive and insensitive anisotropic materials. Int. J. Plasticity 10 (1994) 719-748

194. Van Houtte, P.: Yield loci based on crystallographic texture. In: Lemaitre, J. (ed.): Handbook of Materials Behavior Models, Academic Press, San Diego, CA (2001) 137-154

195. Van Houtte, P.: Anisotropy and formability in sheet metal drawing, , In: Banabic, D. (ed.): Proc. 8th ESAFORM Conference on Material Forming. Cluj-Napoca, April 2005. The Publishing House of the Romanian Academy, Bucharest (2005) 339-342

196. Veerman, C. et al.: Determination of appearing and admissible strains in cold-reduced sheets. Sheet Metal Industries (1971) 687-694

197. Vegter, H., An, Y., Pijlman, H.H., Huetink, J.: Different approaches to describe the plastic material behaviour of steel and aluminium-alloys in sheet forming. In: Covas, J.A. (ed.): Proc. of the 2nd ESAFORM Conference on Material Forming. Guimaraes (1999) 127–132

198. Vegter, H., van den Boogaard, A.H.: A plane stress yield function for anisotropic sheet material by interpolation of biaxial stress states. Int. J. Plasticity 22 (2006) 557–580

199. Viatkina, E.M. et al.: Forming Limit Diagrams for sheet deformation process: a crystal plasticity approach. In: Habraken, A.M. (ed.): Proc. of the 4[nd] ESAFORM Conference on Material Forming, Liege (2001) 465-468

200. Volk, W.: New experimental and numerical approach in the evaluation of the FLD with the FE-method. In: Hora, P. (ed): Numerical and experimental methods in prediction of forming limits in sheet forming and tube hydroforming processes. ETH, Zürich, (2006) 26-30

201. Vos, M., Banabic, D.: The Forming Limit Band – a new tool for increasing the robustness in the simulation of sheet metal forming processes. Proc. of the IDDRG 2007 Conference, Gyor (2007) (in press)

202. Wagoner, R.H., Chan, K.S., Keeler, S.P. (eds): Forming Limit Diagrams: Concepts, Methods, and Applications. TMS, Warrendale (1989)

203. Weinmann, K.J., Rosenberger, A.H., Sanchez, L.R.: The Bauschinger effect of sheet metal under cyclic reverse pure bending. Ann. CIRP 37 (1988) 289-293

204. Wu, P.D., Neale, K.W., Van der Giessen, E.: On crystal plasticity FLD analysis. Proc. R. Soc. London 453(1997) 1831–1848

205. Wu, P.D. et al.: Crystal plasticity FLD analysis of rolled aluminium sheets. Metallurgical Trans. 29A (1998) 527-535

206. Wu, P.D., MacEwen, S.R., Lloyd, D.J., Neale, K.W.: A mesoscopic approach for predicting sheet metal formability. Model. Simul. Mater. Sci. Eng. 12 (2004) 511–527

207. Wu, P.D., Graf, A., MacEwen, S.R., Lloyd, D.J., Jain, M. Neale, K.W.: On forming limit stress diagram analysis. Int. J. Solids Struct. 42 (2005) 2225-2241

208. Xu, Y.: Modern Formability: Measurement, Analysis and Applications. Hanser Gardner Publications (2006)

209. Yao, H., Cao, J.: Prediction of FLC using an anisotropic yield function with prestrain induced prestress. Int. J. Plasticity 18 (2002) 1013-1038

210. Yoon, J.W., Barlat, F., Dick, R.E., Chung, K., Kang, T.J.: Plane stress yield function for aluminum alloy sheets–Part II: FE formulation and its implementation, Int. J. Plasticity 20 (2004) 495-522

211. Yoon, J.W., Barlat, F., Dick, R.E., Karabin, M.E.: Prediction of six or eight ears in a drawn cup based on a new anisotropic yield function. Int. J. Plasticity 22 (2006) 174-193

212. Yoon, J.W., Barlat, F.: Modeling and simulation of the forming of aluminium sheet alloys, In: Semiatin, S.L. (ed): ASM Handbook, Vol 14B, Metalworking: Sheet forming, ASM International, Materials Park, OH (2006) 792-826

213. Yoshida, F., Urabe, M., Toropov, V.V.: Identification of material parameters in constitutive model for sheet metals from cyclic bending tests. Int. J. Mech. Sci. 40 (1998) 237-249

214. Yoshida, F., Uemori, T., Fujiwara, K.: Elastic–plastic behavior of steel sheets under in-plane cyclic tension–compression at large strain. Int. J. Plasticity 18 (2002) 633-659

215. Yoshida, K., Kuwabara, T., Narihara, K., Takahashi S.: Experimental verification of the path-independence of forming limit stresses. Int. J. Forming Processes 8 (SI) (2005) 283-298

216. Yoshida, K, Kuwabara, T., Kuroda, M.: Path-dependence of the forming limit stresses in a sheet metal, Int. J. Plasticity 23 (2007) 361-384

217. Yoshida, K., Kuwabara, T.: Effect of strain hardening behavior on forming limit stresses of steel tube subjected to non-proportional loading paths. Int. J. Plasticity (2007) (in press)

218. Yu, M.H.: Advances in strength theories for materials under complex stress state in the 20th Century. Appl. Mech. Rev. 55 (2002)198-218

219. Zhou, D., Wagoner R.H.: Use of arbitrary yield function in FEM. In: Boehler, J.P., Khan, A.S. (eds): Anisotropy and localization of plastic deformation. Elsevier, Amsterdam (1991) 688-691

220. Zhou, Y., Neale K.W.: Predictions of FLD using a rate sensitive crystal plasticity model. Int. J. Mech. Sciences 37 (1995) 1-20

221. Życzkowski, M.: Combined Loadings in the Theory of Plasticity, Polish Scientific Publisher, Warsaw (1981)

222. Życzkowski, M.: Anisotropic yield condition. In: Handbook of Materials Behaviour Models. Lemaitre, J. (ed.): Academic Press, San Diego CA (2001) 155-165

# Sheet Metal Forming

Torgeir Welo[1]

[1] Norwegian University of Science and Technology (NTNU), Faculty of Engineering Science and Technology, Department of Engineering Design and Materials, Richard Birkelands veg 2B, N-7034 Trondheim, Norway
torgeir.welo@ntnu.no

**Abstract.** Sheet metal forming represents an extensive research area involving numerous topics such as finite element techniques and modeling, material models, material characterization, formability, component and full-scale test methods as well as a variety of industrial processes and technologies. In the present overview, however, attempts have been made to mainly concentrate on recent technology development and innovative sheet metal forming methods, representing a summary of works presented at ESAFORM conferences during the past decade. The first objective is to demonstrate advances in sheet metal forming by specific examples with different degree of industrial readiness—from new process ideas to fully industrialized processes. The second objective is to give a brief discussion on marked and industrial trends, forming the basis and direction for future technology development and associated research within more basic fields in sheet metal forming.

**Keywords:** Sheet metal forming, state-of-the-art, new technology, innovative methods, flexible and reconfigurable methods, adaptive forming, measurement, sensing, market trends, industrial needs, future research focus.

## 1  Introduction

Sheet metal forming represents probably the research area within metal forming that has been given the majority of attention from academia during the past decade. Dealing with the entire research field of advances in metal forming would be impossible within the scope of this chapter. It has, therefore, been chosen to focus on the more technology related developments that have been presented during the history of ESAFORM. The motivation for this is to provide an overview of the more innovative developments, less known to the broad spectrum of readers, rather than covering the majority of works presented at earlier conferences. This means that more basic articles on traditional sheet metal forming research have purposely been omitted, and the reader is strongly recommended to consult the ESAFORM proceedings or other chapters in the book to get a total overview of related topics. Before going more into details on advances in sheet metal forming technologies, some *key* references for traditional sheet metal forming topics presented at ESAFORM conferences in the period 1998-2006 are summarized below.

**Forming Limit Predictions:** Takuda, H, [1]; Xu, F., Lin, Z.Q. [2]; Banabic, D., Comsa, D.S., Jurco, P., Wagner, S., He, S., Van Houtte, P. [3]; Jaspart, O., François, A, Magotte, O., D'Alvise, L.,[4]; Situ, Q., Jain, M., Bruhis, M., [5].

**Material Models and Yield Surface:** Paraianu, L., Comsa, D.S., Cosovici, G., Jurco, P., Banabic, D.,[6]; Pöhlandt, K., Banabic, D., Lange, K., [7]; Van Haaren, L., van den Boogaard, A.H., Huétink, J., [8]; Teodosiu, C., [9]; Racz, G., Lemoine, X., Haddag, B., Abed-Meraim, F., [10]; Cosovici, G.A., Banabic, D., [11]; Cazacu, O., Plunkett, B., Barlat, F.,[12].

**Springback Compensation in Sheet Metal Forming:** Lenoir, H., Boudeau, N., [13]; Cao, J, Cheng, H.S., [14]; Lingbeek, R., Meinders, T., Ohnimus, S., Petzoldt, M., Weiher, J., [15]; Reese, S., Vladimirov, I.N.,[16].

**Friction and Tool Surfaces:** Felder, E, Devine, I., [17]; Zmudzki, A., Papaj, M., Kuziak, R., Kusiak, J., Pieterzyk, M., [18]; Mkaddem, A., Bahloul, R., Potiron, A., [19]; Carrino, L., Di Meo, N., Sorrentino, L., Strano, M. [20]; Liewald, M., Wagner, S., Becker, D., New [21].

**FEA Techniques and Modeling:** Daniel, J.L., Boubakar, L., Boisse, P, Gelein, J.C, [22]; Onate, E., Zarate, F., Rojek, J., Jovicevic, J., [23]; Haddad, A., Balan, F., Abed-Meraim, [24]; Forestier, R., Ben Tahar, M., Chastel, Y., Massoni, E., [25]; Chenot, J.-L., [26].

## 1.1  Scope of Work and Organization

In the present chapter, attention has been placed on recent advances in sheet forming technology, focusing on new and innovative methods with high industrial potential. The chapter is organized as follows:

In the first part, a number of promising new sheet forming methods and associated technology developments are presented. The former includes methods such as flexible rolling of sheet, sheet incremental forming, hydroforming, hydromechanical forming, mechanical calibration, flexible-reconfigurable forming processes and forming methods at elevated temperatures. This section also gives a brief introduction to new process related developments, including adaptive approaches, sensing and measurement technologies.

The second part of this chapter presents a discussion on future needs for development of sheet metal forming technology. With market and industry trends as a starting point, these lead to a certain direction—or, say, road map—for future sheet metal forming technology development, which in turn leads to specific needs within more basic sheet metal research.

# 2  Advances in Sheet Metal Forming Technology

## 2.1  Flexible Rolling of Sheet Metal

A new so-called flexible rolling technology for tailor rolled strips was introduced by R. Kopp and P. Böhlke [27] at ESAFORM in 2004. The process provides the capability of producing multiple thickness sheets across the width of the rolled blanks. This technology is of particular interest to applications where weight savings and functional integration are of primary focus, e.g. in the automotive industry.

An experimental rolling setup was established using a moveable crosshead with a profiled roll integrated into the rolling mill. One of the main challenges associated with this technology is to achieve a well defined material flow across the width of the strip to avoid flatness defects. The main controlling factors are roll geometry and roll system arrangement along with the definition of an optimal rolling strategy in every pass to achieve the desired final shape of the strip.

The results indicated that process is particularly suited for high strength materials, small to medium widths and moderate wall thickness reductions. The experimental trials demonstrated that 50 % wall thickness reductions could be achieved with acceptable dimensional tolerances. The further work in order to develop and industrialize this technology involves process optimization and development of the rolling device to minimize the number of passes.

## 2.2 Adaptive Forming

Controlling springback represents a great challenge in sheet metal forming operations. In this connection, there are two separate issues that have to be dealt with: (1) dimensional targeting, i.e. designing a tool (and process) with the capability of producing a dimensionally correct part under normal working conditions, nominal blank geometry and mechanical properties; (2) dimensional accuracy, i.e. dealing with the variability in the process, blank geometry and mechanical properties in such a way that the dimensional characteristics of the formed product are within the drawing tolerances (GD&T).

Different strategies do exist to reduce the dimensional variability of sheet metal forming parts. One of these strategies is to introduce in-line adaptive control to the process where measurable process characteristics are registered and the process parameters are adjusted accordingly.

At ESAFORM 2004, P. Sun, J.A. Ferreira, J.J. Gracio [28] presented a new approach for real time control of springback in sheet metal forming. By introducing pressure sensors and position sensors directly to the press equipment, real time measurements of springback could be displayed directly on the control interface of the press. The closed loop system consisted of two pressure sensors installed in the press cylinder, a position sensor installed on the punch rod, two real time hardware cards, hydraulic press and a computer.

The basis philosophy of the system is based on the assumption that the stamping process can be divided into three different main steps: (a) drawing and forming, (b) unloading where springback takes place, and (c) release of punch where it is completely removed from work piece. By measuring the net force (compensated for friction, gravity, etc.) at transition stage (a)-(b) and the corresponding position of the punch rod, this was compared to the similar quantities measured at transition stage (b)-(c). Hence, the displacement over which the force drops from the level recorded at the beginning of unloading down to minimum (net) force as deduced from pressure sensors in the press cylinder represent the springback in terms of change of position.

The technology has been implemented in industrial practice to automatically inspect springback in sheet metal forming in real time. The method is being used to adjust die configuration to compensate for variations in mechanical properties and material thickness, hence providing a fully automated monitoring system for improved quality control. Further work includes establishing a more accurate control model for springback than the linear one used in this work.

Computer aided process control is an important feature in the manufacturing of sheet metal parts. Material parameters and the initial thickness of the sheet metal are the two most important factors with respect to force characteristic, final shape of the part and springback upon unloading. Hence, variations in material and thickness between batches and within a single batch greatly influence the quality of the final product.

Several strategies for automatic process control have been developed to compensate for variability, hence improving the accuracy of formed components. The most common methods are:

- Indirect measurement of material characteristics through 'easily' accessible and measurable process parameters such as force and punch position, which are used in combination with results from FEA process simulation programs (steering models) to control the forming process

- Direct, in-process measurements of bend angle used for compensation of variability occurring between parts. The in-line measurements are utilized in combination with repeated loading and unloading cycles until the desired shape of the part is achieved. The advantage of the method is that it requires a minimum of measured parameters from the process; its drawback is the increase in cycle time and hence cost due to the number of time-consuming loading an unloading steps required.

A new adaptive springback compensation method for air bending of sheet metal parts was proposed by Ridane, N., *et al.* at Esaform 2004 [29] to reduce production cycle time. The strategy is based on using measured data from single partial unloading operation in combination with semi-analytical simulation to directly achieve the desired part shape after unloading. A closed loop control system takes over the process control from the DNC of the press brake just before the final bend angle ($\alpha_f$) is reached. At this bend angle ($\alpha_{pu} < \alpha_f$), an unloading procedure is initiated. Within this step the sheet is unloaded until a certain 'residual' force $F_{pu}$ is reached, which is measured along with the associated change in bending angle. The recorded data (unloading slope and force level) are then utilized together with semi analytical calculations (predicted force at instance of final unloading) to estimate what would have been the angle if the unloading had been taken place at $\alpha_{pu}$ instead $\alpha_f$. Combining the measured and predicted values, the final bend angle providing the correct released bend angle after unloading is calculated. After the determination of the expected springback for the final unloading step, a new control angle for the subsequent over-bending is initiated by the control system.

The experimental results obtained for different materials (St37, X5CrNi1810 and AlMg3) and sheet thicknesses showed that the average deviation between the desired bend angle and the released bend angle as deduced by this adaptive compensation procedure was in the range of 0.4°, and less than 0.85° in all cases. It is conclude that the new in-process control system allows a significant improvement of the bend angle accuracy for the investigated air bending process. The procedure provides the capability of reducing the cycle time and hence the cost of time consuming trial and error. The results show that the developed system works well for different thicknesses, bending conditions as well as strain hardening rates.

## 2.3 Sensing and Measurement Technologies

High precision measurements and advanced technology become increasingly important to monitor the deformation process in metal forming operations. Since most metal forming is done within a set of dies, the deformation of the work piece can only be monitored from other accessible information such as force and die displacement. In most of the works reported in the literature, sensing techniques are based on combining information from theoretical models (analytical and FEM) and process monitoring.

At ESAFORM 2005 Yang, M., and Koyama, J. [30] presented a new die-embedded sensing system with micro semi-conductor sensors fabricated using semiconductor process technology. The sensing system was used to measure complex phenomena by monitoring die stresses in a V-bending process. Micro sensors were used to measure force and bending angle during forming. The micro sensor chip was embedded as close as possible to the contact surface of the die. Information received from four strain components at different locations was used in an elastic mechanics model to predict force and bend angle. Comparing the bending angle obtained by the die embedded sensing system with the bending angle obtained by precise bending angle sensors, the agreement was found to be quite good. The slight discrepancy was explained by variations in friction during the bending process as well as the distance between the sensors, providing less accurate measurements of those sensors located far from the contact area between the work piece and the tool. It was concluded that the new sensing system is favorable compared to more conventional sensors.

## 2.4 Sheet Incremental Forming

Sheet incremental forming is a relatively new forming process introduced to produce low volume bathes (e.g. prototypes) at a reasonable cost level and short lead time. Sheet incremental forming is based on a methodology of producing the desired shape of the part through the progressive movement of a punch, following a predetermined tool path.

At Esaform 2004 Giardini, C. *et al.* [31] presented an intensive experimental and numerical study of sheet incremental forming using a general purpose CNC machine equipped with a special tool, forming the sheet with different die geometries. The main challenge associated with this process is to produce parts with sufficient dimensional accuracy and quality. In the present study, surface roughness and minimum thickness of the formed sheet (AISI 304) were considered for different tool velocity, tool path and steps (incremental movement of punch).

The results show that the tool path, rather than tool velocity, is the more important parameter to obtain the desired geometry of the formed part. On the other hand, the spiral step path does not seem to influence the surface roughness. In the tests, four different multistage paths were tested, all following a spiral path. In the first tool path (1), a simple pressure was applied in the centre of a tool designed to form a square cup. The pressure was increased until a maximum depth of 15 mm was obtained, followed by a spiral trajectory. In the other three cases (2-4), the formed depth of the part was gradually increased, followed by a spiral path at each incremental depth. Comparing the cases (1) with (2-4) shows that the latter provides a more optimal condition with respect to uniform thickness distribution and minimum thickness, whereas (1) improves the centripetal material flow, allowing for larger maximum punch depth without fracture.

The suggested further work involves studies of the impact of other parameters, including lubrication, friction and punch dimensions, combining FEA (DEFORM 3D) with experiments to reduce the number of physical tests. It is concluded that incremental forming is a flexible forming method particularly suited for prototypes and production of low volume bathes at a reasonable cost.

Single point incremental forming was also considered by Duflou, J. R. and Tunçkol, Y. at ESAFORM 2006 [32]. The aim of the present study was to study the influence of the boundary conditions and part geometry on forming force required for process planning purposes. The experiments using aluminum alloy Al 3003-O temper sheet were carried out in a 4-axes milling machine with a spherical tool.

Force measurements in single point incremental forming were also considered by Bologa, O., Oleksik, V. and Racz [33] in their paper presented at Esaform 2005. Their primary focus was to determine the influence of various process parameters, such as blank thickness, forming height of part and punch diameter on different force components ($F_x$ and $F_z$). They found that the primary influential parameters for $F_z$ are blank thickness and part height.

Ambrogio, G. *et al.* [34] presented a study of the influence of relevant process and material parameters on formability of the part in sheet incremental forming. In this work they found a positive effect of punch speed rotation on the formability, most likely as a result of local heating due to friction at the interface between the punch and the part. Moreover, they also found that the formability decreases with increasing tool pitch, typically in the range of 20-50 % when going from 1 mm to 3 mm tool pitch, depending on rotation speed. Reduced tool radius seemed to have a positive effect on part formability.

## 2.5 High-Precision Forming Processes

One of the major challenges in classical deep drawing processes is obtaining high drawing limit ratios without the occurrence of fracture or wrinkling of the part being formed. In order to improve the dimensional accuracy and reduce the weight of deep drawn parts, new innovative processes have been proposed. An example of such processes is hydromechanical deep drawing, combining conventional deep drawing with simple hydroforming technology. The process set-up typically consists of a blank holder, die, punch and a cavity with pressurized water (or oil) underneath the blank. Forming is performed by forcing the blank into the cavity as the pressure gradually increases. In addition to improving dimensional accuracy, hydromechanical forming is known to reduce residual stresses and make it possible to form complex parts in one single step. In order to increase the industrial application of hydromechanical forming, it is necessary to improve the understanding of various process parameters on product quality. One of the first attempts of using FEA to improve the understanding of the hydromechanical forming was presented by Chouati, O, Gelin, J.C. and Baudet, C. [35] at the first ESAFORM conference in 1998. In their work, focus was placed on the frictional condition between the tool and the blank using a sliding type Coulomb friction model in the analyses.

Moshfegh, R, *et al.* [36] presented a numerical and experimental study of the impact of material model (Hill and Barlat) on thinning in hydromechanical and conventional deep drawing using the explicit code LS-Dyna. Their results showed good correlation between numerical and experimental results. Reference is made to Jensen, M.R, Olovsson, L, and Danckert, J. [37] for modeling of pressure distribution in hydromechanical deep drawing.

In the automotive industry, hydroforming is widely used to produce hoods, panels, frame members, etc. The main advantage of hydroforming is its capability of producing parts with tight dimensional accuracy and the abilities of providing weight saving by part consolidation and more structural integrity than, say, conventional spot welded sheets. However, hydroforming requires large investments due the press size required and the relatively long cycle time compared with other stamping processes being widely used in the automotive industry. As a consequence double sheet hydroforming is an interesting process to cut the production time considerably.

At Esaform 2004, C. Giardini, E. Ceretti and C. Conti [38] presented a numerical and experimental study on double sheet hydroforming, hence obtaining two parts with good finishing in the same time that would normally be required for one single part. The experimental setup consists of a die, a bottom sheet, an upper sheet and a blank holder combined with pressurized fluid to obtain the desired geometry. Due to the nature of the process, i.e. placing two sheets on top of each other, there will be a nominal difference in geometry between the two parts, but the final part quality can be obtained in subsequent (less expensive) trimming or calibration processes.

Different materials, part geometries and operating conditions were tested and compared with numerical simulations using the commercial software PAM-STAMP. The results showed good correlation between experimental and numerical thickness distributions across the formed parts. The numerical simulations were also used to determine optimal process conditions (e.g. operating pressure), providing the desired part geometry without tendencies for fracture. It was concluded that the double sheet direct hydroforming process is a promising, low cost alternative process.

Aluminum extrusions are widely used in the automotive industry to reduce the weight of individual components. Examples are bumper beams, crash boxes, subframe, cockpit frames, door reinforcement beams, etc. Due to the higher cost of aluminum than conventional steel, it is of particular importance to reduce the manufacturing cost of the former by effective and robust processes, hence minimizing the number of process steps as well as quality costs.

Rotary stretch bending is an effective forming process for open, tube or box type sections, providing low cycle times and tight dimensional tolerances. The basic principle of the process is clamping the profile's ends and obtaining the final part shape by simultaneously applying bending and stretching by rotating two semi-dies about their individual pivot points. Due to the two dimensionality (constant section along the length) of aluminum extrusions, local changes in the cross section have normally to be done in subsequent forming operations.

At Esaform 2004, Bjørkhaug, L. and Welo, T. [39] introduced a new forming process were local reforming of the section was integrated as a part of the bending process. During the bending operation, prior to unloading, imprints were formed along the length of the part by calibration punches. The main purpose of the local forming was to introduce functional surfaces requiring tight dimensional control for mounting features or packaging reasons.

The results show that the introduction of an imprint along the length of the formed part prior to unloading dramatically reduces the subsequent global springback (chord height). Moreover, measuring the local springback of the external flange of the extrusion just beneath the imprint, demonstrated that local springback may in many cases exceed the global springback.


## 2.6 Flexible, Reconfigurable Forming Processes

Production equipment with the required flexibility to produce different part configurations without labor intensive re-tooling is becoming increasingly important. Steadily tougher product design requirements represent the driver, enforcing more dimensional precision along with flexibility in methodology and technology. In the market place, we see increased globalization and growing competition, resulting in increasing pressure to reduce costs and simultaneously reduce lot size and increase the number of variants. The automotive industry represents one typical example of this trend, continuously driving down their investment and hence the tooling cost of the supplier base.

One way of meeting this trend in terms of new bending technologies was presented at Esaform 2006 by Neugebauer, R., Putz, M. and Laux, G. [40] demonstrating the capabilities of the so-called HexaBend technology developed by the Frauenhofer Institute for Machine Tools and Forming Technology in cooperation with industrial partners. The equipment is particularly suited for pipes and profiles e.g. bending prior to hydroforming. The process is of kinematic nature and is equipped with a flexible die with 6 degrees of freedom, allowing three-dimensional parts to be produced in one single step. The tool setup consists of a moveable die relative to a fixed die called the bending head. The moveable die provided the flexibility of performing 3 translations and 3 rotations. Internal mandrel and mandrel balls may be used as optional equipment for tight radii bending of thin walled tubes.

Bending is performed by pushing the tube between the bending head and the moveable die, which was positioned at an offset, providing a bending moment along the tube between the two dies. A pusher is used to push the profile through the die assembly, forming an arc over the length of the profile. Adjustment of the die position relative to the bending head provides different arc configurations along the length of the bent product.

Comparing this bending technology to more conventional bending technology, like press and draw bending, it is obvious that the additional flexibility is obtained at the expense of the introduction of more influential parameters. Using FEA (ABAQUS Explicit), Neugebauer, R., Putz, M. and Laux, G. [40] therefore performed Statistical Test Planning to identify the primary parameters with respect to overall bend accuracy and distortions of the cross section. The results showed that the distance between the bending head and the moveable die (i.e. distance over which bending is applied upon forming) is the main parameter with respect to bend radius accuracy. Considering local geometry discrepancy, on the other hand, the diameter of the tube is the main influential parameter. Based on experience from the study, it is suggested that physical tests combined with more accurate numerical studies are required to obtain improved understanding of the significance of the different parameters involved.

Rotary draw bending is commonly used for forming of pipe and tubes, sometimes prior to hydroforming. The equipment can easily be reconfigured to other part geometries, bend angles and bend radii by relatively simple re-tooling at low costs. However, the main challenges associated with process control in rotary draw bending process are (a) springback compensation, (b) determination of the minimum bending radius without occurrence of fracture, (c) determination of the minimum bending radius associated with the risk of wrinkling (local defects) and (d) calculation of the flatting (ovalization) of the part as a function of part and tool geometry. In order to minimize the cost of reconfiguring the forming equipment from production of one part to another part geometry, it is extremely important to establish a methodology to reduce the trial and error. Process design is therefore a critical activity, requiring the selection of bending radii, bend angles, length of claming die, pressure of wiper die, mandrel geometry, etc. A detailed description of a methodology for retooling of rotary draw bending equipment was presented by Carrino, L and Strano, M. at ESAFORM 2003 [41]. In this work semi-analytical design formulae for calculation of springback, thinning, wrinkling and tube ovalization are presented for industrial use. The relationship between main influential parameters has been implemented into a spreadsheet and linked to process parameters (settings) of the bending equipment.

As mentioned above in the discussion of the rotary draw bending process, one of the main challenges in controlling the final geometry of the formed product is prediction of ovalization of the tube. Tube ovalization directly affects the local geometry of the part which may be important to functional requirements such as strength and fit-up. In addition, ovalization of the tube during bending impact the instantaneous stiffness and bending moment, hence the springback and the global part geometry after bending. It is, therefore, very important to take into effect ovalization in the design of the part as well as the process. Paulsen F. and Welo, T. [42] presented a work on analytical modeling of the tube ovalization at ESAFORM 2003, combining the deformation theory of plasticity with an energy method. The results - presented in terms of a closed form relationship between tube deformation, material and geometry parameters - showed that the mains factor with respect to tube ovalization are (in ranked order) (1) initial tube diameter, (2) tube thickness, (3) the bending radius and (4) the strain hardening coefficient of the material. Here, it is worth noting that, unlike proposed elsewhere in similar works, the diameter-to-thickness ratio ($D/t$) is not a relevant non-dimensional bendability parameter since $D$ is significantly more important to tube ovalization than $t$ in tube bending. The theoretical model was validated on experiments, and the agreement was excellent.

## 2.7 Forming at Elevated Temperatures

The introduction of additional safety and luxury features in modern vehicles has put pressure on auto makers to reduce the vehicle weight at a reasonable cost. The typical strategy is to obtain weight reductions by replacing traditional mild steel by high strength steel, aluminum or magnesium alloys. However, these 'new' materials typically provide significantly lower formability than mild steel, making the forming process more complicated and to some extent limiting the design freedom of the parts. In this connection, forming at elevated temperatures is an attractive process to achieve improved formability.

At ESAFORM 2003, Meinders, T., van den Boogaard, A.H., and Huetink, J., [43] presented a paper focusing on forming behavior of alternative materials for the automotive industry. Experiments showed that the drawing ratio of Al-Mg alloys at elevated temperatures could be improved up to the same levels as that of mild steel. In order to represent the material behavior correctly at elevated temperatures both strain hardening, strain rate and temperature effects as well as the bi-axial stress-strain response (Vegter yield criterion) of the sheet were included in the model. The applicability of the model was validated by experiments at various temperatures using a drawing ratio of 2.09 in all cases. Both the experiments and the numerical simulations clearly showed a more uniform thickness distribution along with less thinning in the middle of the deep drawing parts formed at elevated temperatures. The present work also discussed forming of thin-walled, high strength steel, focusing on the development of a new improved wrinkling indicator base on change in local curvature.

Magnesium alloy is one of the lightest materials for practical (volume) applications in the automotive industry. In addition to relatively high material cost, its relatively moderate formability at room temperature has up to now limited the use of magnesium alloy for high volume applications. Work reported in the literature has reported that magnesium has to be formed at temperatures above 200 °C to provide competitiveness in terms of formability.

An interesting study of magnesium alloyAZ31-0 formed at elevated temperatures was presented at ESAFORM 2004 by Yoshihara, S. *et al.* [44]. In this work they used the principles of local heating and cooling to study the formation of fracture and wrinkling in square cup drawing. A universal testing machine was used for controlling the displacement in the experiments conducted. The die and the blank holder were conducted by heaters to heat the blank. The actual position and load of the water cooled punch were recorded during deformation. A water cooling system was installed to directly cool the part in order to create a temperature profile across the part during forming.

The numerical results showed that the maximum draw height of the cup was 17 mm at room temperature. By providing a constant temperature (up to 300 °C ) across the part, the draw height was increased to 49 mm. Finally, by providing an optimal temperature profile over the formed part, the corresponding draw height was increased to as much as 64 mm, clearly demonstrating the industrial potential of this technology.

Another study of magnesium alloy forming at elevated temperatures was presented at ESAFORM 2006 by Bruni, C. *et al.* [45]. In the present work, the dimensional accuracy of AZ31 Mg alloy formed in V-bending was investigated. The final part geometry was discussed in terms of the impact of the temperature field during forming on springback as well as thermally induced deformations introduced during the cooling process.

Prior to forming, material characterizations were done to determine the stress-strain behavior at different temperatures (up to 300 °C ) and strain rates. In the bend tests, an induction coil was used to heat up the specimen to the desired temperature. A V- shaped punch used to bend the part into a U-shaped die at different bending angles ($\alpha$). The springback was evaluated in terms of the springback temperature ratio $K_T$ (released radius -to-bending ratio at given temperature) and the springback cooling ration $K_R$ (released radius to-bending ratio at a given cooling rate).

The results show that $K_T$ increases with increasing temperature, i.e. relative springback decreases, approaching a value close to 1.0 at 300 °C . The reason for this is that the yield stress of magnesium is relatively more sensitive to the temperature than is the Young's modulus. The results also showed that springback was more or less unaffected of the punch speed, even at elevated temperatures. It was also observed that $K_R$ is somewhat reduced when cooling from low temperatures (100 °C) compared to $K_T$, indicating increased springback from cooling.

The study concludes that both instantaneous material and expansion effects need to be considered for optimizing the design of the bending process for magnesium alloys.

Merklein, M. and Beccari, S., [46] compared the formability of aluminum alloy AA5182 and magnesium alloy AZ31 at temperatures between 100 °C and 300 °C using forming limit diagrams, Erichsen and Nakajima tests, and cup deep drawing tests. From the test results it was concluded that under similar testing conditions, the aluminum alloy always shows equal or better formability that the magnesium alloy, except at 350 °C. In Erichsen tests, AZ31 has to be heated up to temperatures in the range of 100 °C to 150 °C to provide formability similar to AA5182 at room temperature. However, the magnesium alloy shows a great improvement in formability when the temperature exceeds 200 °C.

Comparing the two materials in terms of Forming Limit Diagrams showed that AA5182 had the better formability at all temperatures with positive minor strain ($\varepsilon_2 > 0$); which was also confirmed in the Erichsen test. On the other hand, in case of negative minor strain ($\varepsilon_2 < 0$), the magnesium alloy showed the better formability of the two materials.

## 3. Future Need for Development of Technology and Capabilities within Sheet Metal Forming

### 3.1 Market and Industrial Trends

The future trends in different volume markets for sheet metal applications include increased focus on cost reduction, part consolidation, functional integration, tailor making and dimensional accuracy. The main driver for this is increased competition from low cost countries, enforcing high added value products and more automated production in order to stay competitive with products delivered from, say, Western Europe and US.

Cost reduction of sheet metal-based products can be approached by introducing more lean manufacturing concepts to increase the throughput and equipment efficiency in production. It is, however, also necessary to combine lean concepts by step changes in terms of new technology development and working modes; the latter typically calls for a better integration between (new) product and (new) process development.

In the automotive industry, for example, there is a continuous focus on reducing cost and vehicle weight by part consolidation, along with the use of more geometrically complex parts and larger modules reducing the assembly cost. This involves more outsourcing to system suppliers, which in turn put though requirements on technology know-how of the lower TIER part suppliers. There is also a trend in the automotive industry towards more frequent model changes along with more derivates within each model. One obvious consequence of this is increased need for flexible and reconfigurable production equipment, reducing the change-over cost between models and variations within a vehicle program. Another result of this is a need for reducing the cost of part specific tooling, which again would call for module based tool solutions, using modular inserts, for example.

The use of steadily more automated production lines, in order to increase production throughput and reduce labor cost, requires parts with improved dimensional accuracy. One factor which makes this even stronger is the need for light weighting–i.e., involving parts with more structural integrity (closed sections) in processing sensitive materials such as high strength steels, aluminum and magnesium alloys–which calls for high dimensional precision for, say, joining purposes. In order to keep up with such product requirements, manufacturing companies have to improve their production technology and increase their knowledge on through-processing factors affecting the dimensional accuracy of formed parts.

## 3.2 New Technology Development Needs

As discussed above, there is a strong market need to manufacture sheet metal-based products with more geometrical complexity and improved dimensional accuracy at a lower cost. At the same time, the manufacturing set-up has to be sufficient flexible and easily reconfigurable to accommodate more product variants with shorter life time.

These requirements can only be met by a simultaneous development of lean manufacturing concepts and new technology. Considering development of new technology for improved productivity and higher added value products, new product line concepts, involving fewer manufacturing steps and more flexible individual operations, have to be established. In sheet metal forming, this means that more complex forming has to be performed in each forming operation. In addition, the overall equipment efficiency (OEE) and part quality (PPM) has to be improved, despite the added complexity of the individual forming operation. It is, therefore, time to challenge the traditional way of doing sheet metal forming of conventional materials. New forming processes must provide the capability of robustly producing high quality parts without local (fracture, wrinkles, excessive thinning, buckling, distortions, etc) failures or global (springback) deviations. Again, this means that recently developed technology such as thermal forming, induction forming, hydromechanical forming, (low cost) hydroforming, rotary-stretch bending, etc. have to be developed further according to industrial needs. Forming technology of light weight ('engineered') materials such as high strength steels, aluminum and magnesium alloys has to be developed to the same level, providing the same shaping possibilities as forming methods for conventional mild steel.

Focus has to be placed on developing improved tools for process control to accommodate the needs for improved product quality. Methods such as statistical process control (SPC) to better utilize information from ongoing production in determining root causes for waste and downtime in production has to be established. In this connection, the approach can be divided into two directions: (1) improved measurement and sensor technology, making it possible to continuously and accurately register key production characteristics; (2) establish new tools for utilizing the information in improving running production with a system to feed back the data to product designers working on next generation products.

Adaptive processes have been discussed in the state-of-the-art review previously in this chapter. The ability of using adaptive processes as a mean to reduce production costs is becoming increasingly important in metal forming operations. Adaptive process control can be established at different levels — (i) using SPC data to define sufficiently robust production lines with capability to produce components with required quality; (ii) using recorded production data from previous bathes to run in production of following bathes based on a simple correction factors; (iii) closed loop automated feedback system where production characteristics are continuously recorded and converted, leading to an instantaneous correction of the forming cycle based on advanced steering models. There is a strong need to further develop adaptive forming technology together with statistical/FEA-base steering models and improved sensor technology. In a long term perspective this would put the manufacturing industry in a position to use shorter process routes together with simpler mechanical forming processes to produce components with better quality than before.

The quality (as determined from customer requirements) of a sheet metal component is always an accumulated result of individual processing taking place at different stages throughout the process chain. Here, the output (variations) from the previous operation serves as input (variation) to the next operation, which in turn based on processing parameters (variation) in this operation determine the input (variations) to the next, and so on. A general challenge in the sheet metal manufacturing industry is to determine the impact of variability in upstream operations, say two or three steps up the processing chain, on variability in a specific operation. In this connection, there is a need to further develop through-process models and technology, combining FEA, SPC and statistical tools into new effective tools for process control.

## 3.3 Fundamental Knowledge and Research Needs

The technology needs form the basis and the direction for establishing new basic knowledge. Obviously, FEA modeling is an important tool to gain insight into process parameter and their impact on product quality. Over the past decade, major progress has been made in developing FEA technology for solving industrial sheet forming problems.

In the future, the capabilities in using FEA as an industrial tool has to be strengthened further, in parallel with more basic development of numerical techniques and modeling methodology. In respect to the latter, traditional continuum-based yield surface representation needs to be more directly coupled with texture and microstructure-based approaches in order to improve the accuracy and reduce the need for material testing. More physically based   yield surfaces will be particularly useful in prediction of failure modes such as necking and fracture, especially if adaptive meshing techniques are developed in parallel.

There is also a strong need for coupling statistical methods with current FEA methods to be able to analyze problems related to process variability. Response surface methodology is one example of such coupled approaches currently in use. However, more attention has to be given to treating statistical dependencies in processes where there are coupling between different parameters. Moreover, physically based surrogate models, rather than simple 'curve fitting', have to be established, increasing the applicability of the response surface methodology. In this connection, inverse modeling is an a promising technique for establishing more physically-based surrogate model requiring a minimum of experimental data.

Resulting from an anticipated attention to thermally-based forming in the future, focus must be placed on thermo-mechanical modeling, involving more accurate constitutive models with strain rate and temperature dependency, along with development of numerical techniques with improved effectiveness for coupled problems. As material models are becoming more complex, new material characterization and testing techniques have to be established. This may be illustrated by the complexity of separating temperature effects from strain rate effects in obtaining material data for different metals. Another example is determination of the parameters needed to represent the yield surface and yield direction in complex stress states. Even with major progress in developing better material characterization and testing methods, these have to be combined with inverse modeling (FEA) techniques.

Development of new testing methods for component and prototype testing becomes increasingly important as the industry is striving for more innovative forming methods to accommodate market needs. The same comments apply to the continuously ongoing needs to verify material and FEA models. Experimental techniques will always form the basis for technology development and innovation in industry as well in academia.

# References


1. Takuda, H, Forming Limit Prediction of Sheet Metals by Means of Some Criteria for Ductile Fracture, ESAFORM 2003, pp. 171-174, Salerno, Italy, April 28-30, 2003.
2. Xu, F., Lin, Z.Q., A Fast Formability Evaluation for Al 6111-T4 by a One-Step Approach: Choice of Yield Theory, ESAFORM 2003, pp. 179-182, Salerno, Italy, April 28-30, 2003.
3. Banabic, D., Comsa, D.S., Jurco, P., Wagner, S., he, S., Van Houtte, P., Prediction of Forming Limit Curves from Two Anisotropic Constitutive Models, ESAFORM 2004, pp. 455-458, Trondheim, Norway, April 28-30, 2004
4. Jaspart, O., François, A, Magotte, O., D'Alvise, L., Numerical Simulation of the Stretch Forming Process for Prediction of Fracture and Surface Defect, ESAFORM 2004, pp. 499-502, Trondheim, Norway, April 28-30, 2004
5. Situ, Q., Jain, M., Bruhis, M., A New Approach to Obtain Forming Limits of Sheet Materials, ESAFORM 2006, pp. 299-302, Glasgow, United Kingdom, April 26-28, 2006.
6. Paraianu, L., Comsa, D.S., Cosovici, G., Jurco, P., Banabic, D., An Improvement of the BBC2000 Yield Criterion, ESAFORM 2003, pp. 215-218, Salerno, Italy, April 28-30, 2003.
7. Pöhlandt, K., Banabic, D., Lange, K., Determining Yield Loci of Sheet Metal from Uniaxial and Plane Strain Deformation Data, ESAFORM 2003, pp. 223-226, Salerno, Italy, April 28-30, 2003.
8. Van Haaren, L., van den Boogaard, A.H., Huétink, J., Modelling of Aluminium Sheet Material at Elevated Temperatures, ESAFORM 2004, pp. 551-554, Trondheim, Norway, April 28-30, 2004.
9. Teodosiu, C., Some Basic Aspects of the Constitutive Modelling in Sheet Metal Forming, ESAFORM 2005, pp. 239-243, Cluj-Napoca, Romania, April 27-29, 2005.
10. Racz, G., Lemoine, X., Haddag, B., Abed-Meraim, F., Implementation and Comparison of Advanced Behaviour Laws for Sheet Metal Forming, ESAFORM 2005, pp. 293-296, Cluj-Napoca, Romania, April 27-29, 2005.
11. Cosovici, G.A., Banabic, D., A Deep-Drawing Test Used yo Evacuate the Performances of Different Yield Criteria, ESAFORM 2005, pp. 329-332, Cluj-Napoca, Romania, April 27-29, 2005.
12. Cazacu, O., Plunkett, B., Barlat, F., Orthotropic Yield Criterion fo Mg Alloy Sheets, ESAFORM 2005, pp. 379-382 Cluj-Napoca, Romania, April 27-29, 2005.
13. Lenoir, H., Boudeau, N., An Optimization Procedure for Springback Compensation, ESAFORM 2003, pp. 143-145, Salerno, Italy, April 28-30, 2003.

14. Cao, J, Cheng, H.S., An Accelerated Method for Springback Compensation, ESAFORM 2004, pp. 519-522, Trondheim, Norway, April 28-30, 2004
15. Lingbeek, R., Meinders, T., Ohnimus, S., Petzoldt, M., Weiher, J., Springback Compensation: Fundamental Topics and Practical Application, ESAFORM 2006, pp. 403-406, Glasgow, United Kingdom, April 26-28, 2006.
16. Reese, S., Vladimirov, I.N., Springback Simulation by Means of a Stabilized Solid-Shell Finite Element Technology, ESAFORM 2006, pp. 415-418, Glasgow, United Kingdom, April 26-28, 2006.
17. Felder, E, Devine, I., Friction in Sheet Metal Forming: A Theoretical Model and Its Experimental Validation, ESAFORM 1998, pp. 175-178, Sophia Antipolis, France, March 17-20, 1998.
18. Zmudzki, A., Papaj, M., Kuziak, R., Kusiak, J., Pieterzyk, M., Optimum Die Shape Design for Evaluation of Material Properties, ESAFORM 2003, pp139-142, Salerno, Italy, April 28-30, 2003.
19. Mkaddem, A., Bahloul, R., Potiron, A., Numerical Analysis and Response Surface Method to Investigate Friction Effects in Sheet Bending Application, ESAFORM 2005, pp. 261-264, Cluj-Napoca, Romania, April 27-29, 2005
20. Carrino, L., Di Meo, N., Sorrentino, L., Strano, M. The Influence of Friction in the Negative Dieless Incremental Forming Process, ESAFORM 2006, pp. 203-206, Glasgow, United Kingdom, April 26-28, 2006.
21. Liewald, M., Wagner, S., Becker, D., New Approaches on Coulomb's Friction Model for Anisotropic Sheet Metal Forming Applications, ESAFORM 2006, pp. 283-286, Glasgow, United Kingdom, April 26-28, 2006.
22. Daniel, J.L., Boubakar, L., Boisse, P, Gelein, J.C, Finite Element Constitutive Laws and Contact Modeling for Sheet Forming Simulations, ESAFORM 1998, pp. 183-186, Sophia Antipolis, France, March 17-20, 1998.
23. Onate, E., Zarate, F., Rojek, J., Jovicevic, J., Sheet Stamping Analysis Using Rotational-Free Shell Elements, ESAFORM 1998, pp. 207-210, Sophia Antipolis, France, March 17-20, 1998.
24. Haddad, A., Balan, F., Abed-Meraim, On the Implementation of Hardening Models in Sheet Forming Simulations, ESAFORM 2003, pp. 187-190, Salerno, Italy, April 28-30, 2003.
25. Forestier, R., Ben Tahar, M., Chastel, Y., Massoni, E., Application of an Inverse Method to the Analysis of the Bulge Test, ESAFORM 2003, pp. 343-346, Salerno, Italy, April 28-30, 2003.
26. Chenot, J.-L., State of the Art and Recent Developments in Numerical Modeling of Metal Forming Processes, ESAFORM 2005, pp. 71-76, Cluj-Napoca, Romania, April 27-29, 2005
27. Kopp, R., Böhlke, P. Development of a New Rolling Process for Tailor Rolled Strips, ESAFORM 2003, pp. 151-154, Salerno, Italy, April 28-30, 2003.
28. Sun, P., Ferreira, J.A., Gracio, J.J, The Experiment Device of Controlling Springback in Real Time During Sheet Metal Forming, ESAFORM 2004, pp 459-462, Trondheim, Norway, 2004.
29. Ridane, N., Heller, B., Chatti, S. and Kleiner, M., Adaptive process control of air bending on a press brake using process simulations, pp. 523-526, Esaform 2004, Trondheim, Norway.
30. Yang, M., Koyama, J., Development of die-embedded micro sensing system for bending process, Esaform 2005, pp 289- 292, Cluj-Napoca, Romania, 2005.
31. Giardini, C., Ceretti, E., Attanasio, A., Pasquali, M., Feasibility limits in sheet incremental forming: experimental and simulative analysis, pp. 515- 518, Esaform 2004, Trondheim, Norway.
32. Duflou, J. R., Tunçkol, Y. Force modeling for single point incremental forming, Esaform 2006, pp 287-290, Glasgow, Scotland, 2006.

33. Bologa, O., Oleksik, V. and Racz , Experimental research for determining the forces on incremental sheet forming process, Esaform 2005, pp 317- 320, Cluj-Napoca, Romania.
34. Ambrogio, G., Filice, L., Fratini, L., Micari, F., Some relevant correlations between process parameters and process performance in incremental forming of sheet metals, ESAFORM 2003, pp. 175-178, Salerno, Italy, 2003.
35. Chouati, O, Gelin, J.C. and Baudet, C., Modelling the combined hydroforming of multiple sheet parts, ESAFORM 1998, pp. 199-202, Sophia-Antipolis, France, 1998.
36. Moshfegh, R., Nilsson, L, Jensen, M.R., Danckert J., Fininite element simulation of the hydromechanical deep drawing process, ESAFORM 1998, pp. 203-206, Sophia-Antipolis, France, 1998.
37. Jensen, M.R, Olovsson, L, and Danckert, Numerical model for the oil pressure distribution in the hydromechanical deep drawing process, ESAFORM 1998, pp. 219-222, Sophia-Antipolis, France, 1998.
38. Giardini, C., Ceretti, E., Conti, C., A study of an application of double sheet hydro forming, ESAFORM 2004, pp 563- 566, Trondheim, Norway, 2004.
39. Bjørkhaug, L. and Welo, T. Local calibration of aluminium profiles in rotary stretch bending, Esaform 2004, pp 531- 534, Trondheim, Norway 2004.
40. Neugebauer, R., Putz, M., Laux, G. The influence of die design on HexBend technology, Esaform 2006, pp. 239-242, Glasgow, Scotland, 2006.
41. Carrino, L., Strano, M., A comprehensive process design methodology for rotary draw bending of tubes, ESAFORM 2003, pp. 239-242, April 28-30, Salerno, Italy, 2003.
42. Paulsen, F., Welo, T., An analytical model for prediction of tube ovalization in bending , ESAFORM 2003, pp 775-778, April 28-30, Salerno, Italy, 2003.
43. Meinders, T., van den Boogaard, A.H., Huetink, J., Advanced sheet metal forming, Esaform 2003, pp. 159- 162, Salerno, Italy, 2003.
44. Yoshihara, S., MacDonald, B.J., Koizumi, C, Yamamoto, H., Nishimura, H., FE Analysis and Optimization of Square Cup Stamping Using a Local Heating and Cooling Technique, Esaform 2004, pp 547-550, Trondheim, Norway.
45. Bruni, C., Forcellese, A., Gabrielli, F., Palumbo, G., Simoncini, M., Sorgente, D., Tricarico, L., Bending of aluminium strips at elevated temperatures, Esaform 2006, pp 271-274, Glasgow, Scotland, 2006.
46. Merklein, M and Beccari, S., Formability of the magnesium alloy AZ31 at elevated temperatures, Esaform 2006, pp. 295-298, Glasgow, Scotland, 2006.

# Advances and Progresses in Sheet and Tube Hydroforming Processes

Jean-Claude Gelin
FEMTO-ST Institute, Department of Applied Mechanics
ENSMM Besançon, 26 rue de l'Epitaphe
25030 BESANCON France
jean-claude.gelin@ens2m.fr

**Abstract.** The Sheet Hydroforming (SHF) and Tube Hydroforming (THF) are more and more commonly used in industry to produce complex sheet or tubular metallic components with High Limit Drawing ratio or large diameter expansions. The large variety of metallic materials that can be used and the complexity of the shapes that can be produced, avoiding the use of complex process sequences render the hydroforming processes well suited for numerous application fields, where the complexity of the parts requires the combination of processes to get lightweight components. The paper presents a synthesis and last developments in terms of hydroforming technologies, materials and limits of formability, as well as modeling strategies accounting the pressure and fluid flow effects. One also focuses on the identification, optimization and then control strategies that are now used in process design development to render as efficient as possible the whole hydroforming process chain. The developments are related to the contribution of Esaform research community and more generally international scientific community during the last ten years.

**Keywords:** Sheet Hydroforming, Tube Hydroforming, Fluid flows, material behavior, forming limits, simulation, finite elements, optimization and control.

## 1.   Introduction

The hydroforming technologies are appeared in the beginning of 1980's associated to the fact that numerous authors have remarked that hydraulic pressure could entrance the formability of materials [1, 2]. The research works carried out in Eastern Europe, US, Japan and Asia, and progresses in forming equipments performed by machine tools providers permit the development of hydroforming concepts and equipments in tube hydroforming as in sheet hydroforming. The pioneer theoretical researches by Tirosh and co-authors [3, 4, 5] describe how the liquid flows beneath the flange and the die cavity in flange hydroforming, as well as the pressure grows in hydroforming of tubular stapes. This has been completed by analytical and numerical finite element simulations related in [6, 7] in order to understand and to improve the aquadraw deep drawing processes. Since this period, the flange hydroforming gets on increasing interest in laboratories and industry.
The development of tube hydroforming starts in the middle of 1990's with pioneered works related by Dohmann and co-workers [8, 9].

Since the beginning of 1990's, the development of hydroforming processes for flanges or tubes forming are very important, particularly associated to the manufacturing of complex tubular or sheet metal parts, used for Lightweight components in automotive, aeronautic or other application fields [10, 11].

Among different hydroforming processes, one classically distinguishes two distinct technologies. In tube hydroforming technology, a tube generally with a circular cross section, is first shaped by bending and then clamped in a tooling system, filled by a fluid and then pressurized [9]. It results that the tube takes the inner shape of tools that allow producing complex seamless parts. The flange hydroforming technology consists to take benefit from hydraulic pressure to assist flange deep drawing by replacing the die cavity by a cavity filled by an hydraulic fluid [1, 3] In such processes, the pressure in the die cavity decreases the sliding between the flange and the punch and reduces the friction between the flange and the die cavity and also under the blankholder, leading to the increase of Limit Drawing Ratio comparatively to classical deep drawing processes [6, 7]. The flange hydroforming process permit to get complex components in a single stage that cannot be produced by classical deep drawing processes. It also permits to get small size details due to the fact that the pressure applies the flange on the dies.

In order to perform modeling and simulation of hydroforming processes, it is first necessary to properly identify material properties involved in hydroforming. For tube hydroforming, several authors have proposed biaxial experiments where a tube submitted to an inner pressure can be expanded in an open cavity [12, 13]. The results obtained from these experiments show that the hardening exponent has an important role. In order to characterize material behavior through tube expansion tests, some authors have proposed relations issued from analytical modeling [13, 14] that permit to get material parameters from the values of the inner pressure and axial loads.

Another aspect in modeling and simulation of flange or tube hydroforming is associated to the prediction of the limits of formability, i.e. the range of process parameters leading to a component free of defects (necking, fracture, wrinkling and bursting). In that field, some authors have attempted to apply classical Forming Limit diagrams (FLD) to predict necking, bursting and failure [15], but as the pressure strongly influences the FLD, other approaches have been proposed based on ductile fracture mechanics [16], whereas other authors proposed to extend the linear stability analysis to 3D complex stress states to describe necking and fracture in tube hydroforming [17, 18]. In order to proceed in flange or tube hydroforming in the process window avoiding failure, bursting and buckling, various authors have proposed strategies to determine the limiting axial loads to apply to pistons or counter pistons in order to prevent risks of buckling and failure [14]. Control strategy based on coupling FE simulation of tube hydroforming and optimization has been proposed by the authors that propose a method to incrementally build the inner pressure and axial loads vs. axial feed.

One has to notice that numerous research works associated to sheet or tube hydroforming processes are carried out in research labs and in industries. International Conferences associated to material forming and numerical methods, as

Metal Forming, International Conference Technology of Plasticity, Numiform and Esaform Conferences series, always comprise sessions on Hydroforming since the last 10th years, see the Conference Proceedings. The Esaform members have strong activities in that field and the last five Esaform Conferences comprise approximately 6 papers or more in that field each year.


## 2.  Development of hydroforming technologies

One classically distinguishes between flange hydroforming and tube hydroforming technologies. In the first case, the fluid pressure applied on the flange generally enhance the sheet formability limits and permit to get components with very high drawing ratios, that cannot be obtained with classical deep drawing processes. The fluid pressure could be only obtained through a cavity filled of fluid and eventually with an external pump that increases the fluid pressure on the flange. This is a classical process to enhance formability of thin or thick sheet metal parts. The counter pressure and peripheral pushing effects associated to the fluid in the die cavity enhance the formability and permit to get components with High Limit Drawing Ratios [1, 2, 3].
The aquadraw deep drawing process is in that sense an interesting flange hydroforming process where the fluid inside the die cavity flows beneath the flange and blankholder resulting in the decreasing of the friction effects and in the increasing of the Limit Drawing Ratio. This process has been extensively analyzed in numerous papers [4, 5, 6, 7], and is used in industry to get cylindrical like components with high LDR.
Tube hydroforming technologies become popular in the beginning of 1990's as the necessity to develop structural lightweight components starting from tubular or tubular like components increases, as it is render necessary due to economical and ecological reasons in automotive industry or other ones [10, 11]. In Europe, the pioneer technological works carried in Germany by Dohmann and co-authors [8, 9] have demonstrated that tube hydroforming could be used to get complex structural components by combining tube bending, tube hydroforming and tube joining technologies.
Such technologies are now commonly used in automotive industries to produce fuel tanks [19, 20] as well as tubular parts for exhaust systems. The last technological developments in tubular hydroforming processes consist in combining tube bending, tube hydroforming and tube welding in a flexible manner using reconfigurable machine tool equipments easily adaptable to various production batches.
In terms of hydroforming technology improvements, are had also to notice the progress recently accomplished in terms of process control, in particular feed back control approaches based on experimental measurement systems allowing to adjust the loads and tool displacements by using feed back control loops [21,22].

## 3.  Modeling the fluid/structure interactions in hydroforming

### 3.1  Fluid-Structure Interactions in Sheet Hydroforming

In this paragraph, one focuses on the development of modeling strategies related to the fluid velocity and pressure in the case of flange hydroforming to realize cylindrical or 3D cups. The objective is to set various models for the simulation of the distribution of fluid pressure between the flange and the die. Moreover, one has to determine the variation of the pressure in the closed die where the rise in pressure is obtained through the reduction in volume of fluid caused by the displacement of the punch. In that field, pioneer contributions from Tirosh and co-authors [3, 4, 5] relate the modeling of the hydroforming of a circular flange and analyze the influence of pressure and deep drawing ratio on the punch force to be applied, see figure 1; they also determined the evolution of the pressure between the flange and the die in using an analytical approach. The present author and co-authors [6, 7] have proposed an alternative approach to describe the role of the fluid between the flange and the die and have also designed and realized an experimental device to measure the pressure of the fluid corresponding to various points of the flange. The results obtained indicate that the pressure drop beneath the blankholder permits to completely support the flange that is not in contact with the die, avoiding friction effects and increasing of the Limit Drawing Ratio as noticed in [3, 4, 5, 6, 7].

From such an approach the flow located between the flange and the die could be accurately determined and it results that the flow input in the die cavity could also be determined. The related analytical models described in [3, 4, 5, 6, 7] are associated to axisymmetric cases, and it cannot be applied to full 3D cases. Jensen and al. [24, 25] proposed to extend such an approach in using a Finite Element model to solve the flow equations in 2D planar cases and the axial velocity could be determined from the resulting Finite Element Model for any position under the blankholder (see figure 2). A similar approach has also been used in [29] where one has shown that pressure distribution could be accurately determined.

So accounting such analytical approaches combined with Finite Elements Methods, one could assume that the pressure could be accurately determined for any parts of the flange, and the simulation of the sheet metal deformation, accounting the pressure distribution could be carried out in using standard Finite Elements approaches accounting as well non linear geometrical approaches as well as non linear material elastic-plastic behavior associated to the mechanics of sheet metal deformation.

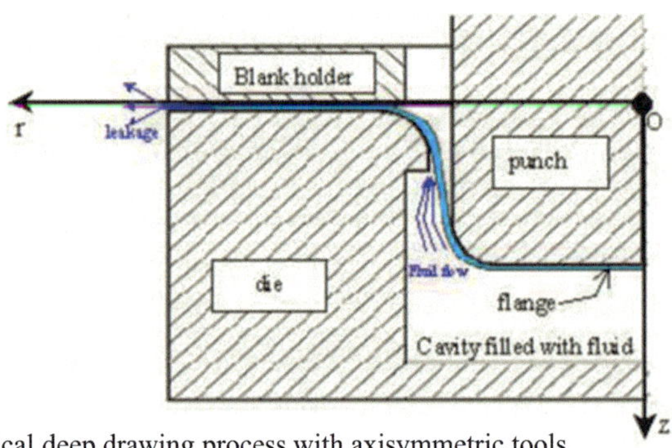


The hydromechanical deep drawing process with axisymmetric tools.

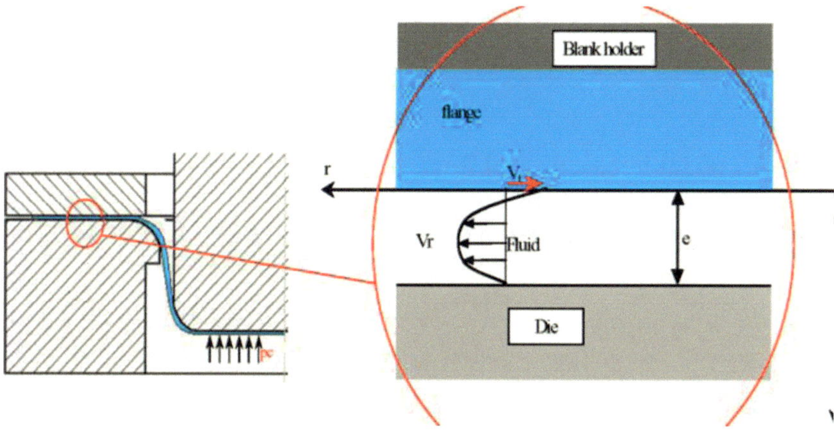


**Fig. 2.** Representation of the flow between the flange and the die.

## 3.2  Fluid-Structure Interactions in tube Hydroforming

The fluid-structure interaction in tube hydroforming where an initially cylindrical tube is pushed in the axial direction by a piston and where the inner pressure inside the tube change the inner shape of the tube, see figure 3, has also be investigated by different authors.. The fluid flow in the cavity has to be accounted in order to describe the process as it is carried out in practice [25,26,27,28].

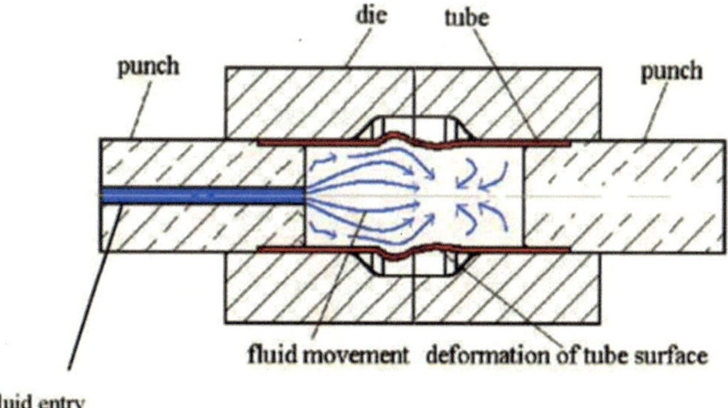


**Fig. 3.** Fluid-Structure interaction in tube hydroforming.

In that field, the common practice for modeling and simulation of tube hydroforming process (THF) consists to account the pressure that is excited by the hydraulic fluid on the tube [29, 30, 31], without accounting the fluid displacement in the inner tube, associated to the fact that tube deformation implies fluid displacements inside the tube. This problem has been addressed in literature by some authors [32] that use standard facilities in FE simulation codes as the Airbag facility in LS Dyna, or that propose computational developments accounting the flow of the fluid inside the tubular cavity in using an Arbitrary Lagrangian-Eulerian approach [32].

The results indicate that the fluid displacements are limited and that the associated phenomena could be neglected in the majority of the industrial cases, apart the case where T shapes imposes an abrupt volume variation in the die cavity.

## 4.  Material behavior and formability limits in SHF and THF

### 4.1 Material behavior in SHF and THF

The materials that can be used as well as in Sheet Hydroforming or Tube Hydroforming concern majority of metallic alloys that can be employed is classical deep drawing or stamping processes as common on special steels, stainless steels, aluminum alloys, copper alloys and also special heavy materials as tantalum alloys. It has to be noticed that new research works are now carried out by prescribing differential temperature to soften the material and thus to facilitate the associated plastic deformation by increasing the temperature forming to enhance the hydroformability of material that imposes strong requirements as magnesium alloys [33]. This is associated to the fact that magnesium alloys are very interesting to reduce part weights as example in automotive industry where the environmental requirements impose to reduce fuel consumption.

Considering the material models to use in hydroforming, there is lot of literature in that field, and the deformation process associated to T-tube expansion problem is considered as a reference case [34, 35, 36, 37]. In the deformation process of the T-tube expansion problem, material points in the expansion zone experience multi-axial loading situations where compression and tension states change as the material moves along the tool surface. There are hoop and radial stresses due to the pressure inside the tube and compression from both ends to feed material. There have been many investigations to optimize the pressure and the material feeding at both ends and a reliable prediction is therefore essential.

It has been shown for this case that the deformation height is not sensitive with respect to the hardening models, considering both isotropic and anisotropic hardening models [34,35].

As the deformation in this problem is strongly restricted by the shape of the die and the displacement controlled feeding of the material from both sides of the tube it may not have been surprising that overall differences in the plastic strain are rather small [36, 37].

### 4.2 Formability Limits in SHF and THF

It is well known that pressure normally increases formability limits. So the formability limits in sheet hydroforming processes where analyzed by different authors as well as in SHF as in THF [4,5].

In SHF, the pioneer work by Tirosh and co-workers [4, 5] relate a standard approach in 2D plane stress case where the fracturing mode is associated to mechanical and physical instabilities. It is shown that the formability limits in SHF are in the same range that in standard deep drawing process apart in the case where the hydraulic counter pressure acts as peripheral pushing effects. It means that the hydroforming process has to be properly designed to benefit from this fact. Extensive analyses of fracture in sheet hydroforming have been also carried in easing the concept associated to linear stability analysis by simply post processing Finite Elements results. It is shown that such an approach could be used to predict limit of formability and bursting occurrence during hydroforming. The study has been completed with the analysis of the risks of failure and bursting that could occur during hydroforming. The author research has carried important developments in that field and one relates on figure 4, the straining paths and the FLDs obtained from the linear stability analysis as related in [17, 18]. It clearly appears no risks of failure or bursting during the hydroforming process.

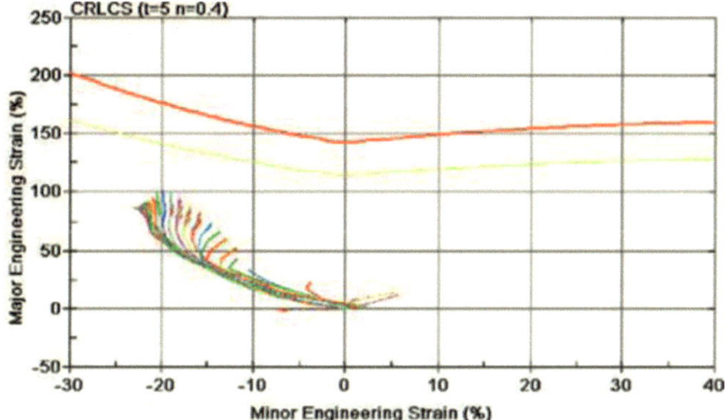


**Fig. 4.** Strain paths and CLFs associated to liner hydroforming.

The analysis of forming limit and fracturing mode in tube hydroforming has been also analyzed in using the concepts of ductile fracture mechanics and fracturing mechanics [35, 36, 37]. The results prove that such approaches could be used to predict the hydroforming limit accounting standard material behaviour combined with a critical material value accounting for ductile fracturing mode. This implies to carefully determine this value, in link with classical plastic properties as plastic yield surface and hardening law.

## 5.  Optimization and control of the hydroforming processes

The optimization procedures are increasingly used in research labs and in industry to determine the optimal process parameters to get the required hydroformed components with the required specifications. The use of the finite element method coupled to optimization algorithms is a solution to obtain the optimum parameters. So, the finite element method is used to evaluate the effect of a set of parameters on the process. The optimization of hydroforming process, in particular, requires the definition of objective functions representing the quality of the components. These functions are used to quantify shape defects or surface distortions (buckling, wrinkling, and bursting). For example, one can use the objective function expressed as:

$$F_{obj}(p) = \frac{1}{n}\sqrt{\sum_{i=1}^{n}\left(\frac{h_i(p)-h_0}{h_0}\right)^2}$$

where $h_0$ is the initial flange or tube thickness $h_i$ stands for the thickness at node i and n is the total nodes number involved in the problem. The optimization method proposed by the author and coworkers [38, 39, 40] consists to combine the simulation of the process and the evaluation of the cost function and then to minimize this cost function with constraints respect in using a SQP algorithm. This method requires computing the gradient of the objective and constraint functions, i.e. to perform sensitivity analyses. Two different methods for sensitivity calculations are used to obtain the gradient vector corresponding to the finite different method (FDM) and to the direct differentiation method (DDM).

One has also proposed a strategy based on a control algorithm to determine the command laws for hydroforming of parts with better quality. It consists to include a control loop inside the incremental loop used in the finite element solution procedure. The developed method is based on a coupling between the response surface method and optimal control [41]. The methodology consists in combining the theory of optimal control with the response surface method (Moving Least Square Approximation). The strategy adopted here consists in building an approximation for the pressure law vs. time. This approximation can be specified in a simple polynomial form or a more complicated one allowing possibilities to get the form of the law. In a first stage, one chose a linear piecewise pressure approximation expressed in the following form:

$$p(t) = \frac{p_{i+1}-p_i}{t_{i+1}-t_i}t + \frac{p_i t_{i+1}-p_{i+1}t_i}{t_{i+1}-t_i}$$

In summary, one thus obtains the following optimization diagram Fig 5.

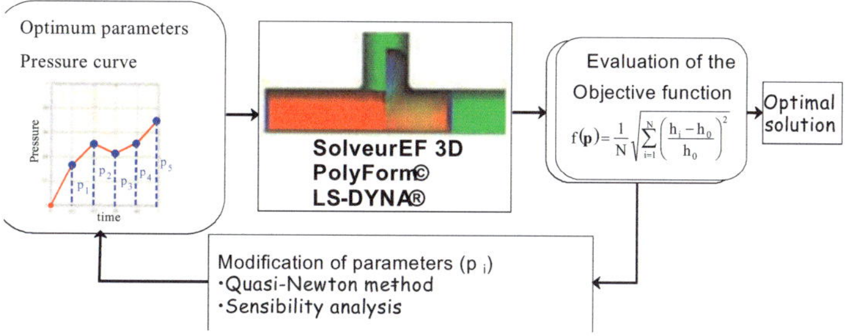


**Fig. 5.** Optimization loop used for process control in hydroforming.

Numerous other contributions for the optimization and virtual design of hydroforming processes, to avoid processing defects and to get the required components with the required shapes, thickness contours and structural properties are appeared in the literature since the last seven years. Among the different approaches, one can distinguish relatively standard approaches that are based on sensitivity analyses that permit to evaluate the effects of geometry, material and processing conditions through on a set of selected design parameters, and then to adjust these parameters in using standard difference methods or through adaptive methods [42]. Other authors mainly concentrate on the loading conditions as axial loads and inner pressure, and propose different methodologies as previously described to adjust loading parameters [43,44], in order to components with the required specifications. The virtual design strategy that is mainly based on the combination of FE simulations and artificial intelligence algorithms or fuzzy logic has also been proposed [45, 46, 47, 48]. It has been proved that these approaches could improve both qualities of the hydroformed parts as well processing conditions. It also appears analyses that intend to analyze hydroforming processes starting from the elaboration of the initial flange or tubular parts to the final products [49]. The last developments are related to the resultant structural properties associated to the in use properties of components obtained through a sequence of forming processes implying bending, hydroforming, blanking and welding sequences [50]. This is particularly important for the design of lightweight structural frames used in automotive or aeronautic industries [50].

## 6. Conclusions

In the last fifteen years, the hydroforming processes have been largely developed and are nowadays used in industries as alternative processes to sheet metal stamping processes, or tube bending and welding ones that are used for producing tubular like parts. The economical and ecological requirements and the necessity to account the ecological directives impose to reduce fuel consumption that results to decrease the weights of automotives or airplanes. It pushes the process designers to develop lightweight parts and components. In that sense, the hydroforming processes, in combination with bending and welding ones are a real alternative to standard processing design. The scientific developments as well as technological ones permit now to understand and to accurately evaluate the effects of processing conditions associated to the inner fluid pressure and inner fluid movements during hydroforming. The development of powerful hydroforming equipments and combined tooling systems able to combine bending, hydroforming and welding render possible the manufacturing of complex parts on the same processing chain. On the other hand, the new incoming material and alloys as austenitic stainless steels, dual phase steels, TRIP steels and structural aluminum or magnesium alloys impose processing conditions that can be accounting through hydroforming as the hydroforming of thermally soften magnesium alloys. The parallel progresses in modeling of material behaviors and limits of hydroforming permit now to clearly define the process windows that can be used to get the parts without defects.

The associated progresses in hydroforming process modeling permits in parallel to define a complete modeling, simulation and optimization chain, starting from the flange or tube up to the evaluation of the resulting mechanical and structural properties after hydroforming, bending, trimming and joining stages, including the crash resistance. The modeling and numerical strategies have been also strongly developed are now able to account all the processing stages involved in the complete hydroforming process chain. The associated optimization and processes design strategies have also largely progressed in the last five years, and permit really to attempt virtual hydroforming process design by combination of optimization tools and FE ones. Nether less it remains lot of developments to carry out in hydroforming associated to the modeling of fluid-structural interactions, to the accurate prediction of hydroforming limits under complex loadings. From the technological viewpoint, the main progress to accomplish are related to the development of low costs, flexible and reconfigurable hydroforming cells including trimming and welding capability easily adaptable to the developments of new incoming lightweight parts.

# References


1.   K. Nakamura, T. Nakagawa, "Reverse Deep Drawing with Hydraulic Counter Pressure using Peripheral Pushing Effect. Ann. CIRP 35/1 (1986), 173-176.

2.   N. Alberti, A. Forcelises, L. Fratini, F. Gabrielli, Sheet Metal Forming of Titanium Blanks using Flexible Media, Annals of the CIRP 47/1 (1998)., 217-220.

3.   J. Tirosh, P. Konvila, "On the Hydrodynamic Deep Drawing Process", Int. J. Mech. Sci., 27 (1985), 595-608.

4.   S. Yossifon, J. Tirosh, "Buckling Prevention by Lateral Pressure in Hydroforming Deep Drawing", Int. J. Mech. Sci., 27 (1985), 177-185.

5.   S. Yossifon, J. Tirosh, "Rupture Instability in Hydroforming Deep Drawing Process", Int. J. Mech. Sci., 27 (1985), 559-570.

6.    J.C. Gelin, P. Delassus, Modeling and simulation of the Aquadraw deep drawing process, Annals of the CIRP, Vol. 42/1 (1993), 305-309.

7.    J.C. Gelin, P. Delassus, J.F. Fontaine, Experimental and numerical modeling of the effects of process parameters in the aquadraw deep drawing process, J. Mater. Process. Technol., 45 (1994), 329-334.

8.    F. Dohmann, C. Hartl, Hydroforming – a method to manufacture lightweight parts, Journal of Materials Processing Technology, 60-1 (1996), 669-676.

9.    F. Dohmann, C. Hartl, Tube hydroforming – research and practical application, Journal of Materials Processing Technology, 71-1 (1997), 174-186.

10.   M. Merklein, M. Geiger, New materials and production technologies for innovative lightweight construction, Manufacturing if lightweight components by metal forming, Journal of Materials Processing Technology, 125-126 (2002), 532-536.

11.   M. Merklein, M. Geiger, A. Klaus, Manufacturing of lightweight components by metal forming, Annals of the CIRP 52/2 (2003), 521-542.

12.   K.I Manabe, M. Amino, Effects of process parameters and material properties on deformation process in tube hydroforming, Journal of Materials Processing Technology, Vol 123 (2002), 285-291.

13.   T. Sokolowski, K. Gerke, M. Ahmetoglu, T. Atlan, Evolution of tube formability and material characteristics by hydraulic bulge testing of tubes Journal of Materials Processing Technology, 92 (2000), 34-40.

14.   N. Asnafi, Analytical modeling of tube hydroforming, Thin-walled Structures 34 (1999), 295-330.

15.   M. Koc, T. Atlan, Prediction of forming limits and parameters in tube hydroforming process, Int. J. Mach. Tool Des. Res., 42 (2002), 123-138.

16.   L.P. Lei, B.S. Kang, S.J. Kang, Prediction of the forming limit in hydroforming processes using the finite element method and a ductile fracture criterion, Journal of Materials Processing Technology, 113 (2001), 673-679.

17.   N. Boudeau, A. Lejeune, J.C. Gelin, Influence of material and process parameters on the development of necking and bursting in flange and tube hydroforming, J. of Materials Processing and Technology, Vol.125-126 (2002), 849-855.

18.   A. Lejeune, N. Boudeau, J.C. Gelin, Influence of material and process parameters on bursting during hydroforming process, Journal of Materials Processing Technology, Vol. 143-144 (2003), 11-17.

19.   E.J. Vinarcik, Automotive light metal advances, Part I, Innovative designs and emerging technologies, Light Metal Age 60 (2002), 38-41.

20.   BS. Kang, B.M. San, J. Kim, A comparative study of stamping and hydroforming processes for an automobile fuel tank using FEM, Int. J. of Machine Tools & Manufacture, 44 (2004), 87-94.

21.   P. Groche, M. Ertugrul, C. Metz, Increase of Process Stability of Sheet Metal Hydroforming due to Feed Back Control, Steel Research International. Vol. 76-12 (2005), 879-883.

22.   P. Groche, R. Steinheimer, D. Schmoeckel, Process Stability in the Tube Hydroforming Process, Annals of the CIRP, Vol. 52/1 (2005), 229-232.

23.  M.R. Jensen, L. Olovsson, J. Danckert, L. Nilsson, Numerical model for axisymmetrical deep drawing processes, Int. J. Forming Processes, Vol 2, n°3-4 (1999), 193-210.

24.  L Lang, J Danckert, K B Nielsen, Investigation into sheet hydroforming based on hydromechanical deep drawing with uniform pressure on the blank, Journal of Engineering Manufacture, 218- 8 (2004), 833-844.

25. M.  Ben  Tahar,  E. Massoni, Numerical and experimental study of sheet metal hydroforming, Proceedings of the 8th International Conference on Numerical Methods in Industrial Forming Processes, AIP Conference Proceedings,  712 (2004) 1160-1165

26.  E. Ceretti, C. Contri, C. Giardini, Tube hydroforming on an AL7003 extracted, Journal of Materials Processing Technology, 177 (2006), 672-675.

27.  L. Filice, L. Fratini, F. Micari, A simple experiment to characterize material formability in tube hydroforming, Annals of the CIRP, 50/1 (2001), 181-184.

28.  Massoni, E; Aliaga, C, 2D finite element simulation of tube hydroforming process  In Simulation of Materials Processing: Theory, Methods and Applications (USA), A.A. Balkema Publishers, (1998), 893-898.

29.  JC. Gelin, C. Labergere, S. Thibaud, Recent advance in process design for sheet and tube hydroforming, in Advances in Material Forming Processes, ed. By D. Banabic, Springer Verlag, to be published (2007)

30.  R. Neugebauer, A Sterzing, M. Siefert, P. Kurka, The potential and application of temperature supported hydroforming of magnesium alloys, In Proc. Of the $8^{th}$ Int. Conf. on Technology of Plasticity, Ed. By PF. BARIANI, Edizioni Progretto Padovo, (2005), 293.

31.  Y.-M. Hwang and W.-C. Chen, Analysis of tube hydroforming in a square cross-sectional die, Int. J. Plast. 21 (2005), 1815–1833.

32.  M. Jansson, L. Nilsson and K. Simonsson, On constitutive modelling of aluminium alloys for tube hydroforming applications, Int. J. Plast. 21, 1041–1058 (2005).

33.  M.-G. Lee, C.-S. Han, K. Chung, J.R. Youn and T.J. Kang, Influence of back stresses in parts forming on crashworthiness, J. Mat. Proc. Tech. 168  (2005), 49–55

34.  Y. Choi, C.S. Han, J.K. Lee, R.H, Wagoner, Modelling multi-axial deformation of planar anisotropic elasto-plastic materials, part II: Applications, International Journal of Plasticity , 22-9 (2006), 1765-1783.

35.  L.P. Lei, B.S. Kang, S.J. Kang, Prediction of forming limit in hydroforming processes using the finite element method and a ductile fracture criterion, Journal of Materials Processing Technology, 113 (2001), 673-679.

36.  L.P. Lei, B.S. Kang, S.J. Kang, Bursting failure prediction in tube hydroforming processes by using rigid-plastic FEM combined with ductile fracture criterion, International Journal of Mechanical Sciences, 44-7, (2002), 1411-1428.

37.  J. Kim, S.W. Kim, W.J. Song, B.S. Kang, Analytical and numerical approach to prediction of forming limit in tube hydroforming, International Journal Mechanical Science, 47-7 (2005), 1023-1037.

38.  J.C. Gelin, C. Labergere, Application of optimal design and control strategies to the hydroforming of thin walled metallic tubes, Int. J. Forming Processes, Vol. 7 n°1-2 (2004), 141-158.

39.  C. Labergere, A. Lejeune, J.C. Gelin, Optimization and control of flange and tube hydroforming processes, J. Steels and Related Materials, Vol, 2  (2004), 221-227.

40.  J.C. Gelin, C. Labergere, S. Thibaud, Modelling and control for the hydroforming of metallic liners used for hydrogen storage, J. of materials Processing Technology, Vol. 177 (2006), 697-700.

41.  J.B. Yang, B.H. Jeon, S.I. Oh, Design sensitivity analysis and optimization of the hydroforming process, Journal of Materials Processing Technology, 113/1-3 (2001), 666-672.

42.  S. Jirathearanat, T. Altan, Optimization of Loading Paths for Tube Hydroforming, in AIP Conference Proceedings, Volume 712 (2001), 1148-1153.

43.  F. Dohmann, Ch. Hartl, Hydroforming-applications of coherent FE-simulations to the development of products and processes, Journal of Materials Processing Technology, Volume 150 / 1-2 (2004) , 18-24.

44.  M. Strano, S. Jirathearavat, S.G. Shr, T. Altan, Virtual process development in tube hydroforming, Journal of Materials Processing Technology, 146 (2004), 130-136.

45.  R. Di Lorenzo, L. Filice, D. Umbrello, and F. Micari, An integrated approach to the design of tube hydroforming processes: artificial intelligence, numerical analysis and experimental investigation, Proc. of Numiform 2004 Conference, AIP Conference Proceedings, 712 (2004), 1118-1123.

46.  R Di Lorenzo, L Filice, D Umbrello, F Micari, Optimal design of tube hydroforming processes: a fuzzy-logic-based approach, Journal of Engineering Manufacture, 218-6 (2004), 599-606.

47.  K.S. Park, B.J. Kim, Y.H. Moon, Optimization of Tube Hydroforming Process by Using Fuzzy Expert System, Materials Science Forum. Vol. 475-479, Part 4, (2005), 3283-3286.

48.  J.C .Gelin, C. Labergere, S. Thibaud, N. Boudeau, Design of hydroforming processes for metallic liners used in high pressure hydrogen storage, AIP Conference Proceedings, 778, (2005), 538-542.

49.  J. W. Yoon, K. Chung, F. Pourboghrat, F. Barlat, Design optimization of extruded preform for hydroforming processes based on ideal forming design theory, International journal of mechanical sciences, 48,1-2 (2006)., 1416-1428.

50.  S. Thibaud, N. Boudeau, J.C. Gelin, TRIP Steel: Plastic behaviour modelling and influence on functional behaviour, J. of Materials Processing Technology, 177 (2006), 433-438.

# Hot Metal Extrusion


Per Thomas Moe[1], Sigurd Stören[1] and Han Huetink[2],

[1] Department of Engineering Design and Materials, Norwegian University of Science and Technology, Richard Birkelands vei 2B, 7491 Trondheim, Norway

[2] Department of Mechanical Engineering, University of Twente, Twente, The Netherlands



**Abstract.** The paper focuses on the main scientific problems related to hot metal extrusion of thin walled tube, strip and sections, covering analytical approach, 2D- and 3D-numerical simulation and experimental validation. The main physical phenomena and parameter estimation related to thermo-mechanical constitutive equations, shear localization, bearing channel, dry hot friction and metal flow stability in extrusion is treated. The scientific, generic insight and knowledge of these extrusion phenomena should be considered as pre-competitive. In industry this knowledge must be integrated with practical experiences and skills in order to be able to predict and control dimensional, surface and microstructure variability during a press cycle, and inspire to innovations in the field of extrusion and its down stream processes such as tube drawing, cold forging, bending and hydroforming.

**Keywords:** Extrusion, thin-walled sections, metals, FEM, experiments.


# 1  Introduction

This Section gives a short overview of extrusion and drawing. Sub-Section 1.1 covers extrusion and drawing as technological forming processes presented in the ESAFORM papers in the Mini-Symposium "Extrusion and Drawing". In Sub-Section 1.2 the focus is on hot metal extrusion of thin-walled strips, tubes and sections, with special attention to direct extrusion of aluminium alloys. This process is of special scientific interest because of the complex and strong interaction between metal flow, temperature, die displacement, interface phenomena, friction, localized shear and microstructural evolution.

## 1.1  Extrusion and drawing

Extrusion and drawing are material forming processes for manufacture of long lengths with constant cross-section. Extrusion transforms cast and homogenized billets to rods, open sections and hollow sections by pressing the billet in the container with a ram through a die in a hydraulic press. Rods are normally a semi-fabricated product, used as feedstock for cold drawing of long lengths or cold or hot forging and machining of components.

In the case of hot aluminium extrusion thin-walled open and hollow sections can be made with very complex cross-sectional area. Hollow sections are produced with a mandrel that shapes the inner surface of the section. The mandrel is fixed to the die by "bridges" so that the metal flow has to be separated and subsequently joined after passing the bridges, but before leaving the die outlet. These longitudinal welds are called "seam welds".

At the ESAFORM Conferences, a mini-symposium Extrusion and Drawing was organized. Papers related to backward and forward extrusion of components, drawing of long lengths, joining of sections along its length by friction stir welding and even cold and hot rolling of steel, copper, aluminium and magnesium and their alloys, including super-alloys, were presented. A significant part of the presentations focused on extrusion of thin-walled open and hollow sections. In this chapter we will therefore treat basic thermo-mechanical problems related to hot semi-continuous extrusion of thin-walled sections of long length. The interaction between alloy microstructure, temperature distribution during the press cycle, metal flow field, shear strain localization, dry friction at high temperature, die distortion, flow stability, joining processes and final properties of the extrudate makes hot extrusion of thin-walled aluminium sections one of the most complex and fascinating processes in the big family of metal forming processes.

## 1.2   The basic scientific challenges of hot metal extrusion

In thin-walled extrusion, the metal has a tendency to flow faster out of the die in one part of the section than in other parts of the section. This is seen when observing the first metal extruded through the die, from the uneven cross-sectional front of the section. When some of the metal has been extruded the potential differences in outlet speed are manifested by building up of longitudinal stresses in the section. These stresses are transmitted backward into the die and die inlet and thereby cause incremental changes to the flow field so that the variation in outlet speed is reduced. This causes local thickening or thinning of some part of the section, cyclic waves or even cracks. Obtaining fundamental insight, knowledge and control of these phenomena is the main target of the generic, pre-competitive scientific research. The research results can be applied by the extruders in combination with their detailed practical experience and skills as a means to control thickness, shape, microstructure, surface properties and bulk properties of the extruded section, design optimal tools and dies with increased efficiency, precision and minimum variability. The vision is an innovative ideal continuous, steady state process with zero variations along the section length may be realized. Work related to the Conform extrusion process [1], the potential of producing curved sections[2], and integration of the  process chain extrusion, quenching, shape calibration and hydroforming [3] may give directions of future innovations.

## 2   Application and validation of numerical methods

During the last thirty years numerical tools have enabled scientists and engineers to study complex physical phenomena and to solve practical engineering problems. In relation to aluminium extrusion and similar forming processes it  is of special interest to further develop numerical tools in order to understand and model complex interactions and to predict and expand the limit of extrudability as function of reduction, initial temperature, process speed, material properties, shape and thickness of thin-walled extruded strips, tubes and shapes, as well as to predict and reduce the variability of dimensions, microstructure and properties of the extruded section during a process cycle.

## 2.1 Finite Element Modeling

The first use of finite element models (FEM) for extrusion dates back to the early 70s. Zienkiewicz, Argyris, Kobayashi and co-workers clearly demonstrated the potential of numerical models compared to analytic approaches for practical cases of extrusion [4], [5], [6], [7]. Still, most early applications were related to 2D geometry and very low extrusion ratios. Simulations were mainly useful for studying extrusion and drawing of steel and similar metals. Aluminium extrusion through flat faced dies could more easily be handled by the conventional models from Computational Fluid Dynamics (CFD) [8], but there was no satisfactory tool yet even for 2D geometries. During the 80s significant efforts were made to establish dedicated commercial and non-commercial finite element codes for metal forming [9]. In the late 80s and early 90s a program dedicated to the study of aluminium extrusion [10] was developed at the Norwegian Institute of Technology. It took into account heat losses to the container, ram and die, ram movement and realistic friction conditions, and was capable of predicting the temperature and forces for two-dimensional cases. In the early 90s commercial simulation programs such as DEFORM, FORGE, ABAQUS, LS-DYNA, FLUENT and MARC were further developed and used by both Academia and the Industry. However, while 3D FEM started to be of significant practical value within the fields of forging and sheet metal forming by the second half of the 90s, the lack of progress with practical 3D aluminium extrusion models was the source of great disillusionment and discontent within the Industry and to some extent within the Academia.

Extruded aluminium profiles are usually thin-walled and of very complex shape. While the challenges related to the establishment of viscoplastic and friction models should not be underestimated, the main obstacles to the practical use of 3D FE models were, and still are, the high computational costs. While the advances of computer power could be expected to contribute to solving this issue in the long run, very important steps could be made by devising more effective and suitable numerical models. Building on the work of Huétink, Mooi and Lof advanced aluminium extrusion modeling significantly in the late 90s [11], [12]. The complex issue of bearing channel mechanics was treated, and elastic die deformations were introduced. Another early effort to establish useful 3D models for extrusion was made by Eikemo et al. [13]. However, the most impressive study in the late 90s was probably that of van Rens [14]. Like the models of Lof and Mooi, van Rens' models treat both bearing channel flow and tool deformations. Van Rens also dedicated much time and effort to the application of suitable and robust meshing algorithms, which will be important if the FE models are to become useful tools for scientists and engineers. Van Rens also ran a series of validation cases, but still significantly more work has to be done within this field before FEM results are to be trusted completely. During the early 2000s researchers have made more widespread use of the commercial software packages mentioned above as well as FE programs dedicated to extrusion of aluminium and polymers. A number of problems have been assessed, and experience is gradually built. In this respect, ESAFORM has had a very important role to play since it has allowed researchers to share experiences at an early stage. Van Rens presented his early work at the first conference in 1998 [15], [16] and Lof presented a part of his work the following year [17]. In 2002 Williams presented the efforts of her group [18] in building a dedicated model for aluminium extrusion.


## 2.2  Application of models

**Container flow and back end effects.** Container flow issues have traditionally been the easiest to study and solve with the finite element software. In the early 90s the combined use of ALMA2 and experiments gave valuable insight into the mechanics of metal flow. Valberg's new and improved grid line technique clearly demonstrated the transient nature of the aluminium extrusion process and how the dead zones in fact evolve significantly during an extrusion run. His results have been systematically compared with the results provided by ALMA2 and DEFORM 2D[19], [20]. Hanssen et al. used ALMA2 to study problems related to the inflow of contaminated material from the end of the profile and from the walls using half-moon dies [21]. In 2004, Hatzenbichler et al. presented numerical studies on the back-end effects using DEFORM 2D [22] [23]. Flitta et al. studied the effect of friction on container flow with Forge2 [24]. There have also been detailed studies of more complex processes such as extrusion through multi-hole dies [25] and feeder dies [26]. The processes of indirect extrusion with and without active friction have been studied numerically [27], [28]. Finally, numerical process design for isothermal extrusion has been the subject of an extensive numerical investigation [29]. The presentations at ESAFORM reflect the work published in scientific journals. During the early 2000s there were a large number of studies treating container flow for relatively low extrusion ratios using mainly commercial Lagrangian software [30], [31], [32], [33], [34], [35], [36], [37]. Through their detailed studies with the program Forge2, Sheppard and co-workers have contributed to a demonstration of the state-of-the-art [38], [39].


**Flow stability and flow control.** Usually, metal flow is controlled by the short bearing channel at the outlet of the extrusion die. The friction and wear of the bearing channel has been carefully studied by a number of researchers. Clode and Sheppard [40] presented very valuable insight into the formation section surfaces and the factors controlling their quality. Lefstad [41], Abtahi [42], Grasmo [43], Tverlid [44] and Moe [45] performed detailed studies, set forth the hypothesis of a Coulomb friction model and carefully tested that hypothesis through finite element modelling and a large number of experiments. Die wear has been further studied experimentally by Björk et al. [46], and numerical studies of shear boundary layers were performed by Aukrust et al. [47], [48]. Lof [49] and van Rens [14] have taken into account various models of friction in their numerical tools and studies of flow.

Future use of FE modelling in relation to aluminium extrusion will include the tuning of metal flow for particularly difficult sections. In this relation the optimization techniques presented at the ESAFORM Conferences are of great value. There have been very few presentations on the optimization of flow in direct relation to extrusion at ESAFORM, but the scientific literature today contains a considerable number of references on the issue [50], [51], [52], [53], [54], [55], [56]. It should be noted that very interesting practical use of their insight into flow control has been made by Müller and Buntoro. They have designed dies that allow bending of profiles during extrusion [57], [58]. In an article from 2006 Müller gives a detailed presentation of the state-of-the-art of a technology [2] that has a significant commercial potential.

**Die manufacture and die deformations.** A practical issue of great importance during extrusion of aluminium is die deformation and durability. Die deformations cause significant distortion of the die outlet, for which an appropriate compensation must be sought. After a number of extrusion runs dies are worn out or break. Wear and breakage can in many cases be the cause of excessive dimensional variability, which can lead to scrapping of extruded products. The potential of FE simulations is great. In the future, simulations will be able to predict the tool deformation with the corresponding outlet distortions and profile dimensions. This may be the starting point for die correction. Furthermore, simulations can be used to provide information of load cycles and the state of stress and temperature in the die. Such information can be used for optimal die design and for the assessment of fatigue. While several papers have been published on the subject of die deformation in relation to the extrusion of steel [59], [60], [61], [62], [63], [64], there are very few examples with the extrusion of aluminium. The possibility of treating the die and work piece is present in most commercial software as well as in the tools of Mooi et al. [65] and van Rens [14]. Wu treats die structure optimization of an aluminium rectangular hollow pipe extrusion process [66]. Lee and Im [67] and Lin [68] have evaluated wear and flow problems.

At ESAFORM 2004 Costa et al. [69] presented a die design for thin tubes extrusion based on 2D simulation. Mousolis et al. [70] focused on fracture of an aluminium extrusion die. Moe et al. [71] presented a set of experiments where the load and the corresponding deformation of a porthole die have been continuously measured during hot extrusion of aluminium. Pressure sensors, strain gauges, mandrel deflection sensors and temperature sensors give a good impression of the immense load to which dies are exposed. Plančak et al. [72] have designed pressure sensors for cold extrusion and compared experimental results with simulation data.

**Seam welds and extrusion welds.** During extrusion of hollow profiles the material flow is split and subsequently welded underneath the bridges. The fact that hollow sections of extremely complex cross-sections can be extruded is an advantage. However, hollow profile extrusion is a much more complex process than open profile extrusion. The die deformation is often greater, and the quality of seam welds needs to be assessed. Generally, seam welds are of high quality since the weld is formed at a high pressure and temperature in the absence of air. However, ductility and fatigue properties have to be considered, and contamination in relation to end effects is a possibility. An issue that has been carefully studied is the effect of low welding pressure on the mechanical properties of welds. The advent of 3D FE simulation has made possible detailed studies of welding conditions. Buffa, Donati et al. [73], [74], [75], [76], Edwards et al. [77], Jo et al., [78], [79], Kim et al., [80] and Kleiner et al. [81] have performed systematic numerical studies of seam welding often in combination with experiments. Donati et al. presented parts of their work at ESAFORM 2005. At the same conference Ghiotti et al. [82] demonstrated a new procedure and test for the investigation of seam weld quality. Loukus et al. [83] have earlier studied plastic properties of seam welds by systematic testing.

**Microstructure evolution and properties.** One of the great advantages of FE models is their ability to provide detailed information about deformation and temperature histories of the material during forming. An important objective of for example the European VIRFORM project was to link mechanical and other properties of the material to processing route (through-process modeling). Both the rolling and extrusion processes were extensively studied at European universities. Such an approach requires detailed micro-macro models taking into consideration all significant mechanisms. In relation to VIRFORM Holmedal et al. [84] studied extrusion of aluminium and the evolution of the microstructure with the ALMA-program. Earlier, Aukrust et al. [85] have performed a study of texture evolution during extrusion. The link between the mechanical properties and the microstructure was one-way,, i.e. the effect of microstructure evolution on mechanical properties was not assessed. Valuable contributions have also been made by van de Langkruis et al. [86], [87], Misiolek et al. [88], Duan et al. [89] and Sheppard [90]. Both Misiolek and Sheppard and co-workers have presented their work related to dynamic grain size evolution or recrystallization at the ESAFORM Conferences [91], [92], [93].

## 2.3 Validation of models

It is well known that the predictions of simulations can not be expected to be better than the information about material and interface conditions fed into the models. As the efficiency and capability of 3D numerical programs are gradually improving, the requirements to material and friction models necessarily increase. However, requirements depend on how the simulation technology is applied. Simple estimates of force and temperature may only require a rough modelling approach and approximate material data. The evolution of microstructure must be assessed with a multi-scale model capable of assessing texture evolution, dislocation densities and both dynamic and static recrystallization. Studies of flow stability most certainly will pose strict requirements to interface modelling and elastic responses of dies and work pieces. Strain localization and constitutive behaviour will also be important issues. During extrusion material particles may experience very different modes of loading as well as strain rates ranging from 0 to 10 000 1/s and temperatures in the range from 400 to 600 °C. Strain rate localization is of interest close to the outlet.

**Materials testing and analysis.** Traditionally, material data for aluminium extrusion have been obtained through torsion and compression tests. Both of the tests methods differ significantly from extrusion with respect to the deformation history and strains and strain rates involved. Djapic Oosterkamp et al. [94] have studied high strain rate properties of selected AA6082 and AA7108 aluminium alloys up to 2000 1/s with Hopkinson Split Pressure Bar and conclude that there appear to be different rate controlling mechanisms at low and high strain rates. If correct, such an observation should be taken into account in the extrusion modelling. It should be mentioned that there are mini-symposia at the ESAFORM Conference related to materials testing and inverse analysis that in the future will be increasingly important for the researchers within the fields of steel and aluminium extrusion.

**Inverse process modeling.** Establishment of material data through representative tests is important, but it is even more important to carefully validate models. The ESAFORM conference has got some examples of careful comparisons of simulated and experimental data. These include the work of Flitta et al. [95] on the evaluation of the use of Norton-Hoff and Zener-Hollomon equations in relation to hot extrusion. A similar study has been performed by Wajda et al. [96] and Moe [45]. The conference has also given examples of creativity related to experimental set-ups and process data extrusion. Logé et al. have devised small scale press equipment for tribology and rheology studies during hot extrusion [97]. A similar idea was presented by Moe et al. where a very long axisymmetric and slightly conical bearing channel was used to evaluate both friction and flow close to the die outlet [98]. Morsi et al. have discussed the consequences of miniaturization of the extrusion process [99].

The extrusion process is undeniably a very complex process, and there are limits to the precision of inverse analyses based on extrusion. Heat losses to the ram, die and container must be registered. Non-homogeneous deformation and heat generation must be taken into account. Attention must be paid to the study of causes and effects. For example, a change of the ram speed usually causes a change in extrusion pressure, but to which extent strain rate hardening is dampened by heating due to plastic dissipation must be assessed. Strain localization and friction behaviour must be taken into account. The studies of profile shape or seam welding are even more complex, and may require advanced multi-scale models. Obviously, FE models and new statistical and optimization techniques are of great value in such analyses. Furthermore, during the last twenty years, new experimental tools have been developed for the analysis of extrusion. Many of these have been presented at the ESAFORM Conferences.

**Grid technique for flow analysis.** Valberg's grid technique for the analysis of metal flow including extrusion [100], [19] is a significant improvement compared to older grid techniques. It renders possible detailed studies of flow in zones with significant shear localization. It was used successfully in the study of the development of dead zones during extrusion, seam welding and the generation of the profile surface in the bearing channel. The technique has been used in reverse studies of extrusion where friction and flow conditions have been modelled.

**Pressure measurement.** Pressure is a parameter of primary interest in the study of extrusion. Pressure differences drive the flow of material, and pressure or force is needed to move the ram. The pressure causes dies to deflect and profile shapes to deviate from nominal. By measuring the pressure at a number of places in the container it is possible to deduce information about flow and friction behaviour as well as about the elastic response of die and work piece. In the future, sheets of nano-pressure sensors will enable an almost continuous description of the pressure on the interface between the tools and work pieces. However, sensors for pressure measurement have been used in relation to metal forming at least since the 50s [101]. One of the most popular types of pressure sensors is the pin sensor, which has been used to study pressure and friction in relation to rolling and forging [102], [103], [104], [105], [106], [107], [108]. The most obvious alternative to the pin sensor is a sensor that constitutes an integral part of the tool and is placed in a cavity in the tool [109], [110], [111], [112]. Plančak et al. [113], [114] and Yoneyama et al. [115] have applied pin sensors in the study of cold extrusion. A new version of Plančak's sensor was presented at ESAFORM 2006 [72]. Yoneyama has developed optical sensors that constitute an integral part of the extrusion die. Although very accurate temperature compensation techniques can be used for fibre optic sensors, Yoneyama has only used the sensors for cold extrusion [116]. Mori et al. and Tan et al. have also used pressure sensors in the detailed study of cold extrusion of aluminium [117], [118]. Moe and Støren developed pressure sensors based on capacitive techniques, which were first presented at ESAFORM 2002 [119]. The sensors have been used at high temperature, and compensation schemes for temperature changes have been developed. The sensors have been used for upsetting and rod, thin-strip and tube extrusion [120], [70], and have proven very useful for the assessment of flow, friction and die deformation.

**Temperature measurement.** Due to plastic heat dissipation the billet is heated during extrusion. Temperature rises that exceed 150 °C are quite common. The outlet temperature must be carefully controlled during extrusion to prevent melting and cracking and to secure a desirable microstructure of the finished product. During isothermal extrusion the ram speed and/or the initial temperature distribution of the billet are controlled in order to obtain an outlet temperature that does not change with time. A simplified analytical and numerical analysis of induction heating of billets was presented by Moe et al. at ESAFORM 2005 [121]. Direct measurement of the outlet temperature has been performed by Lefstad [41] with thermocouples scraping the surface of the profile. Recently, Tercelj et al. presented another method for determining the temperature at the bearing surface [122].

**Shape measurement.** The die outlet shape is important to the shape of the profile, but extremely difficult to determine in-line. It is possible, however, to measure the shape of the profile optically as it exits the die. Moe [45] has presented a technique for high speed in-line profile shape measurement. Measurement is performed with a laser and a high-speed camera capable of collecting several thousand images per second. By projecting the laser line on the profile surface it is possible to determine the height and frequency of buckles while extruding a thin-strip of aluminium. The technique can also be used for thickness measurement, but in such a case measurements on both sides of the profile must be performed, and the results must be compared.

**Thin strip analysis.** In 1993 Støren outlined a new theory of extrusion and proposed a set of extrusion test cases for numerical codes [123]. One of the cases is with a thin- and wide strip with a 2D extrusion ratio equivalent to the 3D extrusion ratio (and consequently a profile width to container diameter ratio of $4/\pi$. The reason for setting this limitation is that the flow in the centre of the container will be in plane strain and can be assessed satisfactorily by 2D simulation codes. The thickness of the extruded profile can be set so that the limits of the mechanism of self-stabilization can be tested. The limiting case is a profile for which buckling or waving can be provoked at certain extrusion speeds and outlet temperatures. Thin strip extrusion has previously been studied by Abtahi [41], Grasmo [42] and Tverlid [43]. In a set of experiments performed in the early 2000s Støren and co-workers set out to critically test the limits of extrusion of a 6060 alloy and at the same time perform measurement of pressure, temperature and profile shape. The results of the work have been presented at ESAFORM 2002, 2003 and 2005 [98], [120], [123], [125]. Buckling was provoked for profiles of width 78.5 mm and thickness 1.1-1.2 mm. By increasing the initial billet temperature a stable flow could be made unstable. Halvorsen and Aukrust [126] have performed similar experiments with dies with various feeder dimensions and subsequently modelled buckling with the finite element software MARC, linking instability to the dimensions of the feeder. Both Lof [127] and van Rens have studied extrusion test cases very similar to the ones proposed by Støren. Zasadziński et al. have discussed the limits on wall thickness during aluminium extrusion [128].

## 3.  The future; the next ten years

The next ten years of scientific investigation of hot metal extrusion depends very much on the industry's acceptance and support of an open network of academics working together in a generic "multi-scale" trans-disciplinary environment, covering metal physics and metallurgy on nano-levels related to interfaces and friction, as well as micro-structural evolution on micro- and meso-level, crystal plasticity as basis for continuum constitutive description, numerical modelling of thermo-mechanical metal flow, interacting with thermo mechanic distortions of  tool stacks and dies. A particular challenge will be to develop sensors and common experimental procedures for the different laboratories around the world.

It is our hope that the mini-symposium Extrusion and Drawing can be the place where academics, industrial scientists and PhD-students can meet and have dialogs about generic problems, scientific methods, approaches and results from different laboratories around the world related to extrusion and its downstream processes as drawing, cold forging, hydro-forming and high speed forming.. In addition, ESAFORM enables the meeting and discussion between all mini-symposia scientists working with theories of plasticity and tribology, inverse modelling, sensors, experimental design and statistics, numerical methods and multi-scale simulation.

## Acknowledgments.

Two of the authors, Sigurd Støren and Per Thomas Moe, acknowledge many years of cooperation with scientists in Hydro Aluminium and SINTEF as well as financial support from Norwegian Research Council and Hydro Aluminium.

## References

1.    Manninen, T., Ramsay, P., Korhonen, A. S.: Flash Formation In Conform Extrusion. In: Proceedings of ESAFORM 2005, pp. 557-560.
2.    Müller, K. B.: Bending of extruded profiles during extrusion process, Int. J. Machine Tools and Manuf. 46 (2006) 1238-1242.
3.    Moe, P. T., Baringbing H. A., Lademo, O.-G., Welo T., Støren S.:  A study of pure bending of hollow extrusions with internal pressure. Proceedings of ESAFORM 2006, pp. 651-654.
4.    Iwata, K., Osakada, K., Fujino, S.: Analysis of hydrostatic extrusion by the finite element method. Transactions of the ASME, J. Engineering for Industry 94 (1972) 697–703.
5.    Zienkiewicz O.C., Godbole P.N.: Flow of plastic and viscoplastic solids with special reference to extrusion and forming processes, Int. J. Num. Meths. Eng. 8 (1974) 3.
6.    Argyris J. H., Doltsinis J. St.: On the large strain inelastic analysis in natural formulation Part I: Quasistatic problems, Computer Methods in Applied Mechanics and Engineering 20 (1979) 213-251.
7.    Kobayashi, S., Oh, S. I., Altan, T.: Metal forming and the finite element method. Oxford University Press, New York (1989).
8.    Shestopal, V. O., Shestopal, O. Ya.: Flow of metal in high-temperature extrusion, Comput. Methods Appl. Mech. Engrg. 25 (1981) 85-99.
9.    Huétink, J.: On the simulation of thermomechanical forming processes, Ph.D. Thesis, University of Twente, The Netherlands, 1986.
10.    Holthe K., Hanssen L., Støren S.: Numerical simulation of the aluminium extrusion process in a series of press cycles, In: Proceedings NUMIFORM'92, Rotterdam 1992, pp. 611-618.
11.    Mooi, H. G.: Finite element simulations of the aluminium extrusion process, PhD-thesis, University of Twente, The Netherlands (1996).
12.    Lof, J.: Developments in finite element simulations of aluminium extrusion, PhD-thesis, University of Twente, The Netherlands (2000).
13.    Eikemo M.S., Espedal M.S., Fladmark G.: On the numerical solution of a three dimensional extrusion model, Comput. Visual. Sci. 1 (1997) 1–14.
14.    van Rens, B. J. E.: Finite element simulations of the aluminium extrusion process, PhD-thesis, Technical University Eindhoven, The Netherlands (1999).
15.    van Rens, B. J. E., Brekelmans W. A. M., Baaijens F. P. T.: Steady, three dimensional flow calculation of aluminum extrusion with complicated die geometries. In: Proceedings of ESAFORM 1998

16. van Rens B. J. E., Brekelmans W. A. M., Baaijens, F. P. T., Modeling friction near sharp edges using an Eulerian approach. In: Proceedings of ESAFORM 1998.

17. Lof, J., Huétink, J.: Numerical simulation of the aluminium extrusion process. In: Proceedings of ESAFORM 1999, pp 29-32.

18. Williams, A. J., Croft, T. N., Cross, M.: Computational Modelling of Metal Forming Processes. In: Proceedings of ESAFORM 2002, pp. 67-70.

19. Valberg, H.: Extrusion welding in aluminium extrusion, Int. J. Mater. Prod. Technol. 17 (2002) 497-556.

20. Valberg, H., Sandbakk, Ø.: Direct non-lubricated Al-extrusion partitioned into the sub-processes shearing, scratching, radial compression and extrusion. In: Proceedings of ESAFORM 2003, pp. 307-310.

21. Hanssen, L., Lefstad, M., Rystad, S., Reiso, O.,Johnsen, V.: Billet surface flow in aluminium extrusion using "half-moon" dies. In: Proceedings of ESAFORM 1998.

22. Hatzenbichler, T., Buchmayr B., Mauerhofer, H.: Evaluation of the main influence parameters on the back-end defect in direct extrusion of aluminium alloys by numerical study with finite elements. In: Proceedings of ESAFORM 2004, pp. 609-612.

23. Hatzenbichler T., Buchmayr B., Umgeher A.: A numerical sensitivity study to determine the main influence parameters on the back-end effect. J. Mater. Process. Technol. 182 (2007) 73-78.

24. Flitta, I., Sheppard, T.: Investigation of friction during the extrusion of Al-alloys using FEM simulation. In: Proceedings of ESAFORM 2002, pp. 435-438.

25. Peng, Z., Sheppard, T.: Simulation of multi-hole die extrusion. Mater. Sci. Eng. A 367 (2004) 329-342.

26. Leśniak, D., Libura, W., Pidvysotskyy V., Milenin, A.: Influence of pre-chamber die geometry on extrusion of solid sections with different wall thickness. In: Proceedings of ESAFORM 2002, pp. 459-462.

27. Valberg, H. and Misiolek, W.: Plastic deformations in indirect hot extrusion of an aluminium alloy characterised by FEM-analysis with experiment. In: Proceedings of ESAFORM 2006, pp. 487-490.

28. Thedja W. W., Müller, K. B.: Indirect Extrusion With Active Friction. In: Proceedings of ESAFORM 2003, pp. 283-286.

29. Li, L., Zhou J., Duszczyk, J.: Numerical process design for isothermal extrusion. In: Proceedings of ESAFORM 2003, pp. 275-278.

30. Chanda T., Zhou J., Kowalski L., Duszczyk J.: 3D Fem Simulation of the Thermal Events During AA6061 Aluminum Extrusion. Scripta Materialia 41 (1999) 195–202.

31. Chanda T., Zhou J., Duszczyk J.: FEM analysis of aluminium extrusion through square and round dies. Materials and Design 21 (2000) 323-335.

32. Chanda T., Zhou J., Duszczyk J.: A comparative study on iso-speed extrusion and isothermal extrusion of 6061 Al alloy using 3D FEM simulation. J. Mater. Process. Technol. 114 (2001) 145-153.

33. Kim Y.-T., Ikeda K., Murakami T.: Metal flow in porthole die extrusion of aluminium. J. Mater. Process. Technol. 121 (2002) 107-115.

34. Kim Y-T., Ikeda K.: Flow Behavior of the Billet Surface Layer in Porthole Die Extrusion of Aluminum, Metall. Mater. Trans. 31A (2000) 1635-1643.

35. Peng, Z., Flitta I., Sheppard, T.: Simulation of Multi-hole Die Extrusion by Finite Element Method. In: Proceedings of ESAFORM 2004, pp. 605-608.

36. Peng, Z., Sheppard T.:, Effect of die pockets on multi-hole die extrusion. Mater. Sci Eng. A 407 (2005) 89-97.

37. Zhou J., Li L., Duszczyk J.: 3D FEM simulation of the whole cycle of aluminium extrusion throughout the transient state and the steady state using the updated Lagrangian approach. J. Mater. Process. Technol. 134 (2003) 383–397.

38. Duan, X., Velay X., Sheppard T.: Application of finite element method in the hot extrusion of aluminium alloys. Mater. Sci. Eng. A 369 (2004) 66-75.

39. Flitta, I., Sheppard, T.: Material flow during the extrusion of simple and complex cross-sections using FEM. Mat. Sci. Tech. 21 (2005) 648-656.

40. Clode M.P., Sheppard T.: Formation of Die Lines during Extrusion of AA6063. Mat. Sci. Tech. 6 (1990) 755-763.

41. Lefstad, M.: Metallurgical Speed Limitations during the Extrusion of AlMgSi-Alloys. Doctoral thesis, Norwegian Institute of Technology, Norway (1993).

42. Abtahi S.: Friction and interface reactions on the die land in thin-walled extrusion. Doctoral thesis, Norwegian Institute of Technology, Norway (1995).

43. Grasmo, G.: Friction and Flow Behaviour in Aluminium Extrusion. Doctoral thesis, Norwegian University of Science and Technology, Norway (1995).

44. Tverlid S.: Modelling of Friction in the Bearing Channel of Dies for Extrusion of Aluminium Sections. Doctoral thesis, Norwegian University of Science and Technology, Norway (1997).

45. Moe, P. T.: Pressure and Strain Measurement During Hot Extrusion of Aluminium. Doctoral thesis, Norwegian University of Science and Technology, Norway (2005).

46. Björk, T., Bergström, J., Hogmark, S.: Tribological simulation of aluminium hot extrusion. Wear 224 (1999) 216–225.

47. Aukrust T., LaZghab S.: Thin shear boundary layers in flow of hot aluminium. Int. J. Plasticity 16 (2000) 59-71.

48. LaZghab S., Aukrust T., Holthe K.: Adaptive exponential finite elements for the shear boundary layer in the bearing channel during extrusion. Comput. Methods Appl. Mech. Engrg. 191, 2002, pp. 1113-1128.

49. Lof, J.: Elasto-viscoplastic FEM simulations of the aluminium flow in the bearing area for extrusion of thin-walled sections. J. Mater. Process. Technol. 114 (2001) 174-183.

50. Lee, G.-A., Kwak D.-Y., Kim S.-Y., Im Y.-T.: Analysis and design of flat-die hot extrusion process 1. Three-dimensional finite element analysis. Int. J. Mech. Sci. 44 (2002) 915-934.

51. Ulysse, P.: Optimal extrusion die design to achieve flow balance. Int. J. Machine Tools & Manufacture (1999) 1047-1064.

52. Ulysse, P.: Extrusion die design for flow balance using FE and optimization methods. Int. J. Mech. Sci. 44 (2002) 319-341.

53. Mehta, B.V., Al-Zkeri I., Gunasekera J.S., Buijk A.: 3D flow analysis inside shear and streamlined extrusion dies for feeder plate design. J. Mater. Process. Technol. 113 (2001) 93-97.

54. Nanhai H., Kezhi, L.: Numerical design of the die land for shape extrusion. J. Mater. Process. Technol. 101 (2000) 81-84.

55. Raj, K. H. Sharma R. S., Srivastava, S., Patvardhan C.: Modeling of manufacturing processes with ANNs for intelligent manufacturing. Int. J. Machine Tools & Manufacture 40 (2000) 851-868.

56. Li Q., Smith C. J., Harris C., Jolly M.R.: Finite element modelling investigations upon the influence of pocket die designs on metal flow in aluminium extrusion Part II. Effect of pocket geometry configuration on metal flow. J. Mater. Process. Technol. 135 (2003) 197-203.

57. Buntoro I. A., Müller, K. B.: Overview of various bending methods directly after extrusion process. In: Proceedings of ESAFORM 2002, pp. 443-446.

58. Buntoro, I. A., Müller, K. B.: Investigation of the material flow during the concurrent bending and extrusion process of Al-profiles. In: Proceedings of ESAFORM 2003, pp. 279-282.

59. Wanheim, T., Balendra R., Qin Y.: Extrusion die for in-process compensation of component-errors due to die-elasticity. J. Mater. Process. Technol. 72 (1997) 177-182.

60. Chung J. S., Hwang S. M.: Application of a genetic algorithm to the optimal design of the die shape in extrusion. J. Mater. Process. Technol. 72 (1997) 69-77.

61. Arif A. F. M.: On the use of non-linear finite element analysis in deformation evaluation and development of design charts for extrusion processes. Finite Elements in Analysis and Design 39 (2003) 1007-1020.

62. Byon S. M., Hwang S. M.: Die shape optimal design in cold and hot extrusion. J. Mater. Process. Technol. 138 (2003) 316-324.

63. Long H.: Dimensional errors of cold formed components using FE simulation. J. Mater. Process. Technol. 151 (2004) 355-366.

64. Yan, H., Xia, J.: An approach to the optimal design of technological parameters in the profile extrusion process. Science and Technology of Advanced Materials 7 (2000) 127-131.

65. Mooi H.G., Koenis P.T.G., Huétink J.: An effective split of flow and die deformation calculations of aluminium extrusion. J. Mater. Process. Technol. 88 (1999) 67-76.

66. Wu, X, Zhao, G., Luan, Y., Ma, X.: Numerical simulation and die structure optimization of an aluminium rectangular hollow pipe extrusion process. Mat. Sci and Eng. A 435-436 (2006) 266-274.

67. Lee G.-A., Im Y.-T.: Finite-element investigation of the wear and elastic deformation of dies in metal forming. J. Mater. Process. Technol. 89–90 (1999) 123–127.

68. Lin, Z.: Optimization of die profile for improving die life in the hot extrusion process. J. Mater. Process. Technol. 142 (2003) 659-664.

69. Costa, A., Fichera, S., Fratini L., Micari, F.: Die design for aluminium thin tubes extrusion: a 2D simulation approach. In: Proceedings of ESAFORM 2004, pp. 601-604.

70. Mousoulis, M., Tseronis, D., Sideris, I. F., Fotopulos, F. A., Medrea, C.: Preliminary examination of the fracture surfaces of an aluminium hot extrusion

die, due to the premature failure. In: Proceedings of ESAFORM 2006, pp. 503-506.

71. Moe, P. T., Lange, H. I., Hansen, A. W., Wajda, W., Støren, S.: Experiments with die deflection during hot extrusion of hollow profiles. In: Proceedings of ESAFORM 2003, pp. 119-122.

72. Plančak, M., Kuzman, K., Vilotić, D., Čupković, D.: Loading of dies in cold extrusion – FE simulation and experimental verification. In: Proceedings of ESAFORM 2006, pp. 491-494.

73. Buffa G., Donati L., Fratini L., Tomesani L.: Solid state bonding in extrusion and FSW: Process mechanics and analogies. J. Mater. Process. Technol. 177 (2006) 344-347.

74. Donati, L., Tomesani, L.: 3D FEM Analysis of Seam Welds Formation in Aluminum Extrusion and Product Characterization. In: Proceedings of ESAFORM 2005, pp. 561-564.

75. Donati L., Tomesani L.: The prediction of seam welds quality in aluminium extrusion. J. Mater. Process. Technol. 153-154 (2004) 366-373.

76. Donati L., Tomesani L.: The effect of die design on the production and seam weld quality of extruded aluminium profiles. J. Mater. Process. Technol. 164-165 (2005) 1025-1031.

77. Edwards S.-P., Den Bakker A. J., Neijenhuis J. L., Kool W. H., Katgerman L.: The Influence of the Solid State Bonding Process on the Mechanical Integrety of Longitudinal Weld Seams. JSME International Journal Series A 49 (2006) 63-68.

78. Jo H. H., Lee S. K., Lee S. B.,Kim B. M.: Prediction of welding pressure in the non-steady state porthole die extrusion of Al7003 tubes. Int. J. of Machine Tools & Manuf. 22 (2002) 753-759.

79. Jo H.H., Jeong C.S., Lee S.K., Kim B.M.: Determination of welding pressure in the non-steady-state porthole die extrusion of improved Al7003 hollow section tubes. J. Mater. Process. Technol. 139 (2003) 428–433.

80. Kim, K. J., Lee C.H., Yang D.Y.: Investigation into the improvement of welding strength in three-dimensional extrusion of tubes using porthole dies. J. Mater. Process. Technol. 130-131 (2002) 426-431.

81. Kleiner M., Schikorra M.: Simulation of welding chamber conditions for composite profile extrusion. J. Mater. Process. Technol. 177 (2006) 587-590.

82. Ghiotti, A., Bruschi S., Dal Negro, T.: An Innovative Experimental Procedure To Investigate Seam Welding In Aluminium Extruded Profiles. In: Proceedings of ESAFORM 2005, pp. 553-556.

83. Loukus A., Subash G., Imaninejad M.: Mechanical properties and microstructural characterization of extrusion welds in AA6082-T4. J. Mat. Sci. 39 (2004) 6561-6569.

84. Holmedal B., Marthinsen, K., Nes E.: A unified microstructural metal plasticity modell applied in testing, processing and forming of aluminium alloys. Zeitschrift für Metallkunde 6 (2005) 532-545.

85. Aukrust T., Tjøtta S., Vatne H.E., Van Houtte P.: Coupled FEM and Texture Modelling of Plane Strain Extrusion of an Aluminium Alloy. Int. J. Plasticity 13 (1997) 111-125.

86. van de Langkruis, J., Kool, W.H., Sellars, C.M., van der Winden, M.R., van der Zwaag, S.: The effect of β, β' and β'' precipitates in a homogenised AA6063 alloy on the hot deformability and the peak hardness. Mater. Sci. Eng. A 299 (2001) 105-115.

87. van de Langkruis, J., Lof, J., Kool, W.H., van der Zwaag, S., Huétink, J.: Comparison of experimental AA6063 extrusion trials to 3D numerical simulations, using a general solute-dependent constitutive model. Comp. Mater. Sci. 18 (2000) 381–392.

88. Misolek, W. Z.: Material physical response in the extrusion process. J. Mater. Process. Technol 60 (1996) 117-124.

89. Duan, X., Sheppard, T.: Simulation and control of microstructure evolution during hot extrusion of hard aluminium alloys. Mater. Sci. Eng. A 351 (2003) 282-292.

90. Sheppard, T.: Prediction of structure during shaped extrusion and subsequent static recrystallisation during the solution soaking operation. J. Mater. Process. Technol. 177 (2006) 26-35.

91. Misiolek, W. Z., van Geertruyden, W. H., Claves, S. R., Bandar, A. R.: Characterization of aluminium extrusions: modeling of metal flow and microstructure response. In: Proceedings of ESAFORM 2002, pp. 431-434.

92. Flitta, I., Sheppard, T.: Prediction and control of substructure evolution during aluminium extrusion using Finite Element Method. In: Proceedings of ESAFORM 2004, pp. 597-600.

93. Flitta, I., Sheppard, T., Peng, Z.: Effects of evolution of structure during and subsequent to extrusion. In: Proceedings of ESAFORM 2006, pp. 479-482.

94. Djapic Oosterkamp, L., Ivankovic, A., Venizelos G.: High strain rate properties of selected aluminium alloys. Materials Science and Engineering A278 (2000) 225–235.

95. Flitta, I., Sheppard, T.: The effects of flow stress on the extrusion parameters, In: Proceedings of ESAFORM 2004, pp. 617-620.

96. Wajda, W., Moe, P. T., Abtahi, S., Støren, S.: An evaluation of material behaviour during extrusion of AA6060 rods. In: Proceedings of ESAFORM 2004, pp. 245-248.

97. Logé, R. E., Minvielle, M., Cini, E.. Le Floc'h, A., Felder, E.: Development of a Laboratory Hot Extrusion Apparatus for Tribology and Rheology Parameters Identification. In: Proceedings of ESAFORM 2005, pp. 569-572.

98. Moe, P. T., Lefstad, M., Flatval, R., Støren S.: Measurement of Temperature and Die Face Pressure during Hot Extrusion of Aluminium. Int. J. Forming Processes 3 (2003) 241-270.

99. Morsi K., Nanayakkara N., McShane H.B., McLean M.: Preliminary evaluation of hot extrusion miniaturization, J. Mat. Sci. 34 (1999) 2801-2806.

100. Valberg, H., Metal flow in the direct axisymmetric extrusion of aluminium, J. Mater. Process. Technol. 31 (1992) 39-55.

101. van Royen G.T., Backofen W.A.: Friction in Cold Rolling. J. Iron Steel Inst. 186 (1957) 235-244.

102. Banerji A., Rice W.B.: Experimental Determination of Normal Pressure and Friction Stress in Roll Gap during Cold Rolling. CIRP Ann. (1972) 53-54.

103. Al-Salehi F.A.R., Firbank T.C., Lancaster P.R.: An experimental determination of the roll pressure distributions in cold rolling. Int. J. Mech. Sci. 15 (1973) 693-710.

104. Hatamura Y., Yoneyama T.: Measurement of Actual Stress and Temperature on a Roll Surface during Rolling. JSME Int. J. 31 (1988) 465-469.

105. Lenard J.G., Malinowski Z.: Measurement of friction during the warm rolling of aluminium. J. Mater. Process. Technol. 3-4 (1993) 357-371.

106. Jeswiet J.: Aspect ratio, friction forces and normal forces in strip rolling. J. Mater. Process. Technol. 53 (1995) 846-856.

107. Hansen A.W., Valberg H., Welo T.: A technique for measuring the pressure on the tool surface. Proceedings 4th Int. Conf. Technology of Plasticity, Bejing (1993) Vol. I pp. 303-308.

108. Hansen A.W., Valberg H.: Accurate Measurements Inside the Tool in Hot Working of Metals. Proceedongs of the 6th Int. Al. Extr. Techn. Sem, Chicago, Illinois (1996) Vol. II pp. 11-15.

109. Daneshi G.H., Hawkyard J.B.: A split-platen pressure cell for the measurement of pressure distribution in upsetting operations. Int. J. Mech. Sci. 13 (1971) 355-371.

110. Hum B., Colquhoun H.W., Lenard J.G.: Measurements of friction during hot rolling of aluminum strips. J. Mater Process. Technol. 60 (1996) 331-338.

111. Dellah A., Wild P.M., Moore T.N., Shalaby M., Jeswiet J.: An Embedded Friction Sensor Based on a Strain-Gauged Diaphragm, J. Manuf. Sci. Eng. 124 (2002) 523-527.

112. Du H., Klamecki B.E.: Force Sensors Embedded in Surfaces for Manufacturing and Other Tribological Process Monitoring. ASME J. Manuf. Sci. Eng. 121 (1999) 739-748.

113. Plančak M., Bramley A.N., Osman F.H.: Some observations on contact measurement by pin load cell in bulk metal forming, J. Mater. Process. Technol. 60 (1996) 339-342.

114. Cvahte P., Dragojevic, V., Fajfar P., Rodič T.: Measurement of forces during extrusion and drawing of Al rods. Kovine Zlitine Tehnologije 33 (3-4) (1999) pp. 249-252

115. Yoneyama T., Kitagawa M.: Measurement of the contacting stress in extrusion, Proceedings of the 4th ICTP, Bejing (1993) Vol II pp. 553-558.

116. Yoneyama T.: Development of a new pressure sensor and its application to the measurement of contacting stress in extrusion. J. Mater. Process. Technol. 95 (1999) 71-77.

117. Mori T., Takatsuji N., Matsuki K., Aida T., Murotani K., Uetoko K.: Measurement of pressure distribution on die Surface in hot extrusion of 1050 aluminium rod. J. Mater. Process. Technol. 130-131 (2002) 421–425.

118. Tan X., Bay N., Zhang W.: Friction measurement and modelling in forward rod extrusion tests. J. Eng. Tribology 217 (2003) 71-82.

119. Moe, P. T., Støren, S.: A technique for measuring pressure on the die face during extrusion. In: Proceedings of ESAFORM 2002, pp. 463-466.

120. Lefstad, M., Moe, P. T., Støren, S., Flatval, R.: Thin-strip aluminium extrusion – pressure, temperature and deflection recordings of the extrusion die. In: Proceedings of ESAFORM 2002, pp. 471-474.

121. Moe, P. T., Wajda, W., Støren, S., Lefstad, M., Flatval, R.: An Experimental and Numerical Study of Induction Heating. In: Proceedings of ESAFORM 2005, pp. 577-580.

122. Tercelj, M., Kugler G., Turk R., Cvathe P., Fajfar P.: Measurement of temperature on the bearing surface of an industrial die and assessment of the heat transfer coefficient in hot extrusion of aluminium: a case study. Int. J. Vehicle Design 1-2 (2005) 93-109.

123. Støren S.: The Theory of Extrusion – Advances and Challenges. Int. J. Mech. Sci. 35 (1993) 1007-1020.

124. Wajda, W., Lefstad, M., Moe, P. T., Abtahi, S., Flatval R., Støren, S.: A study of the limits of self-stabilization during extrusion of thin-strips. In: Proceedings of ESAFORM 2003, pp. 267-270.

125. Wajda, W., Moe, P. T., Støren, S., Lefstad M., Flatval, R.: Measurement of Temperature and Pressure During Thin-Strip Extrusion. In: Proceedings of ESAFORM 2005, pp. 545-548.

126. Halvorsen, F., Aukrust T.: Studies of the mechanisms for buckling and waving in aluminium extrusion by use of a Lagrangian FEM software. Int. J. Plasticity 22 (2006) 158-173.

127. Lof, J., Blokhuis Y.: FEM simulations of the extrusion of complex thin-walled aluminum sections. J. Mater. Process. Technol. 122 (2002) 344-354.

128. Zasadziński J., Libura W., Misiołek W.Z.: Minimal Attainable Wall Thickness in Aluminium Extruded Sections. Proceedings. 7th Int. Al. Extr. Techn. Sem., Chicago (Ill.), 2000, Vol. I pp. 365-370.

Proceedings of ESAFORM 1998 - Proceedings of the 1st International ESAFORM conference on Material Forming, edited by Chenot J. L. et al., Sophia-Antipolis, France, 1998.

Proceedings of ESAFORM 1999 - Proceedings of the 2nd International ESAFORM conference on Material Forming, edited by Covas J. A., Guimaraes, Portugal, 1999,

Proceedings of ESAFORM 2002 - Proceedings of the 5th International ESAFORM conference on Material Forming, edited by Pietrzyk, M. et al., Kraków, Poland, 2002

Proceedings of ESAFORM 2003 - Proceedings of the 6th International ESAFORM conference on Material Forming, edited by Brucato, V., Salerno, Italy, 2003

Proceedings of ESAFORM 2004 - Proceedings of the 7th International ESAFORM conference on Material Forming, edited by Støren, S., Trondheim, Norway, 2004

Proceedings of ESAFORM 2005 - Proceedings of the 8th International ESAFORM Conference on Material Forming, edited by Banabic, D., Cluj-Napoca, Romania, 2005,

Proceedings of ESAFORM 2006 - Proceedings of the 9th International ESAFORM Conference on Material Forming, edited by Juster, N. et al., Glasgow, United Kingdom, 2006.

# Modelling of Cutting and Machining:
# 10 years of ESAFORM activity


Philippe Lorong[1], Fabrizio Micari[2] and Maurice Touratier[1]

[1] LMSP, Lab. for Mechanics of Systems and Processes, ENSAM, 75013 Paris (France)
[2] Dept. of Manufacturing and Management Eng., University of Palermo, 90128 Palermo (Italy)



**Abstract.** This paper reports on the state of the art in the simulation of cutting and machining processes. The contributions provided by researchers all over the world and published on the Proceedings of the European Scientific Association for material FORMing (ESAFORM) Conferences are highlighted. They role played by ESAFORM in this field of research has been quite active, as demonstrated by the number of contributions, their relevant scientific content and finally by the vitality of the minisymposium on Modeling of Machining that has been organized since 2001 with no interruptions.

**Keywords:** Machining and Cutting Modelling.


# Introduction

Machining is a term covering a large collection of manufacturing processes designed to remove material from a work piece. The primary machining processes are turning, shaping, milling, drilling, sawing, abrasive machining and broaching, but several other advanced machining methods are widely utilized nowadays, such as electric discharge machining (EDM), laser cutting, chemical milling, high-pressure water cutting, electrochemical machining and so on.

It is well known that machining operations have a key role in manufacturing. According to a study carried out in 1998 by Merchant [1], about 15% of the value of all the mechanical components manufactured worldwide is achieved through machining. Probably this quota increased in the last decade due to the increase of the machining market and the demands of the manufacturing industries for micro and nano-components which opened new scenarios as micro and nano-machining. Thus, because of its large economic and technical importance, a great number of researches have been carried out in order to optimise machining processes in terms of improved quality, increased productivity and reduced cost. However, according to Armarego et al. [2], in the USA the cutting tool is properly selected less than in the 50% of the cases, the tool is used at the rated cutting speed only in the 58% of the cases, and finally only 38% of the tools are used up to their full tool-life capability. This situation urges the need for developing more scientific approaches in order to improve machining performance.

Several enhancements have been obtained in the last decade thank to the use of Finite Element simulation. Numerous researchers used FEM to predict some typical machining variables, such as cutting forces, chip morphology, surface integrity, etc. These contributions are well recognized in a reviewer paper proposed by Mackerle [3].

In the last ten years also the ESAFORM Community provided an important contribution to improve the knowledge on machining processes and, in particular, to highlight the advantages and the limits of Finite Element simulations. The former proposal of a minisymposium on Machining and Cutting Modelling at the ESAFORM Conference in Liege (2001) and its development with no interruptions at all the following Conferences, have permitted to assess an important forum where innovative and high quality researches have been presented.

In computational mechanics it is usual to define simulation models which depend on the nature of the expected results. It is clear that all the phenomena observed during the cutting process need appropriated reference scales to be properly and consistently identified and analyzed: separation of physical effects having large lengths of variation from those with smaller characteristic lengths must be done.

**Fig. 1.** The macroscopic and the mesoscopic scales.

It is hereafter possible to introduce two different scales (see Figure 1) to consistently study the cutting operation:

- a *macroscopic scale* to analyze the whole work piece-tool-machine (WTM) system; this macroscopic scale will be more appropriated when dealing with either (both) the WTM's dynamics or (and) predicting the geometry of the machined surface,

- a *mesoscopic scale* at the tool tip/work piece level where the material is modeled as a continuum domain. In machining, this scale is the usual one to make the study of chip formation, in order to analyze chip geometry, stresses and temperatures, as well as the repartition of cutting forces, friction and thermo-mechanical characteristics/properties along the tool-tip.

In this paper a wide bibliography of the papers published in the period 1998-2006 on ESAFORM Conferences Proceedings dealing with machining modeling is reported. For sake of clearness, the contributions have been grouped in two main categories, namely the one referring to a macroscopic approach and the ones carrying out a mesoscopic analysis.


# Enhancements in macroscopic modelling

At the macroscopic scale sufficiently accurate mechanical and geometric models have to be built in order to predict :

1.  the behavior of the WTM system during machining (static or dynamic deformations, dynamical instabilities such as chatter),
2.  the cutting forces, and possibly associated thermal aspects,
3.  the geometry of the final surface (roughness, waviness, form).

The very particularity of the machining simulation at the macroscopic scale is the highly evolutive interaction between tool and work piece. This interaction, modeled by a cutting law may be "history" dependent. This is due to the evolution during machining of the work piece domain (material removal). The cutting law gives the resulting forces versus instantaneous cutting conditions.

As the cutting law is a central point in machining simulation it focuses a lot of works. The first model, based on a simple material flow definition under orthogonal cutting assumption, was introduced by Merchant [77]. Many improvements of this model were proposed during this last decades and ESAFORM papers give a good illustration of corresponding researches. The most relevant enhancements are found in [7] and [57] which include the extension of orthogonal cutting taking into account wear and thermal aspects, in [63] where thermal diffusion is incorporated (necessary in high speed machining); [72] for an extension to oblique cutting; [18][19] for ball-end milling; and [9] which includes ploughing forces occurring under high vibrational conditions. Most of these papers are cited in next sections as they can also be seen like analytical mesoscopic approaches.

To accurately compute cutting forces, the instantaneous cutting conditions are needed : the relative position and motion (velocities) between the tool and work piece, the cutting section. To define the latter, geometrical models are necessary. They may be very simple and based on the finite element model of the work piece [8][20][42]. More detailed analyses exist in order to make accurate studies. They include fine description of the machined surface: dexels in [28][39][44] or, in [52], a Monte Carlo integration procedure to calculate the undeformed chip volume.

With the exception of a few purely cinematic simulations, relative motion between work piece and tool is described by static or dynamic models. Some significant enhancements can be found in [7][8][20][19][28][42][44] when the work piece becomes flexible (thin-walled work pieces), and in [10][17][33] when dynamics of the machine is significant.

To study the dynamic behavior of the process, time domain methods [9][10][17][19][28][44] or frequency domain approaches [42] are used. The former attempt to describe a whole machining operation, while the latter deal with the stability of the process (stability lobes).

TABLE 1 gives a synthesis of the previous considerations. On one row, many occurrences of a reference can occur as it includes different aspects defined by columns.

TABLE 1. Macroscopic level.

| | Geom. Aspects | | | Work piece deformation | | Machine behavior | Time dom. | Freq. dom. | Cutting forces |
| | Surf. prediction | | Chip | | | | | | |
| | FE | Other | (undef.) | Statics | Dynamics | (dynamics) | | | enhancements |
|---|---|---|---|---|---|---|---|---|---|
| 2001 | [8] | [10] | | [7][8] | | [9][10] | [9][10] | | [7][9][63] |
| 2002 | | | | | | | | | |
| 2003 | [20] | [19][71] | | [20] | [19] | [17] | [17][19] | | [17][18][57][72] |
| 2004 | | [28] | [28] | | [28] | [33] | [28] | | [29][33] |
| 2005 | | [39] | [39][52] | | | | | | [38][39] |
| 2006 | [42] | [44] | [44] | | [42][44] | | [44] | [42] | |

In order to accurately simulate machining, it is still very difficult to have sufficiently representative mechanical models. At the macroscopic scale, these models are the cutting law with discontinuous cutting condition (case of milling) and dynamic models of the machine, in particular of the spindle.

One way to solve the encountered difficulties to identify a cutting law is to couple the analyses at macroscopic and mesoscopic scales. The definition of a consistent way to go from one scale to another will be of particular interest, and the ESAFORM congress is the ideal place to achieve enhancements in this topic.

## Enhancements in mesoscopic modelling

As far as the papers aimed to a mesoscopic analysis of the machining process are concerned, very relevant enhancements have been registered in the last ten years, mainly as regards the reliability and the effectiveness of the obtained results.

Actually numerical simulation of machining processes still represents a very critical task, since several issues have not been completely established. Among them flow stress modeling at high strain, strain rate and temperatures, friction modeling at the tool chip interface and chip generation modeling are probably the most relevant. Some improvements have been registered in the last ten years as concern flow stress modeling: in the former ESAFORM papers the well known Johnson-Cook law was assumed and high speed shear tests were utilized to identify the material parameters appearing in such law [75,76]. Most recently some innovative models including the effect of material hardness at high temperature have been proposed [30].

Friction modeling is generally based on Coulomb models, constant shear models or finally sticking-sliding models: no great enhancements in this field occurred in the last decade. Finally chip generation is generally taken into account by means of numerical remeshing in machining modeling, while several advanced damage mechanics models have been implemented to simulate blanking [37,73,74].

However it's worth pointing out that the wider improvements have been registered as far as the results provided by the simulation are concerned, as well their reliability and effectiveness. Ten years ago simulation results were mainly limited to the prediction of the cutting forces and to the analysis of simple processes (orthogonal machining, blanking) under heavy hypotheses. Today the amount of results provided by machining modeling is really huge: for sake of comprehensibility the present review is divided into the follow parts:
- prediction of mechanical variables in machining;
- prediction of chip geometry;
- prediction of thermal variables in machining;
- prediction of tool wear in machining;
- prediction of the surface integrity in machining.

In the next paragraphs the most relevant enhancement achieved in the last ten years will be presented, as well as the new challenges for the next years: ESAFORM has to continue to be protagonist in machining modeling and therefore these new challenges will represent the topics of the related minisymposium in the next future.


**Prediction of mechanical variables**

This section deals with the investigation and the prediction of mechanical variables (cutting forces, strain, shear localization problems, normal and shear stresses) related to the metal cutting processes. These mechanical variables are strictly dependent on the work piece parameters (material type, crystallography, temperature, pre-deformation), cutting tool parameters (tool design geometry, material), and cutting parameters (speed, feed, depth of cut, environment). The former studied were focused on the prediction of a few specific topics (mainly the cutting forces) in turning, blanking and milling operations [4-10, 12, 16-19, 22-26, 28, 29, 31-33, 35, 37-40, 42-44]; most recently the attention has been enlarged to the prediction of the pressure distribution on the rake face of the tool and to strain localization problems [14, 15, 19, 20, 40, 41]. Furthermore the latter studies are pursued to a greater comprehension of the complex physical phenomena underlying the specific machining process: in this sense the development of effective friction models and/or the assessment of the existing ones has been pursued by several researchers [20, 21, 34, 41].

Several materials were considered in the ESAFORM papers, the most common were steels and aluminium alloys although, in the last few years, several efforts were oriented on hard steels [30], titanium alloys [65], two-phase alloys and other materials [16, 27].

As far as the type of simulation is concerned, two basic models were taken into account, namely orthogonal cutting, and oblique cutting models. The orthogonal cutting studies were the former to be carried out: they can be useful for understanding the basic mechanics of machining processes. On the contrary, 3D models are still very expensive in terms of CPU time, although they represent most of the real industrial cases.

It's worth outlining that nowadays FE codes permit to obtain proper predictions of the cutting forces and pressure distribution, once the constitutive model of the material and the frictional conditions have been effectively assigned. The next challenge is to make 3-D simulations competitive also in an industrial environment: in order to pursue this objective probably new simulation approaches, for instance the meshless method, must be developed. A few preliminary proposals in this field can be recognized in ESAFORM Proceedings [11, 13, 36].

**Prediction of chip geometry**

Metal cutting is a chip-formation process. The problem of chip formation and its control has been studied trying to define the mechanism of chip formation, chip flow and chip breaking [30, 45-52]. The parameters involved are the same as above and include tool and work piece materials, cutting data, tool geometry and so on. Chip flow along the contact length with the tool is a very important factor, because it influences in a very relevant way energy dissipation by friction and heat transfer conditions: most of the heat generated in machining is removed from the cutting zone by the chip. Chip control is necessary, especially in turning [30, 45-50] and drilling [51]. Milling creates a natural chip length due to the limited length of cutting edge engagement [52].

**Prediction of the thermal aspects**

High temperatures in machining are the cause of unsatisfactory tool life and limitations on cutting speed. Several numerical techniques were proposed to study the thermal problems in machining and to calculate the temperature distributions within both the work piece and the tool [34, 41, 53-62]. However, the prediction of the thermal aspects is probably the most critical task up to now due to implemented numerical formulation.

In fact, most of the numerical analyses of machining are based on the updated-Lagrangian formulation and carry out a coupled thermo-mechanical analysis. In this case only few milliseconds of cutting time can be simulated, even in the case of 2-D simulation of orthogonal cutting conditions. This aspect is a heavy limitation for the effectiveness of the numerical modeling, since in this short time thermal steady state cannot be achieved. Furthermore, temperature prediction influences other important process variables, such as tool wear and residual stresses.

The above drawback was highlighted by several authors [34, 41, 62], which have proposed different numerical strategies to overcame this problem [41, 55, 59, 62].

In particular, among the parameters to be set in the numerical simulation, the global heat transfer coefficient, $h$, at the tool-chip interface plays a relevant role because it directly impacts on the temperature evolution. A few researchers proposed to arbitrarily fix it [34, 62] to a very high value in order to quickly reach steady state thermal conditions in the tool, although this methodology is physically not consistent [41].

**Prediction of tool wear**

Wear prediction in machining has been recently studied by using FEM technique although it still represents a "border" application. A couple of relevant research issues have to be strongly enhanced to achieve an effective prediction [64]: the former, as mentioned cited, is related to the reliability of the calculated temperature field, the latter is connected to the absence of predictive models for coated tool.

Whereas these problems, several authors tried to model tool wear by FEM [57, 63, 65-67] taking into account advanced tool wear models, mainly based on diffusive wear models.

More in detail, Nouari and Molinari [63] proposed a tool wear model considering the diffusion mechanism to predict uncoated tungsten carbide wear when cutting steels. They also investigated the influence of the process variables (cutting speed and feed rate) as well as of the tool geometry on the tool wear. Filice et al. [67] improved the original Nouari and Molinari model providing a consistent physical law. Hua et al. [65] proposed a diffusive wear model able to predict crater wear on tungsten carbide tools cutting Titanium alloys. The obtained numerical results were comparable with the experimental ones.

However, in the last few years the research on tool wear mechanisms appears to be reduced, although tool wear prediction is one of the most critical issues. Probably the reason is strictly related to the unreliability of temperature prediction in machining. Only once this problem is solved, tool wear prediction can be heavily carried out and extended to coated tools, which are the most important industrially.

**Prediction of the surface integrity**

The reliability of mechanical components depends to a large extent on the physical state of their surface layers. This state includes the distribution of residual stresses induced during the machining process.

Residual stresses can enhance or impair the ability of a component to withstand loading conditions in service (fatigue, creep, stress corrosion cracking, etc.), depending on their nature: compressive or tensile, respectively. Also, residual stress distributions on a component may also cause dimensional instability (distortion) after machining [68]. This poses enormous problems in structural assembly as affects the structural integrity of the whole part. Therefore, prediction and control of the residual stresses in machining is absolutely necessary. For these reasons, in the last years, several researches were oriented on this field [30, 63, 68-70] even if more efforts must be done to improve the prediction of the surface integrity in term of advanced flow stress models, temperature prediction and modeling of tool wear when advanced tool materials are used.

Umbrello et al. [30, 70] and Dillon et al. [63] proposed advanced flow stress models for AISI 52100 bearing steel and AISI 316l stainless steel, respectively. The obtained numerical residual stresses predictions were quite similar to those found experimentally even if a 2D model was set in both the cases. Furthermore Umbrello et al. [70] used an hybrid ANN-FEM approach to predict residual stress profiles.

TABLE 2 gives a synthesis of all the previous considerations. ESAFORM contributions for each one of the above listed sectors of research are reported. Table 2 graphically demonstrates the relevant growth of interest associated to mesoscopic analyses of machining processes: ten years ago few papers were published and the results were mainly limited to the prediction of the cutting forces and to the analysis of simple processes (orthogonal machining, blanking) under heavy hypotheses. Today the number of involved laboratories and the amount of results provided by machining modeling is really huge and appears still increasing.

TABLE 2.  Mesoscopic level

|  | Mech. Aspects | Chip Formation | Thermal Aspects | Tool Wear | Surface Integrity |
|---|---|---|---|---|---|
| 1998 | [4, 5] | [45] | - | - | - |
| 1999 | [6] | - | - | - | - |
| 2000 | - | - | - | - | - |
| 2001 | [8-10] | [46-48] | [53-55] | [63] | [68] |
| 2002 | [11-14] | - | [56] | [64] | - |
| 2003 | [15-26] | [49] | [57] | [57, 65, 66] | - |
| 2004 | [27-33] | [30, 50] | [58, 59] | - | [27] |
| 2005 | [34-40] | [51, 52] | [34, 60, 61] | - | [69] |
| 2006 | [41-44] | - | [41, 62] | [67] | [62, 70] |

## Conclusions

Modeling of cutting and machining represents a key issue for the future development of such technologies in the future. Scientific research in this field is facing complex and challenging tasks, both at the macroscopic and mesoscopic level. The role of the scientific community working in this area inside ESAFORM has been relevant in the past and appears very promising for the future.

## References


1. M. E. Merchant, *Mach. Sc. Tech.* **2/2**, 157-163 (1998).
2. E. J. A. Armarego, I. S. Jawahir, V. A. Ostafiev, P. K. Venuvinod, "Modeling of Machining Operations", CIRP Working Group Paper, STC-C Paris, France, 1996.
3. J. Mackerle, *Int. J. Mach. Tools & Manuf.* **43**, 103-114 (2003).

4.  A. M. Goijaerts, Y. W. Stegeman, L. E. Govaert, D. Brokken, W. A. M. Brekelmans and F. P. T. Baaijens, "Modelling of the metal blanking process", Proc. of the 1st ESAFORM Conference, 239-242 (1998).

5.  H. H. Wisselink and J. Huetink, "3-D modelling of sheet metal shearing", Proc. of the 1st ESAFORM Conference, 243-246 (1998).

6.  M. Vaz Jr., "Numerical Simulation of Machining Processes - a Challenge", Proc. of the 2nd ESAFORM Conference, 89-92 (1999).

7.  T. Matsumura, A. Ishii, T. Shirakashi and E. Usui, "Error compensation with simulating turning process" Proc. of the 4th ESAFORM Conference, 623-626 (2001).

8.  L. Masset, J. F. Debongnie and P. Beckers, "Face milling and turning simulation with the finite elements method", Proc. of the 4th ESAFORM Conference, 627-630 (2001).

9.  M. Segreti, A. Moufki, D. Dudzinski and A. Molinari, "Modelling of non-linear vibrations in orthogonal cutting", Proc. of the 4th ESAFORM Conference, 647-650 (2001).

10. A. Dugas, J. J. Lee and J. Y. Hascoet, "Feed rate and tracking errors simulation in high speed milling", Proc. of the 4th ESAFORM Conference, 667-670 (2001).

11. F. Chinesta, P. Lorong, D. Ryckelynck, G. Coffignal, M. Touratier, M. A. Martinez E. Cueto and M. Doblare, "Thermomechanical cutting model discretisation : eulerian or lagrangian , mesh or meshless?", Proc. of the 5th ESAFORM Conference, 567-570 (2002).

12. J. M. A Cesar de Sa, P. M. A. Areias and R. M. Natal George, "Numerical model for cutting using a gradient damage model and element erosion", Proc. of the 5th ESAFORM Conference, 571-574 (2002).

13. P. Lorong, F. Chinesta and D. Ryckelynck, "History-dependent material and the natural element meshless method. study of two approaches in finite transformations", Proc. of the 5th ESAFORM Conference, 579-584 (2002).

14. L. Filice, F. Micari, L. Pagnotta and D. Umbrello, "Prediction and measurement of pressure distribution on the tool in orthogonal cutting", Proc. of the 5th ESAFORM Conference, 595-598 (2002).

15. T. H. C. Childs, "Numerical experiments on material properties and machining parameters influencing normal contact stresses between chip and tool", Proc. of the 6th ESAFORM Conference, 527-530 (2003).

16. T. Shirakashi, "Fem simulation analysis on ductile mode glass machining process", Proc. of the 6th ESAFORM Conference, 539-542 (2003).

17. M. Clausen and K. Tracht, "Dynamic analytical cutting force prediction", Proc. of the 6th ESAFORM Conference, 543-546 (2003).

18. M. Fontaine, A. Moufki, A. Devillez and D. Dudzinski, "Predictive modelling of cutting forces in ball-end milling", Proc. of the 6th ESAFORM Conference, 547-550 (2003).

19. A. Marty, S. Assouline, P. Lorong and G. Coffignal, "Prediction of the workpiece final surface taking into account the workpiece/tool/machne vibrations", Proc. of the 6th ESAFORM Conference, 535-538 (2003).

20. L. Masset, J. F. Debongnie and F. Berger, "Solving problems with contact in machining process simulation", Proc. of the 6th ESAFORM Conference, 555-558 (2003).

21. P. J. Arrazola, F. Meslin and S. Marya, "Tool-chip contact analysis in numerical cutting modelling", Proc. of the 6th ESAFORM Conference, 559-562 (2003).

22. O. W. Dillon, "Fem analysis of machining with a rounded tip cutting tool", Proc. of the 6th ESAFORM Conference, 563-566 (2003).

23. J. F. Mariage, K. Saanouni and P. Lestriez, "Sheet metal deburring and drilling simulation by continuum damage mechanics", Proc. of the 6th ESAFORM Conference, 571-574 (2003).

24. M. T. Santos, P. Lasne and J. L. Romero, "Application of finite elements method to the optimisation of 3d blanking processes", Proc. of the 6th ESAFORM Conference, 583-586 (2003).

25. B. Denkena, V. Jivishov and M. Clausen, "Deviation analysis of fem based cutting simulation", Proc. of the 6th ESAFORM Conference, 587-590 (2003).

26. S. Beccari, L. Filice and L. Fratini, "On the role of superimposed stress states on fracture mechanics in blanking", Proc. of the 6th ESAFORM Conference, 595-598 (2003).

27. T. Shirakashi, "Simulation on cutting process of aluminium single crystal", Proc. of the 7th ESAFORM Conference, 709-712 (2004).

28. S. Assouline, P. Lorong and G. Coffignal, "A method for simulating the machining of thin-walled taking into account their vibrations", Proc. of the 7th ESAFORM Conference, 717-720 (2004).

29. T. Matsumura, T. Shirakashi, E. Usui and T. Furuki, "On the development of a milling process simulator based on energy approach", Proc. of the 7th ESAFORM Conference, 725-728 (2004).

30. D. Umbrello, R. Shivpuri and J. Hua, "Modeling of the flow stress for AISI 52100 during hard machining processes", Proc. of the 7th ESAFORM Conference, 741-744 (2004).

31. G. Ambrogio, L. Filice and D. Umbrello, "Numerical analysis of the fracture surface in thick sheets blanking", Proc. of the 7th ESAFORM Conference, 757-760 (2004).

32. R. Hambli, "Numerical modeling of sheet metal blanking processes", Proc. of the 7th ESAFORM Conference, 761-764 (2004).

33. B. Denkena, K. Tracht and M. Clausen, "Dynamic simulation of oblique turning processes", Proc. of the 7th ESAFORM Conference, 765-768 (2004).

34. L. Filice, D. Umbrello, F. Micari and L. Settineri, "On the finite element simulation of thermal phenomena in machining processes", Proc. of the 8th ESAFORM Conference, 729-732 (2005).

35. W. W. Wenzl and S. Treml, "Simulation of the blanking process of aluminium sheet using the engineering software deform", Proc. of the 8th ESAFORM Conference, 741-744 (2005).

36. Y. Guetari, S. Le Corre and N. Moes, "Study on the X-FEM method possibilities for the simulation of machining", Proc. of the 8th ESAFORM Conference, 745-748 (2005).

37. V. Lemiale, P. Picart and J. Chambert, "Comparison of different fracture models for the metal blanking process", Proc. of the 8th ESAFORM Conference, 749-752 (2005).

38. T. Matsumura, T. Shirakashi, E. Usui and I. Hori, "Simulation of drilling process base on energy approach", Proc. of the 8th ESAFORM Conference, 757-760 (2005).

39. K. Tracht, B. Denkena, M. Clausen and J. H. Yu, "Dexel-based milling simulation", Proc. of the 8th ESAFORM Conference, 765-768 (2005).

40. L. Donati, G. Tani and L. Tomesani, "2D orthogonal cutting simulation of C15 steel", Proc. of the 8th ESAFORM Conference, 769-772 (2005).

41. L. Filice, F. Micari, S. Rizzuti and D. Umbrello, "On the correlations between friction model and predicted temperature distribution in orthogonal cutting", Proc. of the 8th ESAFORM Conference, 599-602 (2005).

42. J. V. Le Lan, L. Masset, A. Marty and J. F. Debongnie, "An efficient simulation tool for predicting chatter during cutting operations", Proc. of the 9th ESAFORM Conference, 607-610 (2006).

43. T. Matsumura, T. Shirakashi and E. Usui, "Prediction of milling process based on 2D FEM simulation", Proc. of the 9th ESAFORM Conference, 623-626 (2006).

44. S. C. Assouline, P. Lorong and G. Coffignal, "Accounting workpiece flexibility in macroscopic machining simulations", Proc. of the 9th ESAFORM Conference, 627-630 (2006).

45. G. Coffignal, F. Lapujoulade, J. L. Lebrun and M. Touratier, "Trends in computational and experimental methods in materials cutting", Proc. of the 1st ESAFORM Conference, 231-234 (1998).

46. E. Ceretti, L. Filice and F. Micari, "Analysis of the chip geometry in orthogonal cutting of mild steel", Proc. of the 4th ESAFORM Conference, 607-610 (2001).

47. J. L. Bacaria, O. Dalverny, O. Pantalè, R. Rakotomalala and S. Caperaa, "Transient numerical models of discontinuous chip formation based on a damage effect", Proc. of the 4th ESAFORM Conference, 611-614 (2001).
48. J. C. Hamann, F. Meslin and S. Donyo, "Phenomenological modeling of chip segmentation using the catastrophe theory", Proc. of the 4th ESAFORM Conference, 659-662 (2001).
49. E. S. Dzidowski and G. Chrushielski, "Mesoscopic-macroscopic model of chip formation during machining in quasi-isothermal conditions", Proc. of the 6th ESAFORM Conference, 523-526 (2003).
50. T. Mabrouki, L. Deshayes and J. F. Rigal, "On the modelling of serrated chip formation", Proc. of the 7th ESAFORM Conference, 713-716 (2004).
51. T. Shirakashi, T.Obikawa and J. Shinozuka, "Simulation analysis on chip breaking process by grooved tool", Proc. of the 8th ESAFORM Conference, 733-736 (2005).
52. M. Ancau, "Numerical modeling of the chip geometry in end milling operation", Proc. of the 8th ESAFORM Conference, 773-776 (2005).
53. R. Rossi, E. Sermet, G. Poulachon, J. L. Lebrun, O. W. Dillon, K. Saito and I. J. Jawahir, "Numerical and experimental studies on tool temperature distribution in machining with flat-faced and grooved tools", Proc. of the 4th ESAFORM Conference, 615-618 (2001).
54. F. Meslin and J.C. Hamann, "Experimental observation and modeling of tool-chip interface phenemena", Proc. of the 4th ESAFORM Conference, 619-622 (2001).
55. J. L. Alcaraz, L. N. Lopez de Lacalle, A. Lamikiz and A. Escudero, "Simulation of thermal enhanced machining", Proc. of the 4th ESAFORM Conference, 663-666 (2001).
56. W. Grzesik, "Thermal and frictional characterization of the tool-chip interface for assessment of the process performance", Proc. of the 5th ESAFORM Conference, 587-590 (2002).
57. T. Matsumura, T. Shirakashi, T. Obikawa and E. Usui, "On the development of cutting process simulator for turnig operation", Proc. of the 6th ESAFORM Conference, 519-522 (2003).
58. G. Germain, P. Robert, J. L. Lebrun, P. Dal Santo and A. Poitou, "Experimental and numerical approaches of laser assisted turning", Proc. of the 7th ESAFORM Conference, 721-724 (2004).
59. W. Grzesik, P. Nieslony and B. Bartoszuk, "A model for the cutting heat partitioning in multilayer coating-stell couples", Proc. of the 7th ESAFORM Conference, 749-752 (2004).
60. P. Carlone, C. Cataldo and G. S. Palazzo, "A finite element three dimensional approach to metal evaporative laser cutting : temperature field non-linear model", Proc. of the 8th ESAFORM Conference, 753-756 (2005).
61. A. Aguiar Vieira, A. Monteiro Baptista, R. Natal George and M. Lages Parente, "Experimental and fem study of the temperature field during superficial grinding", Proc. of the 8th ESAFORM Conference, 761-764 (2005).
62. O. W. Dillon, K. C. Ee, P. C. Wanigarathne, I. S. Jawahir and J. C. Outeiro, "Temperature and residual stresses in orthogonal machining", Proc. of the 9th ESAFORM Conference, 603-606 (2006).
63. M. Nouari and A. Molinari, "Modelling of tool wear and optimization of the cutting process", Proc. of the 4th ESAFORM Conference, 643-646 (2001).
64. T. Shirakashi, "Some difficulties on prediction of tool life and machined surface quality through FEM", Proc. of the 5th ESAFORM Conference, 583-586 (2002).
65. J. Hua, P. Mittal and R. Shivpuri, "An FEM based thermal diffusion model for crater wear in machining", Proc. of the 6th ESAFORM Conference, 531-534 (2003).
66. F. Micari, L. Filice and D. Umbrello, "Preliminary analysis of the predictive capability of a wear model in machining", Proc. of the 6th ESAFORM Conference, 575- 578 (2003).
67. G. Casalino, S. Campanelli, A. D. Ludovico and E. Stasolla, "Numerical simulation of single point turning operation", Proc. of the 9th ESAFORM Conference, 619-622 (2006).

68. T. Shirakashi, T. Obikawa and H. Sasahara, "Estimation of the residual stress distribution in machined surface and its control through fem simulation", Proc. of the 4th ESAFORM Conference, 639-642 (2001).

69. D. Umbrello, J. Hua, R. Shivpuri, F. Hashimoto and X. Cheng, "Numerical investigation of effect of hard turning on residual stress distribution of shot peened surface", Proc. of the 8th ESAFORM Conference, 725-728 (2005).

70. G. Ambrogio, L. Filice, D. Umbrello, R. Shivpuri and J. Hua, "Application of NN technique for predicting the residual stress profiles during hard turning of AISI 52100 steel", Proc. of the 9th ESAFORM Conference, 595-598 (2006).

71. G. Giorleo, L. Carrino, W. Polini, U. Prisco, "Topography of turned surfaces", Proc. of the 6th ESAFORM Conference, 551-554 (2003).

72. A. Moufki, A. Molinari, D. Dudzinski, "An analytical thermomechanical modelling of chip three-dimensional formation with complex tool. Application to bar turning", Proc. of the 6th ESAFORM Conference, 567-570 (2003).

73. K. Saanouni, J.F. Mariage, A. Cherouat, "Sheet metal cutting simulation by continuum damage mechanics", Proc. of the 5th ESAFORM Conference, 575-578 (2002).

74. V. Lemiale, P. Picart, S. Meunier, "Numerical modelling of blanking process for thin sheet metal parts", Proc. of the 5th ESAFORM Conference, 559-562 (2002).

75. B. Changeux, M. Touratier, J.L. Lebrun, T. Tomas, J. Clisson, "High speed shear stress for the identification of the Johnson-Cook law", Proc. of the 4th ESAFORM Conference (2001).

76. P. Bariani, G.A. Berti, S. Corazza, "Analysis of material behaviour at high strain rates for modelling machining processes", Proc. of the 4th ESAFORM Conference (2001).

77. M.E. Merchant, "Mechanics of the metal cutting process", J. of Applied Physics, Vol. 16, N°6,318-324, 1945.

Achevé d'imprimer sur les presses de l'Imprimerie BARNÉOUD
B.P. 44 - 53960 BONCHAMP-LÈS-LAVAL
Dépôt légal : avril 2007 - N° d'imprimeur : 703071

Made in the USA
Monee, IL
07 July 2026

56544402R00150